Springer

Tokyo
Berlin
Heidelberg
New York
Barcelona
Hong Kong
London
Milan
Paris
Singapore

K. Honda (Ed.)

Functionality of Molecular Systems

Volume 2
From Molecular Systems to Molecular Devices

With 172 Figures

 Springer

Kenichi Honda
President
Tokyo Institute of Polytechnics
2-9-5 Honcho, Nakano-ku, Tokyo 164-8678
Japan

ISBN 978-4-431-68552-4   ISBN 978-4-431-68550-0 (eBook)
DOI 10.1007/978-4-431-68550-0

Library of Congress Cataloging-in-Publication Data

Functionality of molecular system / K. Honda, ed.
    p. cm.
  Includes bibliographical references and index.

  ISBN 978-4-431-68552-4
  1. Nanostructure materials—Electric properties. 2. Molecular
electronics—Materials. I. Honda, Ken'ichi.
  TA418.9.N35 F86   1998
  620.1' 1297—ddc21

                                        98-46362
                                        CIP

Printed on acid-free paper

© Springer-Verlag Tokyo 1999
Softcover reprint of the hardcover 1st edition 1999

SPIN: 10498182

Preface

In the period from 1991 to 1995, a research project titled Intelligent Molecular Systems with Controlled Functionality was undertaken by a research group with a Grant-in-Aid for Scientific Research from the Ministry of Education, Science and Culture. Quite a number of research papers have been issued by the group and important contributions to the progress of intelligent molecular systems was achieved.

On the occasion of closing the research project, we agreed that it would be of significant worth to publish a book on the rapidly developing field of molecular systems. It was planned to present an overview of the recent progress in this field, introducing at the same time the principal results obtained by the research project. Publication was planned in two volumes: volume 1 covers fundamentals and volume 2 applications.

Molecular systems is indeed a multidisciplinary and transdisciplinary field. Specialists in a variety of disciplines such as physical chemistry, photochemistry, electrochemistry, surface chemistry, organic chemistry, polymer science, and biochemistry willingly agreed to contribute to volume 2. Most authors were active members of the project. For the areas that were not involved in the project, a few specialists from outside were solicited to become authors.

The editorial coordinators, Professor Takeo Shimidzu (Kyoto University), Professor Masahiro Irie (Kyushu University), Professor Iwao Yamazaki (Hokkaido University), and Professor Kenichi Honda (Tokyo Institute of Polytechnics), discussed the outline of the book and decided to focus on two main areas. The first was how to make a practical molecular system from individual molecular units. Here, emphasis should be laid on the matrix material and its architecture to facilitate the systematization of each component unit. This is described in Chapters 1 and 2. In Chapter 2, two main materials, polymer systems and molecular assemblies, are described in addition to the particular systems prepared in extreme ultrathin or ultrafine states.

The second main area of focus was applications aiming at practical use. The function of a molecular system is divided into information transduction and energy conversion. Within the molecular unit itself, there is no essential difference between the two. However, in order to facilitate understanding the overall status of the application trends, both are treated in separate chapters, Chapters 3 and 4 respectively.

In Chapter 3, information transduction is divided into three types, electroactive, photoactive, and chemoactive, depending on the nature of the input signal to the molecular unit. Chapter 4 deals with energy conversion, with the emphasis laid on light energy conversion systems in nature in the section on photoactive systems (4.1). Photoelectrochemical energy conversion is described separately (4.2) because of its rapid development.

The progress of molecular systems has been most remarkable in organic compounds; consequently, most of the materials treated in the present volume are organic systems, with the exception of Section 2.3.2, Ultrafine Particles, and Section 4.2, Photoelectrochemical Conversion. In these sections, the function of inorganic semiconductors is discussed in terms of light energy conversion. The conversion process is often accompanied by sensitizing systems using organic dyes. During compilation, Professor Shimidzu was principal editor for Chapters 1 and 2, Professor Irie for Chapter 3, and Professor Yamazaki for Chapter 4.

The editor expresses his sincere gratitude to Professor H. Inokuchi, head of the previously described research project, for his efforts in realizing this publication as a milestone of the project. He is very grateful to the authors for their contributions and to Dr. T. Shimidzu, Professor Emeritus at Kyoto University; Professor M. Irie, Kyushu University; and Professor I. Yamazaki, Hokkaido University, for the planning of the content. He expresses his heartfelt thanks to Professor K. Kanoda, University of Tokyo; Professor T. Kitagawa, Institute for Molecular Science; Professor N. Nishi, Institute for Molecular Science; Ms. Y. Suzuki, Institute for Molecular Science; and Professor K. Tanaka, Kyoto University, for their kind assistance in compiling and checking the manuscripts. The editor is indebted to Mr. T. Yonezawa and Ms. Y. Nishimura, Springer-Verlag Tokyo, for their devoted assistance.

KENICHI HONDA

Table of Contents

List of Authors

*Editorial Coordinator

MASUO AIZAWA (3.3.1)
Department of Bioengineering, Faculty of Bioscience and Biotechnology, Tokyo Institute of Technology, 4259 Nagatsuta, Midori-ku, Yokohama 226-8501, Japan

MASAMICHI FUJIHIRA (2.2.2)
Department of Biomolecular Engineering, Tokyo Institute of Technology, 4259 Nagatsuta, Midori-ku, Yokohama 226-8501, Japan

AKIRA FUJISHIMA (2.3.2, 4.2)
Department of Applied Chemistry, Faculty of Engineering, The University of Tokyo, 7-3-1 Hongo, Bunkyo-ku, Tokyo 113-8656, Japan

KENICHI HONDA (1, 5)
Tokyo Institute of Polytechnics, 2-9-5 Honcho, Nakano-ku, Tokyo 164-8678, Japan

KAZUYUKI HORIE (3.2.2)
Department of Chemistry and Biotechnology, Graduate School of Engineering, The University of Tokyo, 7-3-1 Hongo, Bunkyo-ku, Tokyo 113-8656, Japan

MASAHIRO IRIE (3*, 3.2.1, 4.3)
Department of Chemistry and Biochemistry, Graduate School of Engineering, Kyushu University, 6-10-1 Hakozaki, Higashi-ku, Fukuoka 812-8581, Japan

KINGO ITAYA (2.3.3)
Department of Applied Chemistry, Faculty of Engineering, Tohoku University, 4 Aramaki Aza Aoba, Aoba-ku, Sendai 980-8579, Japan

TSUYOSHI KAWAI (3.1.1)
Department of Electronic Engineering, Faculty of Engineering, Osaka University, 2-1 Yamadaoka, Suita, Osaka 565-0871, Japan

MASASHI KUNITAKE (2.3.3)
Department of Applied Chemistry & Biochemistry, Faculty of Engineering, Kumamoto University, 2-39-1 Kurokami, Kumamoto 860-8555, Japan

TOYOKI KUNITAKE (2.2.1)
Department of Chemical Science and Technology, Faculty of Engineering, Kyushu University, 6-10-1 Hakozaki, Higashi-ku, Fukuoka 812-8581, Japan

HACHIRO NAKANISHI (3.2.3)
Institute for Chemical Reaction Science, Tohoku University, 2-1-1 Katahira, Aoba-ku, Sendai 980-8577, Japan

YOSHIO NOSAKA (2.3.2)
Department of Chemistry, Nagaoka University of Technology, 1603-1 Kami-Fukuoka, Nagaoka, Niigata 940-2188, Japan

MAMORU OHASHI (3.3.2)
Department of Materials Science, Faculty of Science, Kanagawa University, 2946 Tsuchiya, Hiratsuka, Kanagawa 259-1293, Japan

SHOGO SAITO (3.1.2)
Department of Materials Science and Technology, Graduate School of Engineering Sciences, Kyushu University, 6-1 Kasuga-Koen, Kasuga, Fukuoka 816-8580, Japan

TOHRU SATO (2.1.1)
Division of Molecular Engineering, Faculty of Engineering, Kyoto University, Sakyo-ku, Kyoto 606-8501, Japan

TAKEO SHIMIDZU (1, 2*, 2.1*, 2.3.1, 5)
KRI International, Kyoto Research Park, Shimogyo-ku, Kyoto 600-8813, Japan

YASUHIKO SHIROTA (2.1.2)
Department of Applied Chemistry, Faculty of Engineering, Osaka University, 2-1 Yamadaoka, Suita, Osaka 565-0871, Japan

KAZUYOSHI TANAKA (2.1.1)
Division of Molecular Engineering, Faculty of Engineering, Kyoto University, Sakyo-ku, Kyoto 606-8501, Japan

DONALD A. TRYK (4.2)
Department of Applied Chemistry, Faculty of Engineering, The University of Tokyo, 7-3-1 Hongo, Bunkyo-ku, Tokyo 113-8656, Japan

TOKIO YAMABE (2.1.1)
Division of Molecular Engineering, Faculty of Engineering, Kyoto University, Sakyo-ku, Kyoto 606-8501, Japan

IWAO YAMAZAKI (3.2.4, 4*, 4.1)
Department of Molecular Chemistry, Faculty of Engineering, Hokkaido University, N13 W8, Kita-ku, Sapporo 060-8628, Japan

KATSUMI YOSHINO (3.1.1)
Department of Electronic Engineering, Faculty of Engineering, Osaka University, 2-1 Yamadaoka, Suita, Osaka 565-0871, Japan

1. Concept of a Molecular System, Molecular Unit, and Molecular Material

TAKEO SHIMIDZU and KENICHI HONDA

All materials have their own intrinsic function which might be useful for human society. In most materials, it is usually their bulk properties which are utilized. The concept of a molecular system is derived from the understanding that the functionality of materials is based on the properties of their component molecules. The overall function of a material is the combined effect of the functions of each component molecule, and these component molecules are defined as molecular units.

Let us call material based upon the above concept "molecular material." While there are a great number of different types of molecules, the large variety of structures and properties can be characterized by the nature of the electrons belonging to each molecule. Such variety suggests that these is great potential for the development of specific functional materials. Each molecule has its own characteristics and, as such, can be regarded as being the smallest piece of functional material whose properties are easily distinguishable. This is in contrast to inorganic and metallic solid materials, which are widely used as advanced electronic and photonic materials. With the latter, the smallest functional units are characterless atoms, so that the variety is much more limited. On the other hand, each molecular unit has its own characteristics, which may not disappear entirely, even when it is observed in a cluster or in the bulk state. The inorganic and metallic types of materials also function primarily via interactions involving electrons or photons, while molecular materials have the ability to function in other modes, for example, translocation.

In principle, the electrons associated with a particular molecule govern its various properties and functions. These functions can involve changes in electronic state or distribution. In order to design an appropriate functional molecule, one should take into account the nature of the electrons intrinsic to that molecule, e.g., σ-, π-, unpaired, conducting, localized, or delocalized, as well as the molecular structure. Every possible type of molecule has its own characteristics and can give rise to many different types of specific behavior, based on changes in electronic structure, states, and the spatial arrangement of the nuclei. The latter aspect is specifically associated with molecular functions such as conformational changes, which can facilitate energy and electron transfers.

Each molecule must necessarily exist in a particular quantum state at a particular time, and the molecular function can be defined in terms of a transition from this state to another one. This molecular function or process can involve not only changes in electronic distribution, energy levels, and spin states, but also changes in conformation, for example, isomerization, based on changes in the spatial positions of the nuclei (transposition). Furthermore, the movement of the entire molecule in space (translocation) is involved. Thus, in addition to electronic, magnetic, and photonic functions, which are also possible for inorganic and metallic materials, molecular functions can involve transposition and translocation. Such molecular transformation induces the transport of intrinsic electrons, both of molecules and ions, giving rise to finely controlled chemical reactions, which are essential processes occurring, for example, in living cells. Together, electronic transitions and molecular translocations induce controlled, successive energy and electron transfers and chemical reactions at the intramolecular and intermolecular level.

There is a big difference between molecular material and the usual types of material used for industry, transportation, communication, construction, and so on. From the variety of functions of molecular materials, the stimulus–respone type of function will be described in detail.

Figure 1.1 shows the concept of the function of a molecular unit. A single molecule receives an input signal and is then subjected to an intramolecular change. This change gives rise to the output or response. The input or output can be any kind of signal or energy, such as a light, electric, magnetic, thermal, chemical, or mechanical signal, as shown in Fig. 1.2.

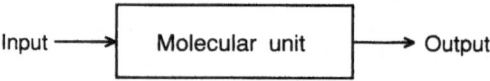

FIG. 1-1. The basic process of a molecular unit

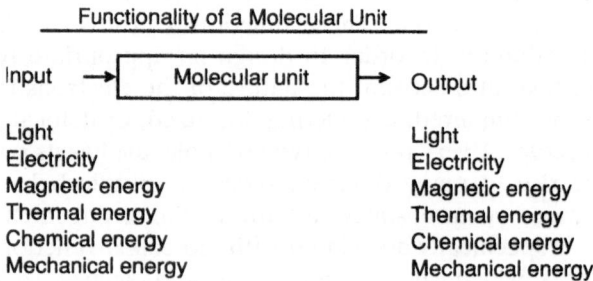

FIG. 1-2. A variety of combinations of input and output energy

Electronic transition between molecular energy levels is a typical case of intermolecular change caused by the input signal or energy. As a result of the intramolecular change, the output or response moves out of the molecule and into its surroundings as some form of energy, as is also shown in Fig. 1.2. Electronic transition will cause successive intra- or intermolecular processes such as reaction, isomerization, molecular strain, emission, internal conversion to thermal energy, polarization, etc. When the input or output takes the form of light, that molecular unit is regarded as having a photoresponsive function. Of course, a single molecular unit on its own will have no discernible effect. These units must be in an assembly to be of any practical use. In an assembly, the geometrical arrangement of molecular units should be optimized. This optimizd arrangement in an assemblage of molecular units is the molecular system. Figure 1.3 shows the concept of a molecular system where each molecular unit is embedded in an appropriate matrix or support. This optimization is essential in any attempt to find practical applications for molecular systems.

For the matrix material, polymers or molecular assemblies are the most hopeful candidates. In either case, control of the spatial or geometrical distribution of the molecular units is indispensable for optimization. We will consider two ways of controlling spatial distribution: (a) intermolecular spatial distribution; (b) intramolecular spatial distribution.

For intermolecular spatial distribution both polymers and molecular assemblies have been the target of intensive research work. Polymers are described in Chap. 2.1 and molecular assemblies in Chap. 2.2. Intermolecular electron and energy transfer is widely studied in terms of the spatial distribution of the donor and acceptor as the functional molecular unit. Typical examples of this process with a molecular assembly as the matrix material are described in Chaps. 2.2.2 and 4.1. The use of inclusion compounds such as cyclodextrin is also interesting, and offers another method of controlling the spatial distribution of functional molecular units as guest species.

For intramolecular spatial distributions, suppose a molecule to contain two independent functional groups. Functional group 1 is connected to functional group 2 by an appropriate spacer, as shown in Fig. 1.4.

In the figure, a circle indicates one functional unit and a square another functional unit. Each of these units exerts an influence on the responsive behavior of the other. This aspect can be understood by the way that the functionality of one unit can be controlled or modulated by operating the second unit. This

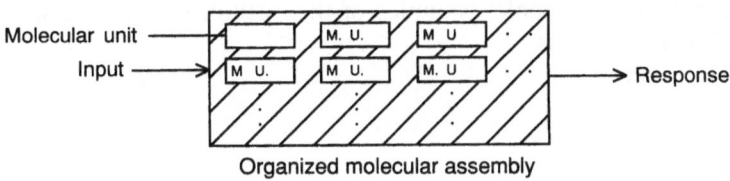

FIG. 1-3. The concept of an optimized molecular system composed of molecular units

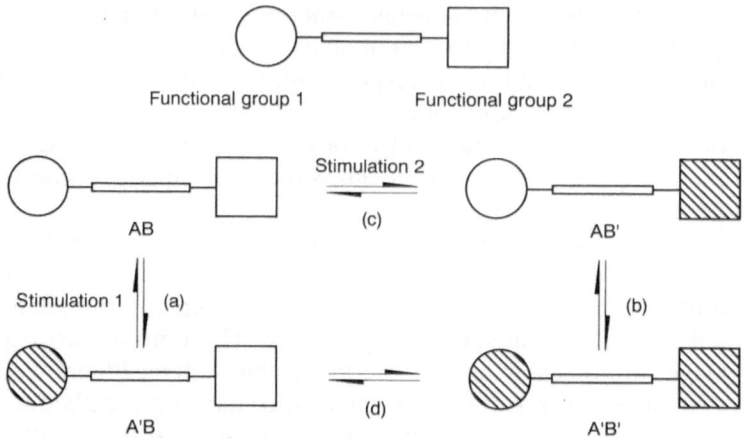

Fig. 1-4. Schematic illustration of the functioning of a bifunctional molecule

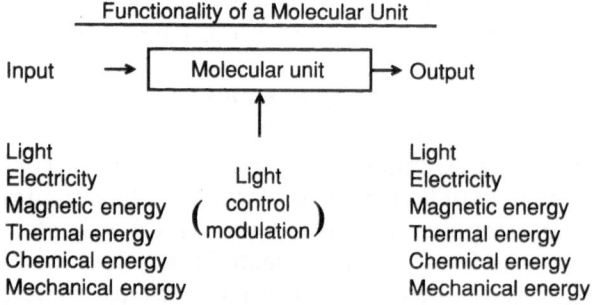

Fig. 1-5. The concept of the modulation of the molecular function

concept is shown in Fig. 1.5. The example shown is intramolecular electron transfer where two functional groups are donor and acceptor, respectively. There has been much research work on the efficiency or intramolecular electron transfer as a function of the relative spatial locations of the donor and acceptor groups, and discussions on "through space" and "through bond" are of continuing interest.

For both intermolecular and intramolecular spatial distributions, natural living systems give us incomparable examples of the optimized spatial distribution of functional molecular units.

2. Architectural Design and Preparation of Molecular Systems

TAKEO SHIMIDZU

The function of a systematized and assembled group of molecules is essentially the same as that of a single molecule. The most fundamental functions of a molecular system are its electron and ion conducting properties. These involve information transduction and energy conversion, although their conduction mechanisms are different. They are seen in polymeric molecules, molecular clusters, and molecular assembly. A polymeric molecule should be considered as a unit repeating material. Bilayer Langmuir–Blodgett (LB) membranes are well-organized molecular assemblies which are prepared utilizing their self-assembling properties. A forced preparative method gives ultrathin material, ultrafine particles, STM prepared molecular materials, etc. They all have their own characteristics and distinguishing functions. In addition, the properties of all molecules remain more or less the same when the molecules are clustered and assembled. The polymers, the molecular assemblies, and the molecular clusters present us not only with their own functions, but also with a stage where the functional molecules can act effectively. In this chapter, the molecular design and fabrication of conducting polymers, molecular assemblies, and molecular clusters are described as one of the most essential, fundamental molecular functional materials.

2.1 Functional Polymers

TAKEO SHIMIDZU

TOKIO YAMABE, TOHRU SATO, and KAZUYOSHI TANAKA (2.1.1)
YASUHIKO SHIROTA (2.1.2)

As typical representatives of functional polymers, electron- and ion-conducting polymers are described in Sect. 2.1. Fundamental theories as well as probable applications for carbon nanotubes and polymer–salt hybrid systems are introduced in Sects. 2.1.1 and 2.1.2, respectively, to explain some of the most recent developments in the major area of functional polymers.

2.1.1 Electron-Conducting Polymers

In 1991 the carbon nanotube (simply called a "nanotube" throughout this section) was discovered as a cathode product in a carbon-arc discharge method which is similar to that used for the preparation of fullerenes [1]. Although thick graphite fibers or carbon needles, with diameters of the order of $10^0 \mu m$, grown during arc-discharge experiments, have been known for over thirty years [2], the nanotube is a completely new type of carbon fiber which comprises coaxial cylinders of graphite sheets, the number of which ranges from 2 to 50 [1]. The most unusual aspect of nanotubes is the helical arrangement of carbon hexagons on the tube surface along its axis. Moreover, it has recently been reported that there exist nanotubes consisting of a single-shell cylinder in the gas-phase product of the carbon-arc discharge method in the presence of iron and methane [3] (Fig. 2.1-1).

A purification process to obtain nanotubes out of cathode products has recently been established, but most include fragments of graphitic sheet and/or amorphous carbon [4, 5]. However, based on such techniques, it is now possible to study the properties of pure nanotubes in the absence of graphite fragments, amorphous carbon, etc. Typical diameters and lengths of the nanotubes obtained [4] are of the order of 10^0–10^2 nm and 10^1–$10^2 \mu m$, respectively.

In this section we discuss theoretical studies on electronic structures, and the results of experiments on the electronic properties of nanotubes. In Sect. 2.1.1.1, we show the complete analytical solution of the one-dimensional π-band structure of single-shell nanotubes without bond-alternation patterns based on the tight-binding crystal orbital method with in the framework of the Hückel ap-

6

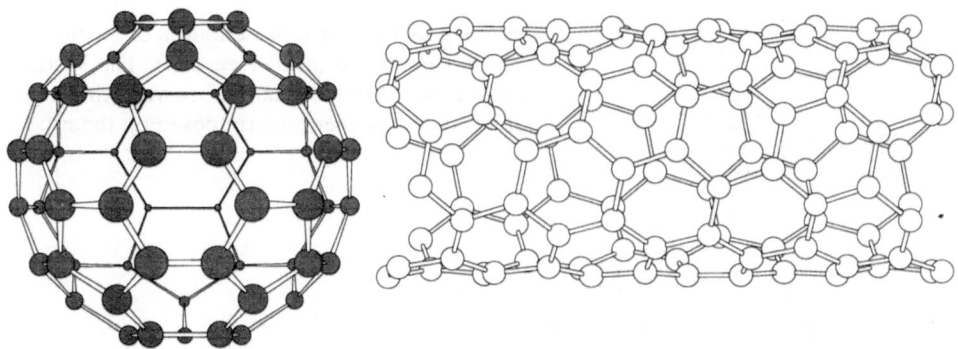

FIG. 2.1-1. *Left*, C_{60}. *Right*, Single-shell nanotube

proximation. In Sect. 2.1.1.2 we present the experimental results of Raman scattering and electron spin resonance (ESR) measurements for purified nanotubes.

2.1.1.1 Electronic Structures of Nanotubes

Nanotubes have such an unusual structure, as described above, that they have been attracting intense interest in their electronic properties and possible uses as a new one-dimensional (1D) conductor. In a study of the electronic structure of nanotubes, it is important to consider the effects of their helical structure and coexisting two-dimensionality as well as one-dimensionality. Here we present some simple analytical results for single-shell nanotubes using the 1D tight-binding crystal orbital method based on the Hückel approximation [6].

Since the geometry of the single-shell nanotube is constructed by rolling up a graphite sheet, any nanotube configurations can be represented by superimposing the hexagon at the origin (O) onto the hexagon (a, b) defined by the vector $\boldsymbol{R} = a\boldsymbol{x} + b\boldsymbol{y}$, as shown in Fig. 2.1-2. Here $\boldsymbol{x}$ and $\boldsymbol{y}$ are primitive vectors with length $\sqrt{3}d_{C-C}$ and (a, b) is the integer set satisfying the conditions $a > b$, $a \geq 1$, and $b \geq 0$. Here d_{C-C} indicates the C–C bond length. The nanotube defined in this manner is called "nanotube (a, b)." The diameter D of nanotube (a, b) is expressed by

$$D = \frac{\sqrt{3}d_{C-C}}{\pi}\sqrt{a^2 + ab + b^2} \qquad (2.1\text{-}1)$$

The conformation angle θ shown in Fig. 2.1-2 is defined as

$$\theta = \arccos\left(\frac{2a+b}{2\sqrt{a^2 + ab + b^2}}\right) \qquad (2.1\text{-}2)$$

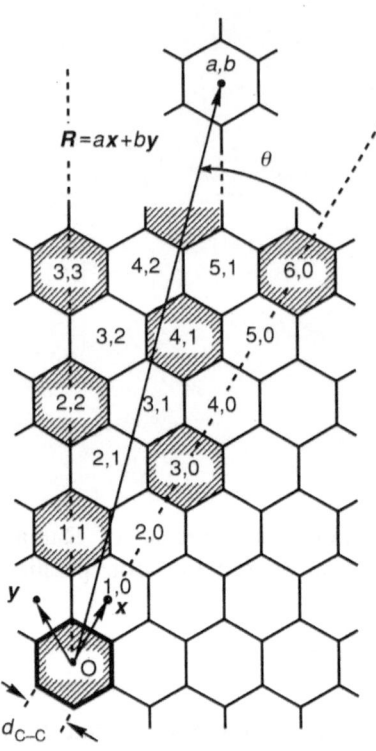

FIG. 2.1-2. Schematic representation of the formation of a nanotube (a, b) (see text). The shaded
hexagons show metallic nanotubes when the original hexagon has been superimposed on them

Under the condition $b = 0$ (when $\theta = 0$) or $a = b$ (when $\theta = \pi/6$), the nanotube
becomes non-helical, while in other cases it is helical. A helical nanotube (a, b) is
chiral, and its enantiomer is expressed by nanotube (b, a).

The electronic structure of two-dimensional (2D) graphite sheet used to study
the 1D electronic structure of nanotubes has been examined. As shown in
Fig. 2.1-3, graphite has the reciprocal lattice vectors K_x and K_y, making $\pi/6$ and
$-\pi/6$ with the lattice vectors x and y, respectively, and their lengths are $2\pi/l$ (here
$l = 3/2\, d_{C-C}$). Then an arbitrary point in the first Brillouin zone can be expressed
as

$$k = k_x \frac{l}{2\pi} K_x + k_y \frac{l}{2\pi} K_y \qquad \left(-\frac{\pi}{l} < k_x, \quad k_y < \frac{\pi}{l} \right) \qquad (2.1\text{-}3)$$

where k_x, k_y are the components of the wave vector.

Since the unit cell of a graphite-sheet lattice consists of two carbon atoms, each
of which supply one π electron, the secular determinant under the Hückel approximation is expressed as

$$\begin{vmatrix} -\lambda & h^* \\ h & -\lambda \end{vmatrix} = 0 \qquad (2.1\text{-}4)$$

where λ and h are given by

FIG. 2.1-3. The crystal lattice and the reciprocal lattice of graphite

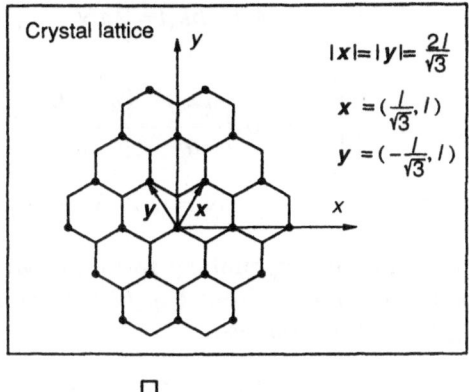

$$|x| = |y| = \frac{2l}{\sqrt{3}}$$

$$x = (\frac{l}{\sqrt{3}}, l)$$

$$y = (-\frac{l}{\sqrt{3}}, l)$$

$$\begin{cases} x \cdot K_x = y \cdot K_y = 2\pi \\ x \cdot K_y = y \cdot K_x = 0 \end{cases}$$

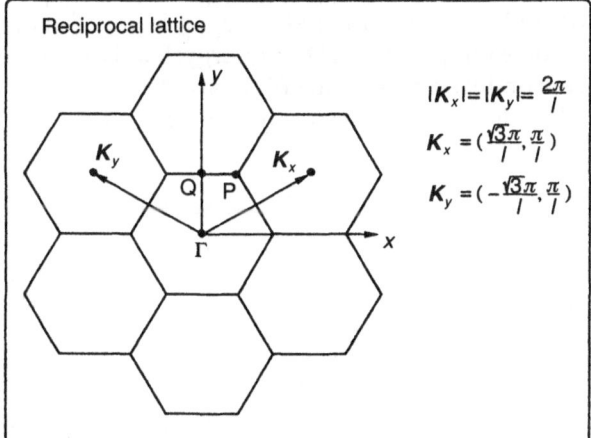

$$|K_x| = |K_y| = \frac{2\pi}{l}$$

$$K_x = (\frac{\sqrt{3}\pi}{l}, \frac{\pi}{l})$$

$$K_y = (-\frac{\sqrt{3}\pi}{l}, \frac{\pi}{l})$$

$$\lambda = \left[E(k) - \alpha \right] / \beta \tag{2.1-5}$$

$$h = 1 + \exp(ik_x l) + \exp(ik_y l) \tag{2.1-6}$$

and h^* is the complex conjugate of h. Here α and β are the Coulomb and transfer integrals, respectively. Equation (2.1-4) has two groups of solutions, i.e., the occupied and the unoccupied energy surfaces $E(k)_{1,2}$, which are described as

$$E(k)_{1,2} = \alpha \pm \beta \left[1 + 4 \cos\left(k_x l - \frac{1}{2} k_y l \right) \cdot \cos\left(\frac{1}{2} k_y l \right) + 4 \cos^2\left(\frac{1}{2} k_y l \right) \right]^{\frac{1}{2}} \tag{2.1-7}$$

To make a nanotube (a, b) by superimposing the original hexagon onto the one at (a, b) in Fig. 2.1-2 is equivalent to imposing the periodic boundary condition

$$ak_x l + bk_y l = 2\pi N \qquad \left(N = 1,\, 2,\, 3, \ldots,\, a\right) \tag{2.1-8}$$

$$E(k)_{1,2} = \alpha \pm \beta \left[1 + 4\cos\left\{\frac{2\pi N}{a} - \frac{(a+2b)}{2a}kl\right\}\cos\left(\frac{kl}{2}\right) + 4\cos^2\left(\frac{kl}{2}\right) \right]^{\frac{1}{2}} \tag{2.1-9}$$

$$\left(N = 1,\, 2,\, 3, \ldots,\, a\right)$$

where k simplifies k_y running from 0 to π/l. As a result, $2 \times a$ numbers of 1D π bands are obtained. From Eq. (2.1-9) we can derive the condition leading to $E(k)_{1,2} = \alpha \pm 0$ as follows:

$$2a + b = 3n \qquad \left(n \text{ is a positive integer}\right) \tag{2.1-10}$$

That is, the nanotube (a, b) becomes metallic with a zero-band gap when $2a + b$ is a multiple of 3. However, when $2a + b$ is not a multiple of 3, the nanotube becomes semiconductive with a finite band gap. Metallic nanotubes can be constructed when the original hexagon is imposed onto the shaded ones in Fig. 2.1-2. For example, the 1D π energy bands of a metallic nanotube $(3, 3)$ and a semiconducting nanotube $(4, 2)$ are shown in Figs. 2.1-4a and 2.1-4b, respectively.

The density of states at the Fermi level $N(E_F)$ is expressed as

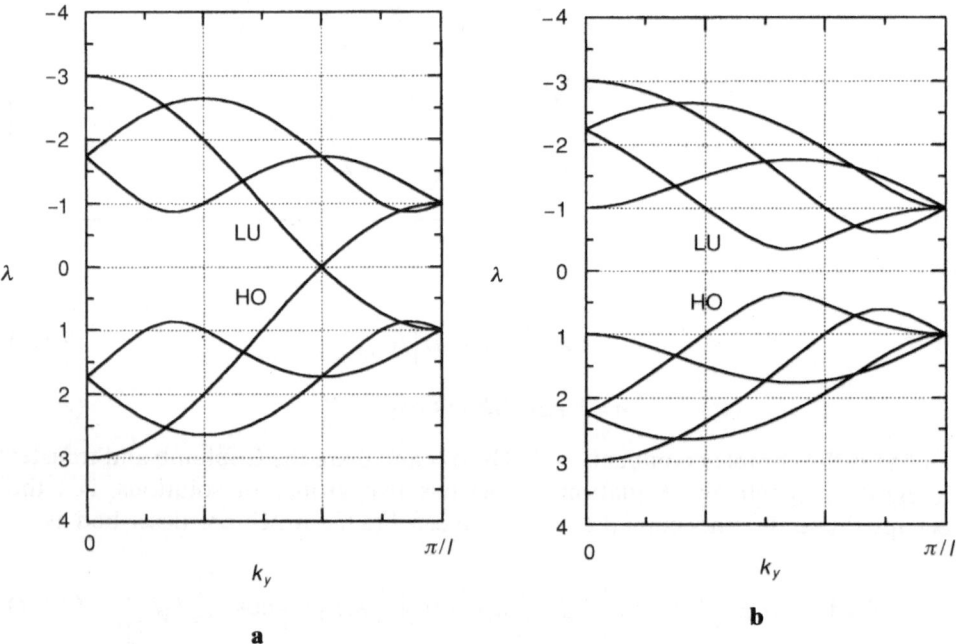

FIG 2.1-4. Energy bands of **a** nanotube $(3, 3)$, and **b** nanotube $(4, 2)$

$$N(E_F) \propto \frac{1}{\left(\dfrac{dE}{dk}\right)_{E=E_F}} = \frac{2a+b}{2\sqrt{a^2+ab+b^2}} \cdot \frac{2a}{2a+b} = \frac{2a}{2a+b} \cdot \cos\theta \quad (2.1\text{-}11)$$

which clearly gives a finite value at $E = E_F$, in contrast to that of one-layer graphite. The effective mass, m^*, derived from Eq. (2.1-9) is expressed as

$$
\begin{aligned}
\frac{1}{m^*} = \frac{1}{\hbar^2}\left(\frac{\partial^2 E}{\partial k^2}\right)_{1,2} = \pm\left(-\frac{\beta}{4\hbar^2}\right)&\left[\frac{3}{2}+2\cos\left(\frac{2\pi N - bkl}{a}\right)+2\cos\left(\frac{2\pi N-(a+b)kl}{a}\right)\right. \\
+\frac{1}{2}\cos(kl)\Bigg]^{\frac{3}{2}} \times &\left\{\left[\frac{2b}{a}l\sin\left(\frac{2\pi N - bkl}{a}\right)+\frac{2(a+b)}{a}l\sin\left(\frac{2\pi N-(a+b)kl}{a}\right)\right.\right. \\
-\frac{l}{2}\sin(kl)\Bigg]+2&\left[\frac{3}{2}+2\cos\left(\frac{2\pi N-bkl}{a}\right)+2\cos\left(\frac{2\pi N-(a+b)kl}{a}\right)\right. \\
+\frac{1}{2}\cos(kl)\Bigg]\times &\left[\frac{2b^2}{a^2}l^2\cos\left(\frac{2\pi N - bkl}{a}\right)+\frac{2(a+b)^2}{a^2}l^2\cos\left(\frac{2\pi N-(a+b)kl}{a}\right)\right. \\
+\frac{l^2}{2}\cos(kl)\Bigg]\Bigg\} & \qquad (2.1\text{-}12)
\end{aligned}
$$

Figure 2.1-5 shows the band gap with respect to diameter D of nanotube (a, b). The nanotubes with $2a + b \neq 3n$ have a finite band gap, where the band gap becomes smaller as the diameter becomes larger. It is reasonable to consider that the nanotube with infinite D resembles graphite. Hence nanotube (a, b) becomes either metallic or semiconductive depending on the combination of (a, b) or, in other words, the combination of helical pitch and diameter.

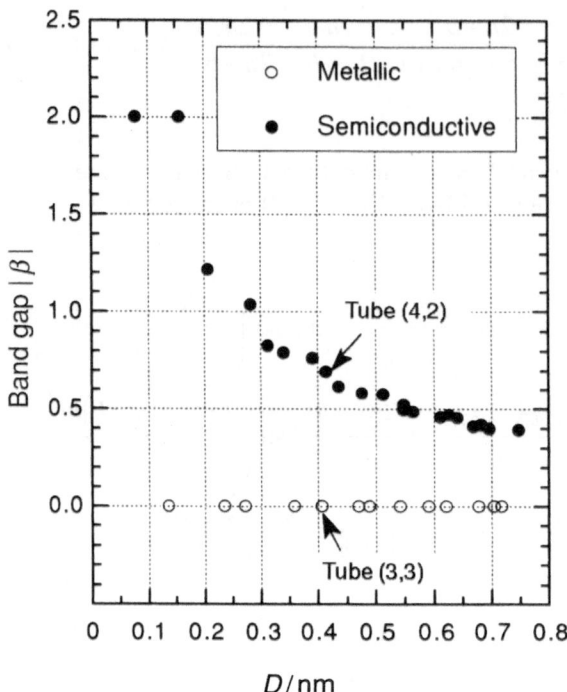

FIG 2.1-5. Band-gap values vs. nanotube diameters (diameter calculated for $d_{C-C} = 1.42\,\text{Å}$)

2.1.1.2 Electronic Properties of Actual Nanotubes

To examine the electronic structure of an actual nanotube, as described above, uncontaminated nanotubes are essential. Recently a purification method for nanotubes has been established [4], and the procedure is given below. The cathode product was obtained by the arc-discharge method in a machine for carbon-cluster preparation under a He pressure of 500 Torr with an applied DC voltage of 18 V. The crude nanotube, mainly consisting of the inner part of this cathode product, was obtained as a float after ultrasonic irradiation for 1 h in methanol as the dispersion medium. The float was milled in ethanol and centrifuged after water and a dispersant were added. Then the float was filtered and centrifuged again. The final float was washed with acid and oxidized at 750°C for 5 min. The purified nanotube thus obtained had an outer diameter of 20 nm and a length of 10 μm on average.

Fourier transform (FT)–Raman scattering and ESR analyses were performed with respect to the purified nanotubes thus obtained [7]. The FT–Raman scattering measurement of the purified nanotube showed only one peak at 1581 cm^{-1}, as shown in Fig. 2.1-6a. This is similar to that of highly oriented pyrolytic graphite (HOPG) shown in Fig. 2.1-6b, which also has only one peak at 1580 cm^{-1} coming from the E_{2g} mode [8]. It has also been deduced theoretically that a single-cylinder nanotube should show an E_{2g} mode at 1601 cm^{-1} [9]. However, there has been a report that a crude nanotube shows another peak around 1346–

FIG. 2.1-6. First-order Raman spectra of **a** a purified nano- tube, **b** highly oriented pyro- lytic graphite (HOPG), **c** a crude tube from the inner part of the cathode product, and **d** that from the outer shell. (**b–d** are taken from [8])

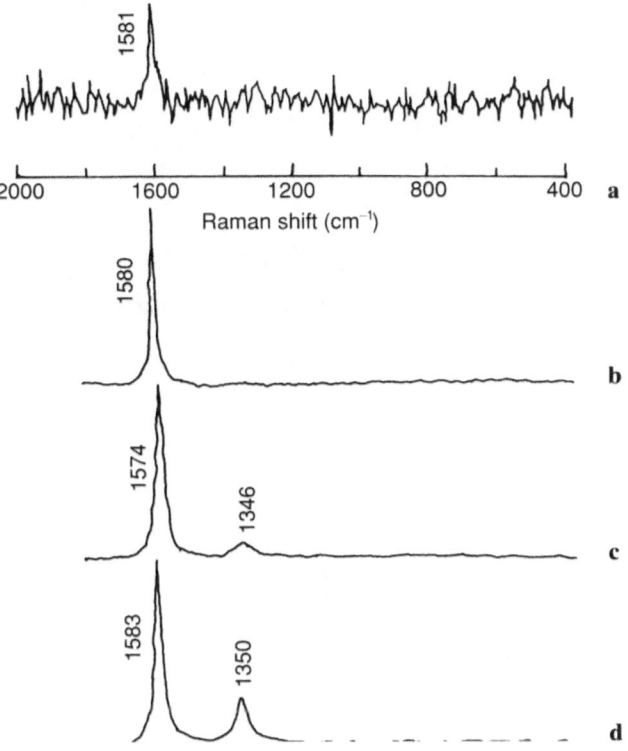

$1350\,\mathrm{cm}^{-1}$ [8], which is characteristic of amorphous carbon, as shown in Figs. 2.1-6c and d. Thus the purified nanotube behaves similarly to graphite from the viewpoint of Raman scattering. ESR measurement has shown that the purified nanotube is almost ESR-silent at room temperature, and only a faint peak was found at low temperature (−150°C). The spin concentrations of this were less than $10^{15}\,\mathrm{spins\,g}^{-1}$. This indicates that a purified nanotube still contains a very small amount of amorphous carbon, which is generally responsible for Curie paramagnetism [10].

Thus it can be concluded that the intrinsic magnetic property of a purified nanotube is almost completely diamagnetic without Pauli or Curie paramagne- tism. In this sense, the purified nanotube cannot be considered metallic from this conclusion. This is consistent with the preliminary electrical conductivity mea- surement [11]. A similar measurement for the "aggregate" of the purified nanotube showed an electrical resistance of ca. $10\,\Omega$ (at room temperature) with an activation energy of $5.0 \times 10^{-3}\,\mathrm{eV}$ [7].

However, the ESR spectrum of the crude nanotube shown in Fig. 2.1-7a comprises a Dysonian shape with a g-value of 2.0171. This value is comparable to that of polycrystalline graphite at 2.0184 [12]. Hence, the crude nanotube is probably contaminated with the small fragments of graphite that were removed from the purified nanotube. The purified nanotube samples after K- and the

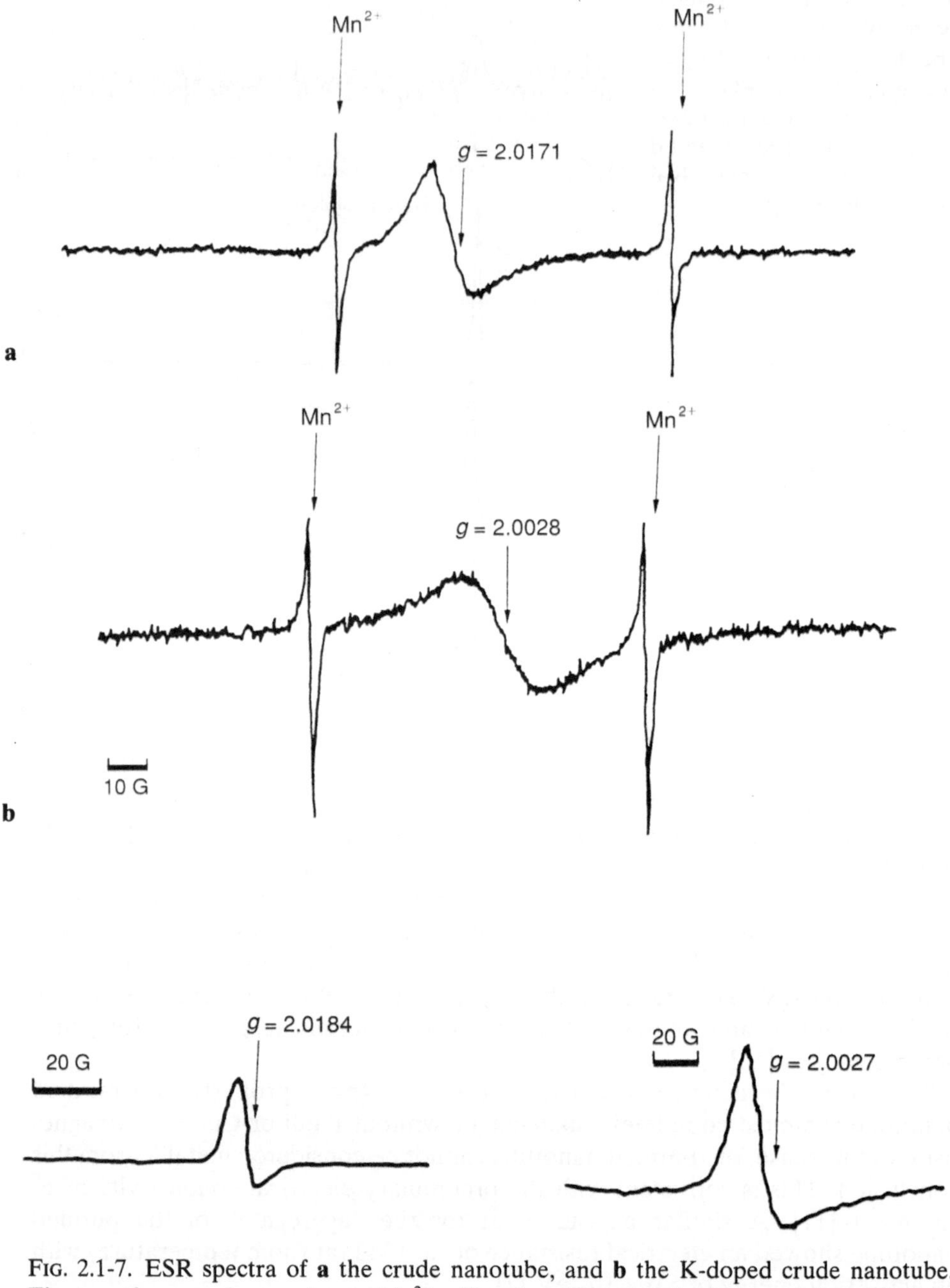

Mn^{2+} Mn^{2+}

$g = 2.0171$

a

Mn^{2+} Mn^{2+}

$g = 2.0028$

10 G

b

$g = 2.0184$

20 G

20 G

$g = 2.0027$

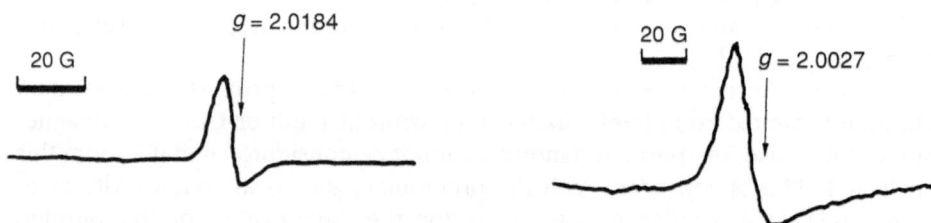

FIG. 2.1-7. ESR spectra of **a** the crude nanotube, and **b** the K-doped crude nanotube. The two sharp peaks are due to Mn^{2+} calibration. Those at the *bottom left* and *bottom right* are for polycrystalline graphite (from [12]) and K-doped graphite (C$_8$K; from [15]), respectively

I_2-dopings again show no ESR spectra, which indicates that neither electron nor hole injection occurs after these dopings. This confirms that the purified nanotube is dopant-inactive, and in this respect is different from graphite [13] and ordinary amorphous carbon materials [14]. However, the K-doped crude nanotube showed the ESR signal in Fig. 2.1-7b with a g-value of 2.0028, compared with that of K-intercalated graphite at 2.0027 [15]. Interestingly, the Dysonian shape of the pristine crude nanotube is completely extinguished with the I_2-doping, indicating the spin-scavenging effect of the electron-accepting property of I_2, which is often observed in the early doping stage of amorphous carbon [16].

Recently, however, Kosaka et al. [17] have reported an ESR spectra of a purified nanotube corresponding to the conduction electron spin resonance (CESR) in their nanotube sample.

2.1.1.3 Conclusion

The intrinsic electronic properties of nanotubes are believed to be semiconductive and not metallic, at least for the purified sample prepared for the present study. However, it is not quite obvious at this stage whether a metallic nanotube cannot be produced in the cathode product, or whether it cannot remain in the final purified sample without, e.g., being burned up in the purification process. Nonetheless, it can be conjectured from theoretical predictions for 1D graphite series [18, 19], which can roughly be compared to nanotubes, that a metallic nanotube would not be thermodynamically stable compared with a semiconductive nanotube. If metallic nanotubes really exist, it is possible that there should be appropriate conditions for their formation from the results in [17].

Moreover, in the same way, we would like to emphasize a metallic nanotube, if obtainable, could be separated by the application of a poling technique often employed for ferroelectric polymers. This involves applying an external electric field, or the application of an electromagnetic interaction under an external magnetic field, since it has been predicted that a metallic nanotube is theoretically equivalent to a molecular solenoid [20, 21].

2.1.2 Ion-Conducting Polymers

Electrical conduction is divided into electronic conduction and ionic conduction, based on whether the charge carriers are electrons or ions. They are essentially different in that electronic conduction is associated with the transport of electrons, while ionic conduction is caused by the transport of matter, i.e., ions. Therefore, electronic conduction is generally enhanced by an increase in pressure, while the opposite is the case for ionic conduction. The discussion in this section is focused on ionic conduction, and describes the general concepts and characteristic features of ionic conduction in ion-conducting polymers, in particular polymer—alkali metal salt systems. Examples of polymer ionic conductors and their conductivities are given, and there is a discussion of the guiding principles for molecular design of host polymers in polymer–salt systems. Potential technological applications of polymer ionic conductors are also described.

As in the case of electronic conduction, the magnitude of ionic conductivity is represented in terms of specific conductivity σ ($S\,cm^{-1}$), or its reciprocal specific resistivity ρ ($\Omega\,cm$). The specific conductivity σ is expressed by Eq. 2.1-13, where n_i represents the concentration of charge-carrier ions of i species per unit volume [cm^{-3}], e_i is the electric charge of the i-species (coulomb), and μ_i is the drift mobility of charge-carrier ions of the i species. For the material to show high ionic conductivity, the values of both n_i and μ_i should be large.

$$\sigma_i = n_i e_i \mu_i \qquad (2.1\text{-}13)$$

It is well known that liquid electrolytes, where free ions produced from electrolytes dissolved in polar solvents move toward the electrode of the opposite sign under an external electric field, have been widely used not only in a variety of electrochemical reactions, but also in practical devices, e.g., batteries. However, materials also exist which show ionic conductivities as high as those of liquid electrolytes at temperatures below their melting points, e.g., β-alumina, AgI, Li_3N, etc. Such materials are usually termed superionic conductors. It was in the 1970s that the new phenomenon that ions can move even in polymer solids was reported. Wright [22, 23] reported that the complexes of alkali metal salts such as NaSCN with poly(ethylene oxide) (PEO) showed high electrical conductivities of over $10^{-5}\,S\,cm^{-1}$ at 80°C. Armand et al. [24] suggested a potential application of polymer—salt complexes as solid electrolytes in solid-state secondary batteries. Since then, there have been many studies on polymer ionic conductors. The social need to develope solid electrolytes for use in solid-state electronic devices has stimulated research interest in ion-conducting polymers. The use of solid electrolytes is expected to lead to the manufacture of thin-film devices as well as to overcome the problems encountered in devices using liquid electrolytes, e.g., solvent leaks. Polymer ionic conductors have received attention not only from the standpoint of technological applications as solid electrolytes, but also from the academic world. This interest focuses on how inorganic salts dissociate into charge carrier ions, and how the ions are transported in solid polymer electrolytes.

Ion-conducting polymers, where ions move in solids as they do in solutions, are referred to as polymer-based solid electrolytes, or solid polymer electrolytes. Compared with inorganic materials such as β-alumina, silver iodide, and lithium iodide, which show high ionic conductivities in the range from 10^{-5} to $1\,S\,cm^{-1}$ at temperatures below their melting points, polymer electrolytes have been found to exhibit much smaller ionic conductivities; the highest values of ionic conductivity achieved to date are of the order of 10^{-4}–$10^{-3}\,S\,cm^{-1}$ in a temperature range from room temperature to ca. 100°C. However, developing polymer electrolytes with high-performance characteristics is an interesting and important subject, and is worth pursuing from both the academic and the practical point of view. Books and reviews articles on ion-conducting polymers are included in the reference list [24–26].

2.1.2.1 General Concepts of Ionic Conduction in Ion-Conducting Polymers

Ion-conducting polymers are classified into single-component systems and composite systems. Single-component polymer ionic conductors include polymers containing covalently bonded salt moieties, e.g., sulfonic and carboxylic acid salts as pendant groups, where cation transport takes place. In composite systems, an inorganic salt, generally an alkali metal salt, is dissolved in a host polymer. Such a composite system is referred to as a polymer—salt hybrid system or a polymer—salt complex, since the formation of complexes of PEO with alkali metal salts has been known for some time.

For the material to exhibit ionic conductivity, charge-carrier ions that move under an external electric field should be present in the system. Mobile charge-carrier ions are generated by the dissociation of salts in both single-component and composite systems. In single-component ion-conducting polymers, the pendant sulfonic acid or carboxylic acid salts dissociate into free ions in the presence of plasticizers or a small amount of water, producing a sulfonic or carboxylic anion bound to the polymer backbone and a counter cation, usually an alkali metal cation, which acts as a charge carrier. Plasticizers or small amounts of water enhance the dissociation of the pendant salt bound to the polymer into free ions. In polymer—salt hybrid systems, host polymers serve as solvents for inorganic salts to dissociate them into charge-carrier ions. That is, charge-carrier ions are generated through the dissociation of inorganic salts caused by the coordination of a host polymer to a salt or by a complex formation between the salt and the host polymer.

The transport of charge-carrier ions in polymer electrolytes contrasts with that in inorganic ionic conductors. Whereas alkali metal cations such as Na^+, Li^+, or Ag^+ move along a specific path which has a tunnel or canal structure resulting from the disorder of the crystal, the transport of charge-carrier ions in polymer ionic conductors is thought to be associated with the segmental motion of host polymers. That is, the motion of charge-carrier ions in polymer electrolytes is strongly coupled to the local structural relaxation of the host polymer. Therefore, high ionic conductivity in polymer electrolytes has been observed only at temperatures above the glass-transition temperature (T_g) of the system, and drops abruptly to a very small value below T_g.

The species and nature of the charge carrier ions and the mechanism of conduction are the subjects of considerable discussion.

2.1.2.2 Characteristic Features of Ionic Conduction in Polymer—Salt Hybrid Systems

A number of recent studies on polymer ionic conductors have dealt with polymer—salt hybrid systems. The hybrid system includes "salt-in-polymer" and "polymer-in-salt" systems. The latter system has been reported very recently and will be described later. The "salt-in-polymer" system has been very extensively studied. The "salt-in-polymer" system can be viewed as a solid solution of an alkali metal salt in the polymer. The formation of charge carrier ions takes place by the dissociation of the salt owing to the coordination of the host polymer to the salt. This section describes some characteristic features of ionic conduction in polymer—salt hybrid systems, particularly "salt-in-polymer" systems.

Species of Charge Carriers. Whereas only a particular ion, either a cation or an anion, is mobile in single-component systems, both cations and anions produced from inorganic salts participate in transport in polymer—salt hybrid systems. In the latter case, the fraction of the ionic current caused by a particular ion in relation to the total ionic current is referred to as the ionic transference number, which is designated by t_+ and t_-. The conductivity is related to the diffusion coefficient (D) by the Nernst–Einstein relation, and t_+ and t_- are expressed as shown in Eq. (2.1-14). There is considerable variation in the reported data on cationic transference numbers. The transference number of a lithium ion in a polymer—lithium salt system has been reported to be less than 0.5. It is therefore suggested that anion transport plays an essential part in ionic conduction in polymer—salt systems where alkali metal cations interact with the ether oxygen of the polymer.

$$t_+ = \frac{\sigma_+}{\sigma_+ + \sigma_-} = \frac{D_+}{D_+ + D_-}, \qquad t_- = \frac{\sigma_-}{\sigma_+ + \sigma_-} = \frac{D_-}{D_+ + D_-} \qquad (2.1\text{-}14)$$

Effect of Morphology on Ionic Conductivity. It is known that PEO is crystalline in nature and forms crystalline complexes with alkali metals such as NaSCN, NaI, LiSCN, and $LiCF_3SO_3$. Extensive studies have been made of the structural analysis of crystalline complexes of PEO with alkali metal salts by infrared, Raman, and NMR spectroscopies, differential scanning calorimetry, and X-ray diffraction. It has been revealed that PEO—alkali metal salt hybrid systems are not homogeneous, and include both the crystalline and elastomeric phases of the polymer itself and of the complex, and phase diagrams have been published [27].

It has been shown that the ionic conductivity of PEO—alkali metal salt complexes is greatly affected by their morphology [28], and that the transport of ions takes place preferentially in the amorphous phase of the complex [29].

Ionic Conductivity as a Function of Salt Concentration. The ionic conductivity of a polymer—salt hybrid system as a function of the concentration of the salt has been shown to increase with increasing concentrations of the salt, reaching a maximum at a particular mole percentage of the salt, and then tending to decrease with further increases in the salt concentration. These results are understood to be a consequence of the combined effects of an increase in the concentration of charge-carrier ions and hindered segmental motions of the host polymer with an increasing concentration of the salt. That is, the increase in the salt concentration leads to an increase in the concentration of charge-carrier ions, but at the same time tends to hinder segmental motions of the host polymer due to the complexation with, or coordination to, the inorganic salt, as suggested from the fact that the T_g of the complexed polymer increases with increasing concentrations of the inorganic salt. Thus, a problem arises from the fact that any attempt to increase the concentration of charge-carrier ions is accompanied by the hindrance of the ion motion in the polymer—salt hybrid system.

Temperature Dependence of Ionic Conductivity.   The temperature dependence of the ionic conductivity of polymer—salt hybrid systems shows characteristic features associated with the segmental motion of host polymers. In general, the ionic conductivity of polymer—salt hybrid systems is greatly affected by temperature, and the ionic conductivity increases strongly with temperature. This is mainly because the chain segmental mobility of host polymers is affected by temperature. In some systems consisting of PEO and inorganic salts such as NaI, where crystalline complexes are formed, the temperature dependence of the ionic conductivities has been shown to follow an Arrhenius-type relation [23]: there exist two temperature regions with different activation energies, and a transition from a higher activation energy ($25–32\,\mathrm{kcal\,mol^{-1}}$) to a lower one ($12–14\,\mathrm{kcal\,mol^{-1}}$) takes place above T_g. However, in many other systems consisting of a fully amorphous phase, the plots of log σ vs. T^{-1} are usually represented by a continuous curve instead of a straight line. This is understood to indicate that the transport of ions is associated with the segmental motions of the amorphous host polymers at temperatures above T_g.

The behavior of continuous curves in the plots of log σ vs. T^{-1}, which has been observed for many amorphous polymer—salt systems, is described by the Vogel–Tamman–Fulcher (VTF) equation or by the Williams–Landel–Ferry (WLF) equation. The theoretical treatment of the temperature dependence of ionic conductivities is fully discussed in the literature [24, 25].

The VTF equation, which was developed to deal with the temperature—viscosity relationship for supercooled liquids, is expressed as Eq. 2.1-15, using the Stokes–Einstein relationship and the Nernst–Einstein relationship.

$$\sigma = A T^{-1/2} \exp\left(-\frac{E_a}{T-T_0}\right) \tag{2.1-15}$$

where A is a preexponential term, T_0 is a temperature which can be associated with the T_g of the polymer electrolyte, and E_a is a pseudo-activation energy related to the configurational entropy of the polymer chain.

The WLF equation, which relates any mechanical relaxation process of polymers to the T_g of the polymer, is expressed as Eq. 2.1-16.

$$\log\left\{\frac{\eta(T)}{\eta(T_S)}\right\} = -\frac{C_1(T-T_S)}{C_2+(T-T_S)} \qquad (2.1\text{-}16)$$

where T_s is an arbitrary reference temperature, and C_1 and C_2 are "universal" constants. By using the Stokes–Einstein relationship and the Nernst–Einstein relationship, Eq. 2.1-16 is transformed into Eq. 2.1-17.

$$\log\left\{\frac{\sigma(T)}{\sigma(T_g)}\right\} = -\frac{C_1(T-T_g)}{C_2+(T-T_g)} \qquad (2.1\text{-}17)$$

Both the VTF and WLF equations are empirical equations, and they are essentially the same. The curved behavior of the plots of log. σ vs. T^{-1} for polymer electrolytes fits well with the VTF or the WLF equation. The temperature dependence of the ionic conductivity of polymer—salt hybrid systems has been shown to be the same as the temperature dependence of the free volume determined from polymer dynamics.

2.1.2.3 Host Polymers in Polymer—Salt Hybrid Systems

To develop polymer—salt hybrid ionic conductors with good performance characteristics, molecular design of host polymers are of crucial importance. In order to achieve high ionic conductivity in polymer—salt hybrid systems, it is necessary to design host polymers that enhance both the generation of charge-carrier ions and their transport.

There are five main requirements for host polymers in a hybrid system. (i) The host polymer should dissolve as much of the inorganic salts as possible. Since charge-carrier ions are generated from the salt by interaction with the host polymer, the host polymer should contain a high concentration of coordinating groups capable of solvating or complexing with inorganic salts to dissociate them into charge-carrier ions. (ii) The host polymer should be flexible in order to be able to solvate inorganic salts efficiently. (iii) Since charge-carrier ions are transported by the aid of the segmental motions of the host polymer, high ionic conductivity is observed only in an elastomeric phase above T_g. Therefore, the host polymer should have a low glass-transition temperature in order to facilitate the transport of charge-carrier ions over a wide temperature range. (iv) It has been revealed that ionic conduction is favored in the amorphous region of the host polymer; therefore, amorphous polymers are preferable to crystalline polymers as host polymers. (v) When polymer electrolytes are used in devices, they should meet the requirements of both electrochemical and dimensional stability.

A number of recent studies on polymer—salt hybrid systems have shown that the ionic conductivity of organic polymer—inorganic salt hybrid systems is en-

$\text{---}(\text{CH}_2\text{CH}_2\text{O})_n\text{---}$

Poly(ethylene oxide)
(PEO)

$\text{---}(\text{CH}_2\overset{\overset{\text{CH}_3}{|}}{\text{CHO}})_n\text{---}$

Poly(propylene oxide)
(PPO)

$\text{---}(\text{OC}_2\text{H}_4\text{O}-\underset{\overset{\|}{\text{O}}}{\text{C}}-\text{C}_2\text{H}_4-\underset{\overset{\|}{\text{O}}}{\text{C}})_n\text{---}$

Poly(ethylene succinate)
(PESc)

$\text{---}(\text{CH}_2\text{CH}_2\underset{\overset{\|}{\text{O}}}{\text{CO}})_n\text{---}$

Poly(β-propiolactone)
(PPL)

$\text{---}(\text{CH}_2\text{CH}_2\text{S})_n\text{---}$

Poly(ethylene sulfide)
(PES)

$\text{---}(\text{CH}_2\text{CH}_2\text{NH})_n\text{---}$

Poly(ethylene imine)
(PEI)

$\text{---}(\text{CH}_2-\text{CH})_n\text{---}$
$|$
$\text{C}=\text{O}$
$|$
O
$|$
$(\text{CH}_2)_2$

Poly(tetrahydrofurfryl acrylate)
(PTHFA)

$\text{---}(\overset{|}{\underset{|}{\text{Si}}}-\text{O})_n\text{---}$

Poly{3-[2,3-(carboxyldioxy)popoxy]-
propylmethylsiloxane}
(PCPMS)

hanced by the use of inorganic salts with low lattice energy, and also by the use of an amorphous polymer with a low glass-transition temperature and with a high concentration of coordinating groups.

A number of polymers containing coordinating groups have been studied as candidates to be host polymers in polymer—salt hybrid ionic conductors [26, 30, 31]. Oxygen, sulfur, or nitrogen atoms function as coordinating atoms. Host polymers studied include polyethers such as poly(ethylene oxide) (PEO) and poly(propylene oxide) (PPO), polyesters such as poly(ethylene succinate) (PESc) and poly(β-propiolactone) (PPL), poly(alkylene sulfide)s, poly(ethylene imine) (PEI), and polymers having pendant organic solvent moieties such as poly(tetrahydrofurfryl acrylate) (PTHFA) [30], and polysiloxanes bearing cyclic carbonate pendant groups (PCPMS) [31].

Since PEO and polyesters are crystalline in nature, comb-shaped polymers having short-chain ether or ester oligomers as the side chain have been prepared to make amorphous polymers, and the ionic conductivities of the polymer–salt systems have been investigated [26, 32]. A variety of comb-shaped polymers have been synthesized, which include, e.g., polyacrylates and their analogues, poly-phosphazenes, and polysiloxanes containing ether or ester groups as side chains.

$$\left(CH_2-\underset{\underset{C=O}{\overset{CH_3}{|}}}{C}\right)_n$$
$$O-\left(CH_2CH_2O\right)_m-CH_3$$

Polymethacrylate bearing oligo(oxyethylene) side chain
(PEOM)

$$\left(CH_2-\underset{\underset{C=O}{\overset{CH_3}{|}}}{C}\right)_n$$
$$O-\left(C_2H_4O-\overset{\overset{O}{\|}}{C}-C_2H_4-\overset{\overset{O}{\|}}{C}-O\right)_m-CH_3$$

m=1: poly(2-methacryloxyethyl methyl succinate)
(P3EMA)

m=2: poly(3,8,11,16-tetraoxa-4,7,12,15-tetraoxo-
heptadecanyl methacrylate)
(P5EMA)

$$O=C-O-\left(CH_2CH_2O\right)_m-CH_3$$
$$\left(CH_2-\underset{\underset{O=C}{|}}{C}\right)_n$$
$$O-\left(CH_2CH_2O\right)_m-CH_3$$

Polyitaconate bearing oligo(oxyethylene) side chain
(PEOI)

$$OCH_2CH_2OCH_2CH_2OCH_3$$
$$\left(N=\underset{|}{\overset{|}{P}}\right)_n$$
$$OCH_2CH_2OCH_2CH_2OCH_3$$

Polyphosphazene bearing oligo(oxyethylene) side chain
(PEOP)

$$\left(\underset{|}{\overset{CH_3}{Si-O}}\right)_n$$
$$O-\left(CH_2CH_2O\right)_m-CH_3$$

Polysiloxane bearing oligo(oxyethylene) side chain
(PEOS)

When using solid polymer electrolytes in solid-state electronic devices, long-term dimensional stability is required. To achieve this aim, the cross-linking of linear and comb polymers has been investigated [33, 34]. The degree of cross-linking is kept low to allow segmental chain motions in the host polymer.

2.1.2.4 Polymer—Salt Hybrid Ionic Conductors and Their Ionic Conductivities

Polyethers, Polyesters, Poly(alkylene sulfide)s, and Poly(ethylene imine) as Host Polymers. Poly(ethylene oxide) (PEO)-based complexes were the first known solvent-free polymer electrolytes and have been studied intensively. PEO–lithium salt complexes have been reported to exhibit high ionic conductivities ($\sigma \geq 10^{-5} \, \mathrm{S\,cm^{-1}}$ at ca. 100°C) compared with other polymer–salt hybrid systems; however, their conductivities at ambient temperatures are relatively low ($\sigma \approx 10^{-7} \, \mathrm{S\,cm^{-1}}$). This is because PEO forms both crystalline and amorphous complexes with alkali metal salts, showing complicated morphologies [24], and because ionic conduction in the amorphous region is mainly responsible for the high conductivity [29].

Linear polyesters such as poly(ethylene succinate) are also crystalline in nature, but can form amorphous complexes with alkali metal salts, showing complicated morphologies. The complexes of linear polyesters such as poly(ethylene succinate) and poly(β-propiolactone) with alkali metal salts have also been reported to show relatively high conductivities [35]. The ionic conductivities reported for some polymer—salt complexes are listed in Table 2.1-1.

TABLE 2.1-1. Ionic conductivities of some polymer–salt complexes

Polymer	Salt	[Salt]/[repeat unit]	Conductivity (S cm^{-1})
Poly(ethylene succinate)			
$M = 1700$	LiBF$_4$	0.33	3×10^{-6} at 65°C
$M_n = 23\,000$	LiClO$_4$	0.12	10^{-5} at 90°C
Poly(β-propiolactone)			
$M_n = 19\,000$			
Annealed	LiClO$_4$	0.10	4×10^{-5} at 70°C
Quenched	LiClO$_4$	0.10	4×10^{-4} at 70°C
Poly(pentamethylene sulfide)			
$M = 10\,000$	AgNO$_3$	0.25	9×10^{-8} at 45°C
			8×10^{-6} at 63°C
Poly(hexamethylene sulfide)			
$M = 7000$	AgNO$_3$	0.25	5×10^{-5} at 66°C
Poly(ethylene imine)			
$M = 2000$	NaI	0.10	2×10^{-6} at 60°C
$M = 100\,000$	NaSO$_3$CF$_3$	0.17	2×10^{-7} at 41°C
			5×10^{-5} at 94°C

Comb Polymers as Host Polymers. One way of producing amorphous complexes of polymers containing ether or ester units with alkali metal salts is the use of comb polymers containing oligo(oxyethylene)s and oligo(ester)s as the side-chain group. A variety of such polymers have been synthesized and found to be a new type of host polymer that forms amorphous complexes with a variety of metal salts.

Polymethacrylates bearing oligo(oxyethylene) side chains have been prepared. When the side-chain length increases, the polymers tend to crystallize. The ionic conductivities of $LiSO_3CF_3$ complexes with poly(methoxy polyethyleneglycol methacrylate) ($n = 22$) (PEM22) are $\sim 6 \times 10^{-4}\,S\,cm^{-1}$ at 100°C and $\sim 2 \times 10^{-5}\,S\,cm^{-1}$ at 20°C [36]. When these values are compared with those obtained for PEO–$LiSO_3CF_3$ complexes with similar salt concentrations, the conductivity data are similar at temperatures higher than 70°C, but at 20°C the values of the conductivities for the PEM22–$LiSO_3CF_3$ complexes are approximately two orders of magnitude higher than those for the PEO–$LiSO_3CF_3$ complex. The complexes of $LiClO_4$ with polymethacrylates having a side chain of oligo(oxyethylene) also exhibit ionic conductivities of $2.2 \times 10^{-5}\,S\,cm^{-1}$ at 25°C in a composition of polymer/$LiClO_4$ (70/30) [37].

A series of poly(itaconic acid) esters with oligo(ethylene oxide) side chains have been synthesized, where the side chains contain from one to seven ethylene oxide units. The polymer with seven ethylene oxide units per side chain produces amorphous complexes with $LiClO_4$ and $NaClO_4$ with the highest conductivities between 10^{-3} and $10^{-6}\,S\,cm^{-1}$ in the temperature range from 400 K to ambient [38].

A new family of comb polymers with varying lengths of side chains containing ester groups have been synthesized, and the ionic conductivities of hybrid films of these polymers with $LiSO_3CF_3$ have been studied as a function of salt concentration and temperature [32]. These polymers are amorphous and show good compatibility with $LiSO_3CF_3$. Ionic conductivities of the hybrid films have been found to increase with increasing side-chain length, and the $LiSO_3CF_3$ complexes of poly(3,8,11-trioxa-4,7,12-trioxotridecanyl methacrylate) and poly(3,8,11,16-tetraoxa-4,7,12,15-tetraoxoheptadecanyl methacrylate) exhibit higher conductivities than a linear polyester, poly(ethylene succinate). The temperature dependence of ionic conductivities has been shown to be in accord with that of the dielectric relaxation of host polymers.

It is known that some polyphosphazenes exhibit very low T_g. In addition, different organic side groups are readily incorporated into the macromolecular structure of polyphosphazenes. In view of these properties, poly(phosphazene) (MEEP) has been synthesized by the reaction of poly(dichlorophosphazene) with the sodium salt of 2-(2-methoxyethyl)ethanol. The $(AgSO_3CF_3)_x$MEEP complexes with T_g lower than −35°C show high ionic conductivities of ca. 10^{-2}–$10^{-3}\,S\,cm^{-1}$ at 70°C. The transference number for Ag^+ is 0.03 or less at 50°C [39]. The complex $(LiSO_3CF_3)_{0.25}$MEEP exhibits a conductivity 2.5 orders of magnitude larger than the corresponding PEO complex at room temperature.

The $LiClO_4$ complexes of polysiloxane comb polymers bearing ethylene oxide oligomers as side chains have been reported to exhibit room-temperature ionic conductivities greater than $10^{-4}\,S\,cm^{-1}$.

Cross-Linked Polymers as Host Polymers. For applications of solid polymer electrolytes, dimensional stability is required as well as high ionic conductivity. Cross linking is one method of improving the mechanical properties of host polymers. In addition, cross linking helps to reduce the degree of crystallinity of polymers.

Chemically cross-linked polyphosphazenes have been reported to show a marked increase in dimensional stability compared with the linear analogue. The polymers form complexes with $LiSO_3CF_3$ which exhibit high ionic conductivities of ca. $4.0 \times 10^{-5} S cm^{-1}$ at 30°C and ca. $1.0 \times 10^{-4} S cm^{-1}$ at 70°C [33].

A cross-linked siloxane-based polymer which has been prepared by the reaction of poly(methylhydrosiloxane), poly(ethylene glycol) monomethyl ether, and poly(ethylene glycol) forms complexes with $LiSO_3CF_3$ that exhibit good ionic conductivities up to $7.3 \times 10^{-5} S cm^{-1}$ at 40°C for a $LiSO_3CF_3$ 15 wt% complex [40].

Cross-linked comb-branch polymers based on polyacrylates with oligo(oxyethylene) side chains using divinyl ether as a cross-linking agent have been prepared. They exhibit conductivities comparable to those of the linear analogues when the cross-link density is kept at ~5%. Maximum conductivities of $\sigma = 10^{-4} \sim 1.58 \times 10^{-3} S cm^{-1}$ at $T = 30–100$°C have been achieved when $n = 5$ for a 5% cross-linked sample complexed with $LiClO_4$ at the composition $[Li^+]/[EO] = 0.05$ [34].

Plasticized Polymer—Salt Systems. The ionic conductivity of polymer–salt hybrid systems can be significantly improved by the addition of plasticizing agents with high dielectric constants for the following reasons. The addition of plasticizing agents decreases the crystallinity, hence increasing the mobility of the ions at ambient temperature and enhancing salt dissociation. The plasticizers need to have good miscibility with polymer—salt systems, low volatility, and stability toward electrode materials.

When a plasticizing salt $LiN[CF_3SO_2]_2$ and the solvent diethyl phthalate were incorporated into free-standing solution-cast films of $(PEO)_x(LiSO_3CF_3)$, the conductivity increased from $7.7 \times 10^{-7} S cm^{-1}$ at 20°C in the absence of the plasticizers to $4.6 \times 10^{-5} S cm^{-1}$ in the presence of both plasticizers [41]. Adding 50 wt% of a modified carbonate as a plasticizer to the $PEO–LiSO_3CF_3$ complex yielded an ionic conductivity of $5 \times 10^{-5} S cm^{-1}$ at 25°C, which is two orders of magnitude higher than that found for a $PEO–LiSO_3CF_3$ electrolyte without a plasticizer [42]. A conductivity of ca. $10^{-3} S cm^{-1}$ at room temperature has been achieved for other systems in the presence of plasticizers.

2.1.2.5 Potential Applications of Ion—Conducting Polymers

Polymers have excellent mechanical properties and allow the manufacture of thin films. Thus, ion-conducting polymers are expected to find potential applications as solid electrolytes for use in miniaturized, light-weight, thin-film devices such as primary and secondary batteries, electrochromic devices, sensors, etc. In these devices, solid polymer electrolytes, which are sandwiched between two electrodes, function as both a separator and an electrolyte. Only ionic conduction should take place in the polymer electrolyte without the participation of electronic conduction.

Solid-state secondary batteries using various electrode materials, including electronically conducting polymers such as polyacetylene and polypyrrole, and solid polymer electrolytes such as PEO–alkali metal salt complexes have been made and their performance characteristics have been examined.

Solid electrolytes for use in batteries should meet the requirements of electrochemical stability, a high conductivity to allow a reasonable current density, compatibility with electrode materials and other components with which they are in contact, and good thermal stability [43]. The complexes of PEO with alkali metal salts have proven to function as useful electrolytes above 80°C.

Solid-state electrochromic devices, using a variety of electrochromic materials and solid polymer electrolytes, which are of considerable interest in view of their application in display devices and smart windows, have been made and their performances examined. For smart window applications, highly conductive solid polymer electrolytes should meet the requirements of transparency, a wide potential window, and fast and reversible responses [44].

2.1.2.6 Towards Development of New Types of Solid Polymer Electrolytes

As a result of extensive studies of "salt-in-polymer" systems, values of ionic conductivities as high as ca. $10^{-3}\,S\,cm^{-1}$ at 70–100°C and ca. $10^{-4}\,S\,cm^{-1}$ at room temperature have been obtained for some solvent-free solid polymer electrolytes without the use of plasticizers. However, the maximum conductivity is still inadequate for most practical purposes. Developing solvent-free solid polymer electrolytes with a conductivity higher than $10^{-3}\,S\,cm^{-1}$ at room temperature will be a target for future studies. New types of host polymers based on novel molecular designs, and new concepts for developing solvent-free polymer electrolytes with high conductivities should be developed. A few examples of recent studies on the development of solvent-free polymer electrolytes with high conductivities are given here.

Polymer blends, e.g., blends of PEO with polyacrylamide (PAAM), have been used as host polymers, where the high molecular weight PAAM inhibits the crystallization of PEO without impeding the segmental motion. It has been shown that the ionic conductivity is enhanced in the blends compared with that in the PEO–LiClO$_4$ complex. Conductivities exceeding $10^{-4}\,S\,cm^{-1}$ at room temperature have been obtained for electrolytes prepared by the in situ polymerization of acrylamide in the polymer [45].

A new type of polymer, referred to as a polycascade polymer, e.g., polystyrene with pendant penta[oligo(oxyethylene)]cyclotriphosphazenes (poly(STEP) n = 3), has been synthesized. The maximum conductivities of poly(STEP)–Li salt complexes are $1.1 \times 10^{-4}\,S\,cm^{-1}$ at 60°C and $5.1 \times 10^{-4}\,S\,cm^{-1}$ at 100°C at [Li$^+$]/O = 0.05. It has been suggested that the density of oxygen atoms around the backbone is one of the important factors for obtaining fast ion transport [46].

Polymer complexes consisting of poly(pyridinium) salt, pyridinium salt, and aluminum chloride have been found to be a new class of highly conductive polymer electrolytes [47].

To date, studies of solvent-free solid polymer electrolytes have focused on "salt-in-polymer" systems, where alkali metal salts are dissolved in host polymers. However, these systems have a problem, because any attempt to increase the concentration of charge-carrier ions by increasing the content of alkali metal salts is accompanied by an increase in the T_g of the complex; this leads to a hindrance of ion motion. Recently, a new concept, "salt-in-polymer" complexes, has been presented to develop solid polymer electrolytes. Lithium salts in appropriate mixtures are combined with small quantities of the polymers PEO and propylene oxide (PPO). These materials have a T_g low enough to remain rubbery at room temperature while preserving good lithium-ion conductivity and high electrochemical stability. For example, a conductivity of ~$10^{-4}\,S\,cm^{-1}$ at ambient temperature has been obtained for a LiI–LiClO$_3$–LiClO$_4$–PPO system [48]. It is expected that future studies will lead to the development of high-performance solid polymer electrolytes that have the desired electrical properties, electrochemical stability, and mechanical properties such as dimensional stability.

In contrast to solvent-free solid polymer electrolytes, gels that contain a large amount of solvent also constitute a new class of promising electrolyte material. For example, the electrolytes are composed of Li salt-solvates of certain organic solvents immobilized in a polymer network of polyacrylonitrile. These electrolytes are dimensionally stable and have a high ionic conductivity of around $10^{-3}\,S\,cm^{-1}$ at room temperature [49, 50].

The gel system is not confined to polymers, but can also be extended to low molecular-weight organic systems. 4,4′,4″-Tri(stearoylamino)triphenylamine (TSATA) and N,N′,N″-tristearyltrimesamide (TSTA) have been found to form gels with a variety of organic solvents, immobilizing solvent at a very low concentrations of TSATA and TSTA [51, 52].

These molecular gels containing tetra-n-butylammonium perchlorate (n-Bu$_4$NClO$_4$) ($1.0 \times 10^{-1}\,mol\,dm^{-3}$) have been found to exhibit high ionic conductivities of around ~$10^{-3}\,S\,cm^{-1}$, which are comparable to those of the corresponding solutions of n-Bu$_4$NClO$_4$ [51, 52]. The TSATA molecular gels containing n-Bu$_4$NClO$_4$ function as electrochromic materials, showing a clear reversible color change on electrochemical oxidation and reduction, and turning green and colorless, respectively [51].

4,4',4''-Tri(stearoylamino)triphenylamine
(TSATA)

N,N',N''-tristearyltrimesamide
(TSTA)

References

1. Iijima S (1991) Helical microtubules of graphitic carbon. Nature 354:56–58
2. Bacon R (1960) Growth, structure, and properties of graphite whiskers. J Appl Phys 31:283–290
3. Iijima S, Ichihashi T (1993) Single-shell carbon nanotubes of 1 nm-diameter. Nature 363:603–615
4. Uchida K, Yumura M, Ohshima S, Kuriki Y, Yase K, Ikazaki F. Proceedings of the 5th General Symposium on C_{60}, 5–6, August 1993, Hachioji
5. Olk CH, Heremans JP (1994) Scanning tunneling spectroscopy of carbon nanotubes. J Mater Res 9:259–262
6. Okahara K, Tanaka K, Aoki H, Sato T, Yamabe T (1994) Band structures of carbon nanotubes with bond-alternation patterns. Chem Phys Lett 219:462–468
7. Tanaka K, Sato T, Yamabe T, Okahara K, Uchida K, Yumura M, Niino H, Ohshima S, Kuriki Y, Yase K, Ikazaki F (1994) Electronic properties of carbon nanotube. Chem Phys Lett 223:65–68
8. Hiura H, Ebbesen TW, Tanigaki T, Takahashi H (1993) Raman studies of carbon nanotubes. Chem Phys Lett 202:509–512
9. Jishi RA, Venkataraman L, Dresselhaus MS, Dresselhaus G (1993) Phonon modes in carbon nanotubules. Chem Phys Lett 209:77–82
10. Tanaka K, Kobashi M, Sanekata H, Yamabe T, Yamauchi J, Yata S (1991) Magnetic susceptibility and magnetization measurements of polyacenic semiconductive materials. Phys Rev B43:8277–8281
11. Langer L, Stockman L, Heremans JP, Bayot V, Olk CH, Van Haesendonck C, Bruynseraede Y, Issi JP (1994) Electrical resistance of a carbon nanotube bundle. J Mater Res 9:927–932
12. Singer LS, Wagoner G (1962) Electron spin resonance in polycrystalline graphite. J Chem Phys 37:1812–1817
13. Pietronero L, Tosatti E (eds) (1981) Physics of intercalation compounds. Springer, Berlin (Solid-state physics, vol 38)
14. Tanaka K, Ohzeki K, Yamabe T, Yata S (1984) A study on the pristine and the doped polyacenic semiconductive materials. Synth Met 9:41–52

15. Lauginie P, Estrade H, Conard J, Guerard D, Lagrange P, Makrini ME (1980) Graphite lamellar compounds EPR studies. Physica 99B:514–520
16. Tanaka K, Koike T, Nishino H, Yamabe T, Yamauchi J, Deguchi Y, Yata S (1987) ESR study of in situ doped polyacenic semiconductive material with iodine and bromine. Synth Met 18:521–526
17. Kosaka M, Ebbesen TW, Hiura H, Tanigaki K (1994) Electron spin resonance of carbon nanotubes. Chem Phys Lett 225:161–164
18. Tanaka K, Ohzeki K, Nankai S, Yamabe T, Shirakawa H (1983) The electronic structures of polyacene and polyphenanthrene. J Phys Chem Solids 44:1069–1075
19. Tanaka K, Koike T, Ueda K, Ohzeki K, Yamabe T, Yata S (1985) Electronic structures of polyacene and polyphenanthrene. Design of one-dimensional graphite. Synth Met 11:61–73
20. Tanaka K, Okahara K, Okada M, Yamabe T (1993) Why some bucky tubes would be metallic? Fullerene Sci Tech 1:137–144
21. Yamabe T, Okahara K, Okada M, Tanaka K (1993) Electronic properties of buckytube model. Synth Met 55–57:3142–3147
22. Wright PV (1975) Electrical conductivity in ionic complexes of poly(ethylene oxide). Br Polym J 7:319–327
23. Wright PV (1976) An anomalous transition to a lower activation energy for dc electrical conduction above the glass-transition temperature. J Polym Sci, Polym Phys Ed 14:955–957
24. Armand MB, Chabagno JM, Duclot M (1979) Polyether as solid electrolytes. In: Vashisha P, Mundy JN, Shenoy GK (eds) Fast ion transport in solids. Elsevier North-Holland, Amsterdam, pp 131–136
25. MacCallum JR, Vincent CA (eds) (1987) Polymer electrolyte reviews, vol 1. Elsevier Applied Science, London
26. Ratner MA, Shriver DF (1988) Ion transport in solvent-free polymers. Chem Rev 88:109–124
27. Robitaille CD, Fauteux D (1986) Phase diagrams and conductivity characterization of some PEO–LiX electrolytes. J Electrochem Soc 133:315–325
28. Payne DR, Wright PV (1982) Morphology and ionic conductivity of some lithium ion complexes with poly(ethylene oxide). Polymer 23:690–693
29. Berthier C, Gorecki W, Minier M, Armand MB, Chabagno JM, Rigaud P (1983) Microscopic investigation of ionic conductivity in alkali metal salts—poly(ethylene oxide) adducts. Solid State Ionics 11:91–95
30. Takebe Y, Shirota Y (1994) Poly(tetrahydrofurfryl acrylate) as a new host polymer for polymer–salt hybrid ionic conductors. Solid State Ionics 68:1–4
31. Zhu Z, Einset AG, Yang CY, Chen WX, Wnek GE (1994) Synthesis of polysiloxanes bearing cyclic carbonate side chains. Dielectric properties and ionic conductivities of lithium triflate complexes. Macromolecules 27:4076–4079
32. Takebe Y, Hochi K, Matsuba T, Shirota Y (1994) Synthesis of a new family of comb polymers with side-chain esters and ionic conductivities of their films containing lithium trifluoromethane sulfonate. J Mater Chem 4:599–604
33. Tonge JS, Shriver DF (1987) Increased dimensional stability in ionically conducting polyphosphazenes systems. J Electrochem Soc 134:269–270
34. Cowie JMG, Sadaghianizadeh K (1990) Effect of side chain length and crosslinking on the ac conductivity of oligo(ethylene oxide) comb-branch polymer–salt mixture. Solid State Ionics 42:243–249

35. Watanabe M, Togo M, Sanui K, Ogata N, Kobayashi T, Ohtaki Z (1984) Ionic conductivity of polymer complexes formed by poly(β-propiolactone) and lithium perchlorate. Macromolecules 17:2908–2912
36. Bannister DJ, Davies GR, Ward IM, McIntyre JE (1984) Ionic conductivities of poly(methoxy polyethylene glycol monomethacrylate) complexes with LiSO₃CH₃. Polymer 25:1600–1602
37. Kobayashi N, Uchiyama M, Shigehara K, Tsuchida E (1985) Ionically high conductive solid electrolytes composed of graft copolymer–lithium salt hybrid. J Phys Chem 89:987–991
38. Cowie JMG, Martin ACS (1991) Ionic conductivity in oligo(ethylene oxide) esters of poly(itaconic acis)—salt mixtures: effect of side-chain length. Polymer 32:2411–2417
39. Blonsky PM, Shriver DF, Austin P, Allcock HR (1984) Polyphosphazene solid electrolytes. J Am Chem Soc 106:6854–6855
40. Spindler R, Shriever DF (1988) Synthesis, NMR characterization, and electrical properties of siloxane-based polymer electrolytes. Macromolecules 21:648–654
41. Walker Jr CW, Salomon M (1993) Improvement of ionic conductivity in plasticized PEO-based solid polymer electrolytes. J Electrochem Soc 140:3409–3412
42. Lee HS, Yang XQ, McBreen J, Xu ZS, Skotheim TA, Okamoto Y (1994) Ionic conductivity of a polymer electrolyte with modified carbonate as a plasticizer for poly(ethylene oxide). J Electrochem Soc 141:886–889
43. Koksbang R, Olsen II, Shackle D (1994) Review of hybrid polymer electrolytes and rechargeable lithium batteries. Solid State Ionics 69:320–335
44. Andrei M, Roggero A, Marchese L, Passerini S (1994) Highly conductive solid polymer electrolyte for smart windows. Polymer 35:3592–3597
45. Wieczorek W, Such K, Florjanczyk Z, Stevens JR (1994) Polyether, polyacrylamide, LiClO₄ composite electrolytes with enhanced conductivity. J Phys Chem 98:6840–6850
46. Inoue K, Nishikawa Y, Tanigaki T (1991) High-conductivity electrolytes composed of polystyrene carrying pendant oligo(oxyethylene)cyclotriphosphazenes and LiClO₄. J Am Chem Soc 113:7609–7613
47. Watanabe M, Yamada S, Sanui K, Ogata N (1993) High ionic conductivity of new polymer electrolytes consisting of polypyridinium, pyridinium and aluminium chloride. J Chem Soc Chem Commun 929–931
48. Angell CA, Liu C, Sanchez E (1993) Rubbery solid electrolytes with dominant cationic transport and high ambient conductivity. Nature 362:137–139
49. Abraham KM, Alamgir M (1990) Li⁺-conductive solid polymer electrolytes with liquid-like conductivity. J Electrochem Soc 137:1657–1658
50 Croce F, Gerace F, Dautzemberg G, Passerini S, Appetecchi GB, Scrosati B (1994) Synthesis and characterization of highly conducting gel electrolytes. Electrochim Acta 39:2187–2194
51. Yasuda Y, Takebe Y, Fukumoto M, Inada H, Shirota Y (1996) 4,4′,4″-Tris(stearoylamino)triphenylamine as a novel material for functional molecular gels. Adv Mater 8:740–741
52. Yasuda Y, Iishi E, Inada H, Shirota Y (1996) Novel low-molecular-weight organic gel: N,N′,N″-tristearyltrimesamide/organic solvent system. Chem Lett 575–576

2.2 Organized Molecular Assemblies

Toyoki Kunitake (2.2.1)
Masamichi Fujihira (2.2.2)

2.2.1 Bilayer Membranes

2.2.1.1 Biological Membrane and Phospholipid Bilayer

Biological membranes are a supramolecular system that is one of the most important structural units in the organization of the biological cell. Their biological role is not limited to being a physical barrier to maintain intracellular environments. Many fundamental biological functions such as selective materials transport, energy conversion, and processing of biological information are performed on the biomembrane. Therefore, the biomembrane is a most useful example in attempts to construct artificial molecular systems.

Attempts to mimic the organization of biomembranes have been conducted using biolipids and related compounds [1]. In 1964, Bangham and Horne demonstrated that a bilayer structure is formed spontaneously from aqueous dispersions of egg yolk lecithin [2]. This finding established that among the constituents of biomembranes, the lipid bilayer plays a central role in maintaining biological molecular organization, and led to extensive efforts to mimic biomembranes. A typical example is that of Gebicki and Hicks [3], who observed the formation of globular aggregates, which they called ufasomes, from dispersed thin films of oleic acid and linoleic acid. Vesicular aggregates have also been described for 1:1 mixtures of saturated fatty acids of C_{12} to C_{20} chains and single-chain lysolecithin [4]. However, these aggregates were not sufficiently stable, and definite evidence for the existence of an isolated bilayer structure has not yet been obtained.

2.2.1.2 Synthetic Bilayer Membranes

The physicochemical basis of why some polar lipid molecules derived from biomembranes form two-dimensionally aligned assemblies has been intensively invertigated [5, 6]. Brockerhoff considered the structural features of membrane lipids in terms of a tripartite structure: a hydrophobic alkyl chain, a hydrophilic head group, and the so-called hydrogen belt region that connects these two moieties [7]. He stressed the importance of the hydrogen belt, which would stabilize the two-dimensional molecular packing. According to the proposition of Tanford, the bilayer formation is explicable by geometric consideration of the volume of the hydrophobic core and the surface area occupied by the head group in membrane lipids [8]. In the same vein, Israelachvili, Mitchell, and Ninham emphasized the role of molecular geometry, asserting that the packing constraints of component molecules decide the surface curvature that determines the aggregate morphology [9]. These theories are useful for a rationalization of the formation of biolipid bilayers. Unfortunately, they cannot provide guiding rules that enable us to design novel bilayer systems.

In initiating the molecular design of bilayer-forming compounds, we made the assumption that the unique structure of the polar head group of biolipid molecules is determined by biosynthetic and physiological requirements rather than by the physical chemistry of membrane formation. It was also important to make a clear distinction between liquid crystalline dispersion and the bilayer membrane. As pointed out earlier by Gray and Winsor [10], the physical properties (such as viscosity and stability) of the mesophases are determined by the intermiceller forces, i.e., the lattice forces, rather than by forces resulting from joining or close packing. A bilayer membrane should be able to exist as an isolated entity without relying on the lattice force to keep its structural integrity. Thus, the formation of bilayer membranes requires a self-assembling tendency greater than that of liquid crystalline dispersion.

The first substance we chose to test was didodecyldimethylammonium bromide $2C_nN^+$ [11]. This compound and related double-chain ammonium halides have been known to show exceptional lyotropic liquid crystalline behavior in water. The didodecyl derivative $2C_nN^+$ gives clear aqueous dispersions by sonication or by heating. When this dispersion was negatively stained by uranyl acetate and observed by electron microscopy, single-walled and multiwalled vesicles with a layer thickness of 30–50 Å were found. Their aggregation characteristics, such as critical aggregate concentration and molecular weight, were consistent with the existence of bilayer vesicles. Other physicochemical measurements pointed to the same conclusion. This finding was the first example of a totally synthetic bilayer membrane, and it became clear that two-dimensional molecular alignment was not restricted to biolipid molecules.

$$CH_3(CH_2)_{n-1} \overset{+}{\underset{CH_3(CH_2)_{n-1}}{N}} \overset{CH_3}{\underset{CH_3}{}} \quad Br^-$$

$$2C_nN^+$$

2.2.1.3 Molecular Design

Subsequent investigations revealed that bilayer formation can be observed in a variety of synthetic amphiphiles. Their (now conventional) molecular structures are derived from combinations of molecular modules, as illustrated in Fig. 2.2-1.

$$CH_3(CH_2)_{n-1}-O$$
$$CH_3(CH_2)_{n-1}-O \overset{O}{\underset{O^-}{\overset{\|}{P}}} O^- \; Na^+$$

$2C_nPO_4^-$

$$CH_3(CH_2)_{n-1}O\overset{O}{\overset{\|}{C}}-\overset{}{\underset{CH_2}{C}}H-\overset{H}{N}-\overset{O}{\underset{\|}{C}}(CH_2)_{m-1} \; -\overset{CH_3}{\underset{CH_3}{\overset{+}{N}}}-CH_3$$
$$CH_3(CH_2)_{n-1}O\cdot\overset{O}{\underset{\|}{C}}-CH_2$$
$$X^-$$

$2C_nGluC_mN^+$

$$CH_3(CH_2)_{n-1}OCH_2$$
$$\phantom{CH_3(CH_2)_{n-1}OC}CH(OCH_2CH_2)_xOH$$
$$CH_3(CH_2)_{n-1}OCH_2$$

$2C_nGlu(ED)_x$

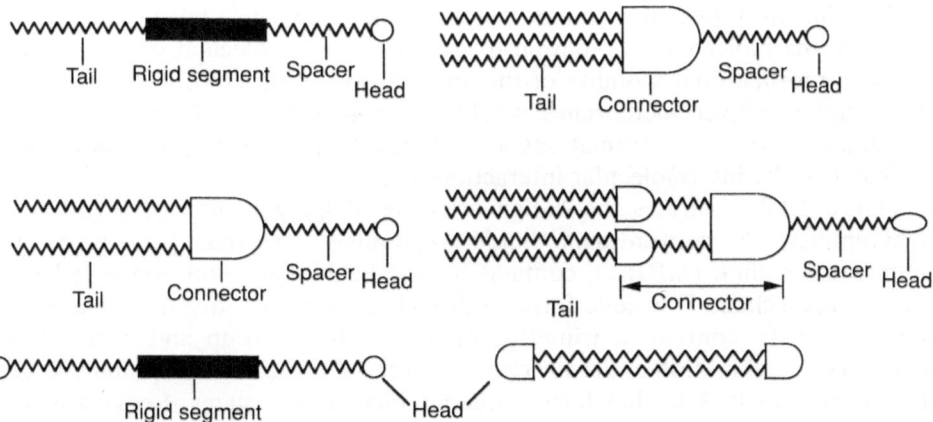

FIG. 2.2-1. Modular structures of amphiphiles which self-assemble to bilayer (and related monolayer) membranes

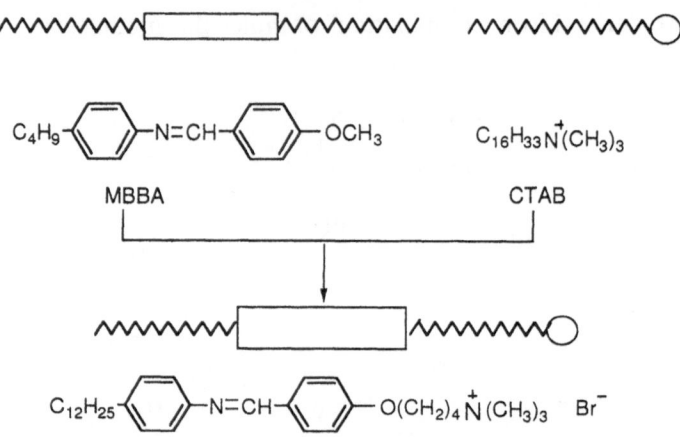

FIG. 2.2-2. Design of bilayer-forming, single-chain compounds

Many double-chain amphiphiles have been synthesized as analogues of natural lecithin molecules, and their bilayer formation has been examined. The molecular modules required are tail, connector, spacer, and head. In the series of simple dialkylammonium salts, the tail may consist of normal alkyl chains of C_{10} to C_{20}. The dispersions may contain vesicles with an inner aqueous phase or multilayer lamellae. The polar head group may be cationic, anionic, nonionic, or zwitterionic. The connector portion that links the hydrophobic alkyl chains and the hydrophilic head group helps to promote alignment of the alkyl chains. A spacer unit between the head group and the connector unit exerts a significant influence on the molecular orientation within the bilayer.

Attachment of two alkyl chains, instead of one, to a single head group should enhance the molecular orientation to give a regular molecular alignment, because conformational mobility of the alkyl chain is restricted. If this were the case, stable bilayer membranes would be produced even from single-chain amphiphiles whose conformations are restricted by the incorporation of rigid segments or by intermolecular interactions [12].

Figure 2.2-2 illustrates the design principle of bilayer-forming, single-chain amphiphiles. A conventional liquid crystalline material, p-methoxy, p'-butylbenzalaniline (MBBA), contains a rigid Schiff base unit connected to a flexible alkyl chain. A micelle-forming surfactant, cetyltrimethylammonium bromide (CTAB), contains a trimethylammonium head group and a hexadecyl chain. The combination of the structural features of these two compounds results in a novel amphiphile that forms aqueous bilayer aggregates. Lengths of the spacer methylene chain and of the hydrocarbon tail exert crucial effects on the aggregation behavior within this structural framework (as in C_nBBN^+ and $C_nBBC_mN^+$ (n, m)). Molecular aggregation detected by surface tension measurements occurs at concentrations of 10^{-4}–10^{-5} M except for $C_n BBN^+$ ($n = 0$) and $C_n BBC_m N^+$ (0,10). The critical aggregate concentration is lowered by lengthening of the hydrocarbon tail (C_n portion), but is little affected by the length of the methylene spacer (C_m portion). It is interesting that a lowering of the surface tension is found for $C_n BBN^+$ ($n = 12$) but not for $C_n BBC_m N^+$ (0, 10) or $C_n BBC_m N^+$ (4,10). In terms of the hydrophile–lipophile balance, these compounds should show similar results. Since this is not the case, the location of the rigid segment must determine, to a large extent, the tendency of these amphiphiles to align at the air–water interface. The development of the bilayer structure is improved with increasing chain lengths of the flexible tail (C_n portion). It is worth emphasizing that $C_n BBN^+$ ($n = 4$) and $C_n BBC_m N^+$ (4,10) fail to form well-organized assemblies in spite of their formation of very large aggregates. A certain tail length (probably $n > 4$) is required for an amphiphile to align in aqueous aggregates as well as at the air–water interface.

Subsequent studies on the molecular design of a single-chain amphiphile established that the kind of head group could be varied as much as that of its double-chain counterpart (cationic, anionic, nonionic, and zwitterionic), and that the structure of the rigid segment is extremely variable. An extensive list of this class of single-chain compounds has been compiled [13]. Cho and co-workers recently reported that an ammonium derivative of cholesterol formed bilayer vesicles [14]. It should be noted that "the rigid segment" implies the structural unit which promotes molecular alignment by aromatic stacking and other intermolecular interactions. Menger and Yamasaki recently reported that a combination of a very long alkyl chain ($C > 20$) and a multicharged oligoethylenimine moiety gave bilayer aggregates [15].

Since the hydrophile–lipophile balance is not the most crucial factor that determines bilayer formation, certain degrees of its imbalance will be accommodated by the stablization gained by molecular alignment. Simple triple-chain compounds $3C_n N^+$ do not seem to give well-developed bilayer structures. In contrast, amphiphiles $3C_n teN^+$ and $3C_n trisC_m N^+$ form stable bilayer membranes in water: [16] (Fig. 2.2-3). The Corey–Pauling–Koltun (CPK) molecular models of these amphiphiles with different connector structures endorse these experimental observations. In the case of $3C_n N^+$ ($n = 12$), the long alkyl chains cannot be aligned at locations close to the ammonuim group because of its tetrahedral configuration. The alkyl chains of $3C_n teN^+$ ($n = 12$) and $3C_n trisC_m N^+$ (12, 2) can

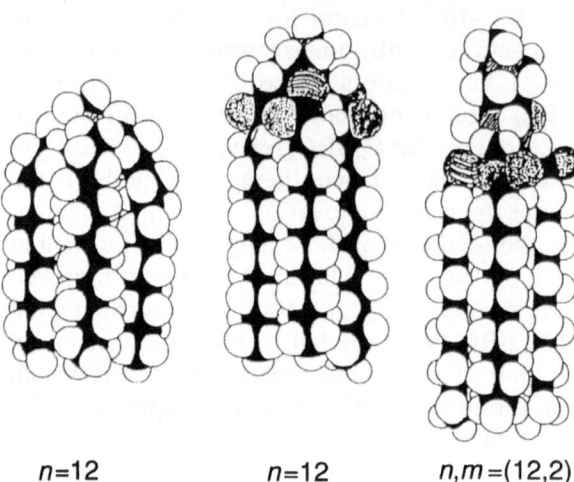

FIG. 2.2-3. Triple-chain ammonium amphiphiles and their CPK molecular models

$n=12$ $n=12$ $n,m=(12,2)$

be well aligned. The compact chain packing is produced by the presence of the ester unit, and the ammonium head can protrude without conformational constraint.

$3C_nN^+$

$3C_nteN^+$

$3C_ntrisC_mN^+$

Even quadruple-chain ammonium amphiphiles such as $4C_{14}N^+$ are suitable for two-dimensional molecular alignment [17]. These compounds give transparent aqueous dispersions at 20 mM, and display typical bilayer characteristics. For example, electron microscopy indicates the formation of vesicles and tape-like aggregates. The chain alignment is strongly dependent on the kind of connector. The native counterpart of these amphiphiles is four-chained cardiolipin, although this cannot form a stable bilayer assembly by itself.

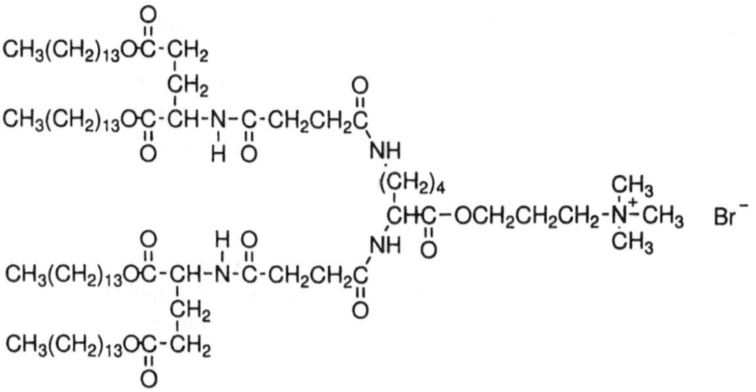

$$
\begin{array}{l}
\text{O} \\
\| \\
CH_3(CH_2)_{13}OC\text{-}CH_2 \\
\qquad\qquad\quad | \\
\qquad\qquad\quad CH_2 \\
CH_3(CH_2)_{13}OC\text{-}CH\text{-}N\text{-}C\text{-}CH_2CH_2C \\
\qquad\qquad\quad \| \quad | \ \| \\
\qquad\qquad\quad O \ \ H \ O
\end{array}
$$

(chemical structure of the tetra-chain ammonium amphiphile labeled $4C_{14}N^+$, with Br$^-$ counterion)

$4C_{14}N^+$

Some archeobacteria contain another class of membrane lipids, such as a macrocyclic amphiphile, that span the bilayer thickness. Covalent bonding of alkyl chain ends of single-chain bilayer components would produce two-headed amphiphiles which may form monolayer membranes in water, e.g., $N^+C_{10}BBC_{10}N^+$ [18]. This is dispersed in water as stable aggregates with molecular weights of 10^6–10^7. Electron microscopy shows that $N^+C_{10}BBC_{10}N^+$ (10 mM) gives highly developed lamellae with a layer thickness of 30–40 Å. This thickness corresponds to the extended molecular length; therefore, the bisammonium molecule is aligned normal to the layer plane.

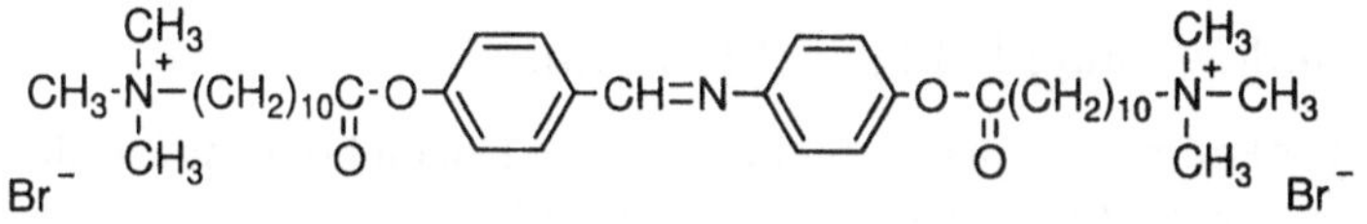

(chemical structure labeled $N^+C_{10}BBC_{10}N^+$, with two Br$^-$ counterions)

$N^+C_{10}BBC_{10}N^+$

(chemical structure labeled $V^+C_nV^+$)

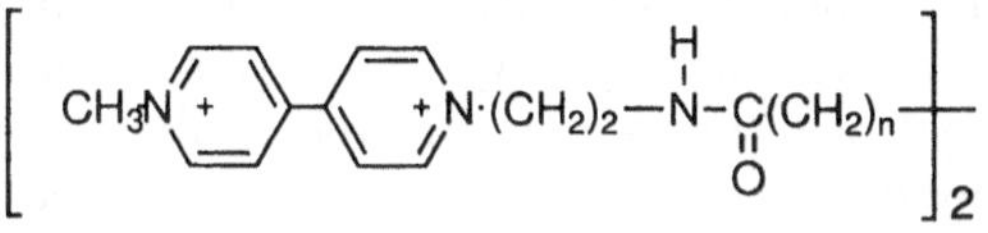

$V^+C_nV^+$

The rigid lamella can be transformed into single-walled vesicles by cosonication with cholesterol in the ratio 3:1. The vesicles are 1000–3000 Å in diameter and 60–70 Å in layer thickness. We suggest that it is preferable to insert the cholesterol on the outer side of the membrane, thus creating a suitable surface curvature for vesicle formation. Remarkable transformations of the morphology of lecithin vesicles by the addition of lysolecithin and cholesterol were reported in a landmark paper by Bangham and Horne [2].

The formation of monolayer vesicles from single-chain amphiphiles with two viologen head groups ($V^+C_nV^+$) was reported by Fuhrhop and co-workers [6]. They conducted extensive research on this type of compound, i.e., the bolaamphiphiles. For example, they prepared a series of macrocyclic compounds with two polar head groups and studied their vesicle formation.

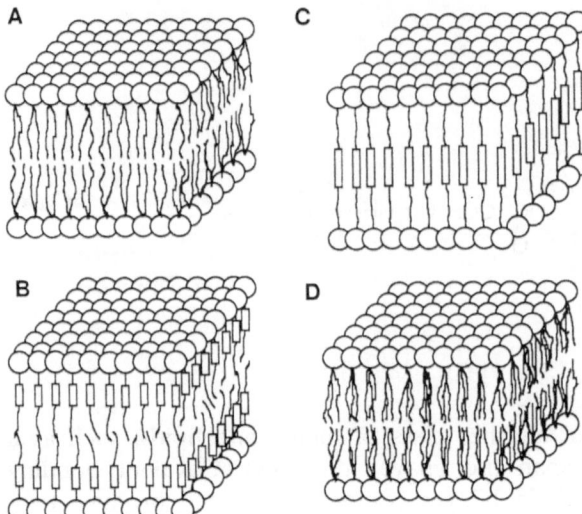

FIG. 2.2-4. Bilayer (and related monolayer) membranes composed of different types of amphiphiles. **A** Bilayer membrane of double-chain amphiphiles. **B** Bilayer membrane of single-chain amphiphiles. **C** Monolayer membrane of single-chain amphiphiles. **D** Bilayer membrane of triple-chain amphiphiles

The results described in this section endorse the general principles of self-assembly of bilayer (and related monolayer) membranes. These molecular membranes are illustrated in Fig. 2.2-4.

2.2.1.4 Hydrogen Bond-Mediated Bilayer Membrane [19]

Construction of bilayer membranes via complementary hydrogen bonding would provide a new class of self-assembly in which novel molecular control is achieved. In general, hydrogen bonding in artificial molecular systems is most effective in solid states or in noncompetitive (aprotic) organic media. However, our recent finding that molecular recognition via hydrogen bonding is effective at the air–water interface implies that hydrogen bond-directed bilayer assembly is feasible in bulk water.

Among the four combinations of equimolar complexes of $2C_{12}OC_2Mela$, $2C_{16}Mela$, and $CAPhC_{10}N^+$, $CAC_{10}N^+$, only $2C_{12}OC_2Mela/CAPhC_{10}N^+$ gave a transparent dispersion in water (ca. 30 mM) upon ultrasonication. The aqueous

$$CH_3(CH_2)_{11}O(CH_2)_2-N\begin{matrix} H \\ \end{matrix}$$

2C₁₂OC₂Mela

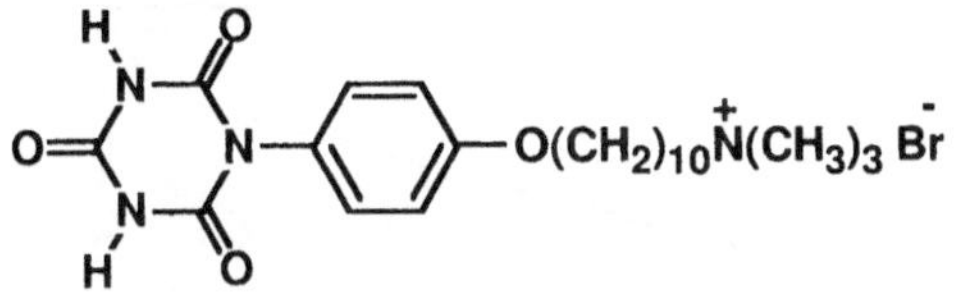

CAPhC$_{10}$N$^+$

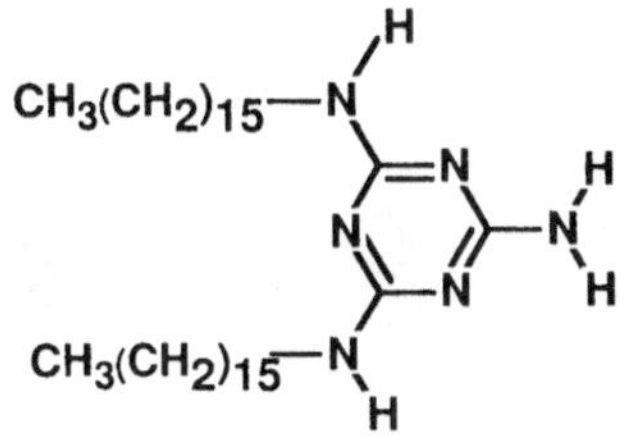

2C$_{16}$Mela

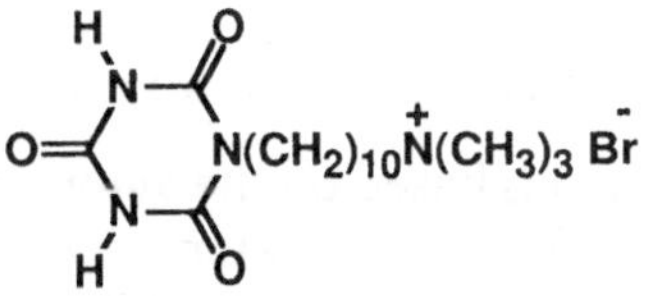

CAC$_{10}$N$^+$

dispersion was stable over a period of 1 month. The other complexes and individual components, except for CAC$_{10}$N$^+$, displayed poor solubilities in water even at lower concentrations. Improved molecular orientation by the phenyl group in CAPhC$_{10}$N$^+$ and facilitated alkyl chain alignment by the ether linkage in 2C$_{12}$OC$_2$Mela appears crucial to give an ordered stable-assembly. Transmission electron microscopy indicated the presence of disk-like aggregates with diameters of several hundred Å and a thickness of ca. 100 Å. Differential scanning calorimetry (DSC) measurement and the spectral properties of membrane probes clearly showed the presence of gel-to-liquid crystal phase transition similar to that of a conventional bilayer membrane. The bilayer assembling proceeds spontaneously even from the two amphiphilic components that are separately dissolved in water or ethanol.

Figure 2.2-5 illustrates schematically the formation of a hydrogen-bond-mediated bilayer membrane. The complementary hydrogen bonding is effective even in water thanks to the hydrophobic molecular association. Self-assembly assisted by the interplay of the hydrogen bond and hydrophobic force produces many exciting possibilities.

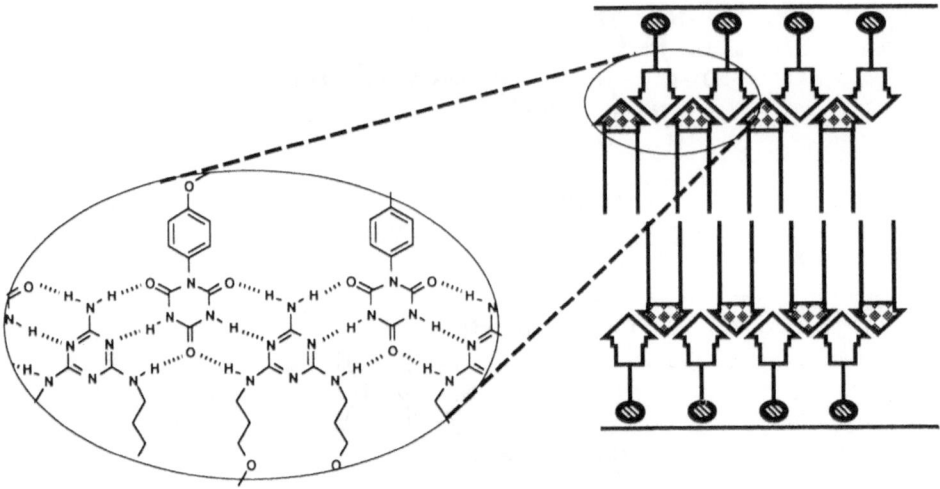

FIG. 2.2-5. Bilayer formation via complementary hydrogen bonding

2.2.1.5 Self-Assembly of Bilayers in Nonaqueous Media

The preceding molecular design is based on the amphiphilic nature of the bilayer component. A molecule is amphiphilic in water when a hydrophobic moiety and a hydrophilic moiety are covalently bonded within the molecule. If this design principle is modified in order to encompass aprotic media, a rich field of novel molecular assemblies is expected to emerge.

Kim and Kunitake [20] prepared a Ca^{2+} complex of phosphate-bearing bilayer membranes by precipitation after mixing separate aqueous solutions. The equimolar complex gave a reversed bilayer structure in $CHCl_3$, as supported by chiroptical measurement and electron microscopic observation. In this case, the bilayer core is made of the Ca^{2+}/phosphate bridged complex, and the alkyl tail is exposed to the $CHCl_3$ media. Kunieda et al. [21] obtained multilayer vesicles of micron size by dispersing tetraethyleneglycol dodecyl ether in a dodecane/water mixture. A reversed bilayer structure with hydrated palyethylene glycol (PEG) chains pointing inward was assumed on the basis of spectroscopic evidence.

We have now developed a more general strategy for the preparation of nonaqueous bilayer assemblies. Long perfluoroalkyl compounds generally display low solubilities in hydrocarbon media due to their small cohesive forces. Ammonium amphiphiles possessing perfluoroalkyl chains in place of hydrocarbon chains can form stable, well-aligned aqueous bilayers. These fluorocarbon bilayers tend to phase-separate (cluster formation) when mixed with hydrocarbon bilayers. The limited miscibility has also been observed at the air/hydrocarbon interface. Long perfluoroalkyl chains linked with the hydrocarbon moiety form insoluble monolayer films and reduce surface tension. On the basis of these

$$CF_3(CF_2)_7CH_2CH_2O\overset{O}{\overset{||}{C}}\text{-}\underset{\underset{\underset{\underset{O}{||}}{\overset{|}{C}}}{\overset{|}{CH_2}}}{CH}\text{-}N\text{-}\overset{H}{\overset{|}{C}}(CH_2)_7\text{-}CH=CH\cdot(CH_2)_7CH_3$$

$$CF_3(CH_2)_7CH_2CH_2O\underset{\underset{O}{||}}{C}\text{-}CH_2$$

2 $C_8^F C_2 GluC_{18}$

$$CF_3(CF_2)_7CH_2CH_2O\overset{O}{\overset{||}{C}}\text{-}\underset{\underset{\underset{\underset{O}{||}}{\overset{|}{C}}}{\overset{|}{CH_2}}}{CH}\text{-}N\overset{H}{\overset{|}{C}}\text{-}\langle\,\rangle\text{-}O(CH_2)_8\text{-}CH=CH\,(CH_2)_7CH_3$$

$$CF_3(CH_2)_7CH_2CH_2O\underset{\underset{O}{||}}{C}\text{-}CH_2$$

2 $C_8^F C_2 GluPhC_{18}$

precedents, we designed new fluorocarbon "amphiphiles" $2C_8^F C_2 GluC_{18}$ and $2C_8^F C_2 GluPhC_{18}$, that undergo spontaneous bilayer formation in organic media [22]. They consist of double fluorocarbon chains and a single oleyl chain as solvophobic and solovophilic moieties, respectively. These two molecular modules are linked through the chiral L-glutamate residue. When dispersed in chlorocyclohexane or benzene, these amphiphiles form stable bilayer assemblies that give rise to morphologies such as tubes, tapes, rods, and particles, as confirmed by electron microscopy. These dispersions show bilayer/monomer phase transition, as inferred by differential scanning calorimetry. This phase transition is accompanied by changes of the helical pitch of the fluorocarbon conformation and of intermolecular hydrogen bonding among the glutamate moiety. The surface tension analysis of adsorbed monolayers at the air/liquid interface indicates that the bilayer formation in bulk chlorocyclohexane is driven by the enthalpic change. Limited miscibilities between fluorocarbon chains and hydrocarbon solvents produce the main enthalpic driving force of the molecular association, and the hydrogen bonding and van der Waals force of the aromatic moieties serve as secondary driving forces. Molecular ordering in the novel bilayers is not inferior to those of conventional aqueous bilayers, although their bilayer stability does not surpass that of aqueous bilayers.

$$CH_3(CH_2)_{11}O\text{-}\overset{O}{\overset{||}{C}}\text{-}\underset{\underset{\underset{\underset{O}{||}}{\overset{|}{C}}}{\overset{|}{CH_2}}}{CH}\text{-}N\text{-}\overset{H}{\overset{|}{C}}\text{-}\overset{CF_3}{\overset{|}{CF}}(OCF_2CF)_2O(CF_2)_2CF_3$$

$$CH_3(CH_2)_{11}O\text{-}\underset{\underset{O}{||}}{C}\text{-}CH_2$$

$2C_{12}Glu(C_3^F O)_3$

This solvophilic/solvophobic relationship may be reversed, since bilayer assemblies are formed in an aprotic fluorocarbon medium [23]. In this case, amphiphilic molecules, such as $2C_{12}Glu(C_3^F O)_3$, are composed of double hydrocarbon chains as the solvophobic moiety and flexible perfluoroalkyl chains as the solvophilic moiety. These particular compounds possess "reversed" amphiphilicity. The as-

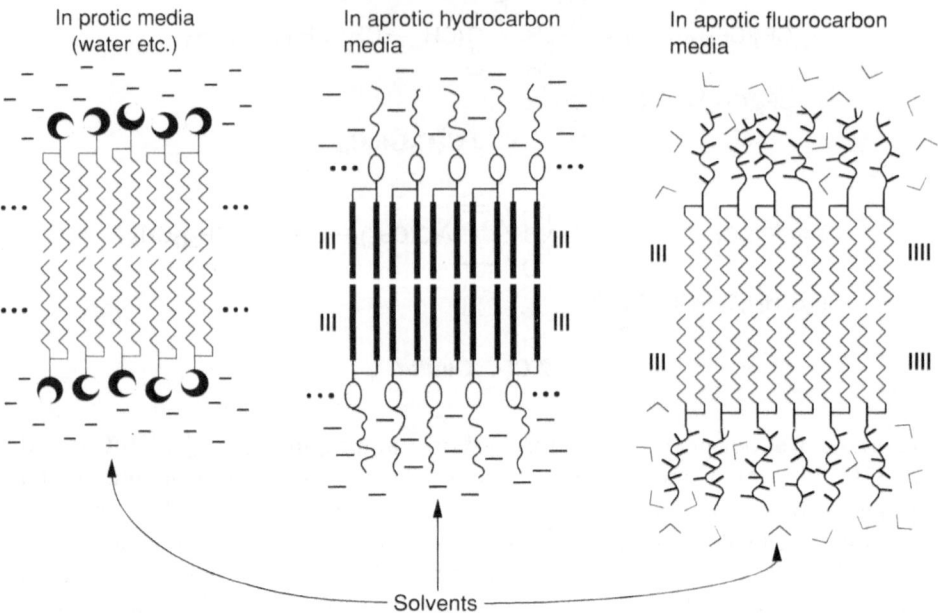

In protic media
(water etc.)

In aprotic hydrocarbon
media

In aprotic fluorocarbon
media

Solvents

FIG. 2.2-6. Generalized bilayer concept

sembly of hydrocarbon chains in fluorocarbon media is more likely than that of fluorocarbon chains in hydrocarbon media in terms of the cohesive energy difference.

These results establish that the assembly of molecular bilayers is a widely observable phenomenon. *Organized molecular assemblies are produced in any medium if the "amphiphilic/amphiphobic" nature of the solute molecules is properly designed.* Figure 2.2-6 represents the generalized bilayer concept. Self-assembling bilayers can be formed in protic media, in aprotic hydrocarbon media, or in aprotic fluorocaron media. The common driving force of the molecular assembly is the solute/solvent immiscibilies (enthapic force) that arise from differences in cohesive energy between solute (amphiphile) and solvent. The magnitude of the cohesive energy of the solute per se is not relevant for promoting effective molecular assembly. Further expansion of the generalized concept is possible by employing appropriate solute/solvent combinations. The concept should become a powerful, strategic instrument to open up new possibilities for a broad area of molecular self-organization.

2.2.2 Langmuir–Blodgett (LB) Membranes

2.2.2.1 History

By 1919, under Langmuir's guidance, Miss Katharine Blodgett had succeeded in transferring fatty acid monolayers from water surfaces to solid supports. Such built-up monolayer assemblies are now referred to as Langmuir–Blodgett (LB) films, while floating monolayers at the air–water interface are called Langmuir films. The first formal report describing the preparation of built-up films did not appear until 1935. A series of their pioneering works is introduced in the book by Gaines [24] together with the contributions of other early investigators, including the earliest scientific paper on the effect of oily films on water by Benjamin Franklin in 1774, and the preparation of the first monolayers at the air–water interface by Agnes Pockels in 1891. The LB method was the first technique to provide a chemist with the practical (and still the most extensively used) capability to construct ordered molecular assemblies. In the mid-1960s, Hans Kuhn et al. began their stimulating experiments on monolayer organization [25]. A self-assembled (SA) monolayer was introduced as a possible alternative to the LB films. The LB and SA films have attracted much attention recently, because optoelectronics and molecular electronics have become areas at the frontier of materials science [26]. For more details of the historical aspects, the reader is referred to a relevant text book [27] as well as Gaines' book mentioned above.

2.2.2.2 Preparation Methods

Monolayers at the Air–Water Interface. Langmuir demonstrated that the long chain fatty acids form films in which the molecules occupy the same cross sectional area, about 0.20 nm^2 per molecule, whatever the chain length. He also concluded not only that the films were indeed one molecule thick, but also that the molecules were oriented at the water surface, with the polar functional group immersed in the water and the long nonpolar chain directed nearly vertically up from the surface. For the measurements, he devised the surface balance with which his name is associated, i.e., a device in which a movable float separates clean water surface from the film-covered area, and forces are measured directly from the deflection of the float. Today an electronic microbalance with Wilhelmy plate is also widely used to measure the surface pressure. The reader will find a discussion on surface pressure, and on the Langmuir, Wilhelmy, and other methods of surface pressure measurements, in Gaines' book [24].

The characteristic description of an insoluble monolayer is usually in terms of its pressure–area (π–A) curve, i.e., the relationship between the surface pressure observed and the area occupied on the water surface by the molecules of the film. In some π–A isotherms, the presence of condensed, expanded, and gaseous films is seen. This is analogous to the pressure–volume behavior of ordinary three-dimensional phases. For more details, see the books mentioned above [24–27].

Langmuir–Blodgett Deposition. A wealth of useful information about molecular sizes and intermolecular forces can be obtained from studies of monolayers on the water surface, but the great resurgence of interest in this area of science has been largely due to the fact that films can be transferred from the water surface onto a solid substrate by the LB technique.

In the most commonly used method, the substrate (e.g., a glass slide) is first lowered through the monolayer so that it dips into the subphase and then withdrawn under a constant surface pressure. The value of surface pressure which gives the best results depends on the nature of the monolayer, and is established empirically. If the slide used has a hydrophilic surface, deposition follows the sequence of events now described. The water wets the slide's surface and the meniscus turns up; there is no mechanism for deposition at the first dipping. As the slide is withdrawn the meniscus is wiped over the slide's surface and leaves behind a monolayer in which the hydrophilic groups are turned toward the hydrophilic surface of the slide. It will be apparent that there must initially be a liquid film between the slide and the deposited monolayer, and that bonding of the monolayer to the slide will only be complete after the intervening layer of water has drained away or evaporated. The second dipping into the subphase differs from the first in that the slide is now hydrophobic; the meniscus turns down and a second monolayer is deposited with its tail groups in contact with the exposed tail groups on the slide. The second withdrawal exactly resembles the first except that the new monolayer is now being deposited onto the hydrophilic head groups of the monolayer already present. This type of deposition, in which layers are laid down each time the substrate moves across the phase boundary, is known as Y-type deposition. It is also possible for deposition to occur only when the substrate enters the subphase (X-type) or leaves it (Z-type), depending on the nature of the monolayer, substrate, and subphase, and the surface pressure [24–27].

The question arises as to whether the manner of depostion determines the final structure of the film, or whether some reorganization takes place after the multilayer has been dipped. At first sight one would expect multilayers deposited in X and Y modes to have the structure shown in Fig. 2.2-7. However, it seems that multilayers of stearic acid always have Y-type structure irrespective of whether they have been deposited in an X- or Y-type sequence, so that simple molecules of this type are able to invert at some stage during the dipping process. Related topics studied by atomic force microscopy (AFM) will be described later.

The transfer ratio is often used as a measure of the quality of deposition. It is defined as the ratio of the area of monolayer removed from the water surface to the area of substrate coated by the monolayer. The area removed from the water surface is easily measured by the mechanism used to maintain a constant surface pressure, and there is then a direct electrical readout of surface area. Under most circumstances a transfer ratio of unity is taken as a criterion for good deposition, and one would then expect the orientation of molecules on the slide to be very similar to their orientation on the water. Occasionally, there is a large but consistent deviation from a value of unity; this points to a situation in which the molecular orientation is changing during transfer. Variable transfer ratios are almost always a sign of unsatisfactory film deposition.

The need for a continuous fabrication of thick, noncentrosymmetric organic films for nonlinear optical (NLO) applications was the driving force for the

FIG. 2.2-7. Structures of X, Y, and Z multilayers. X- and Z-type depositions do not normally guarantee that the multilayers will have the corresponding structure

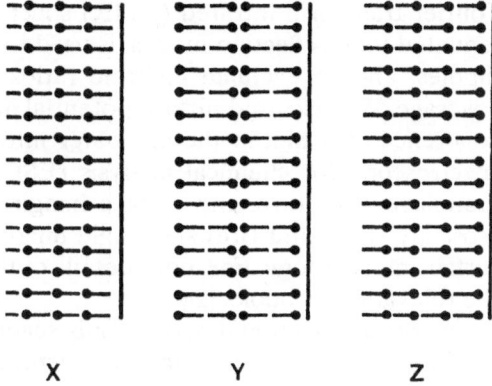

X   Y   Z

development of troughs with two baths in which a film of molecule A is formed on one bath, and a film of molecule B on the other. Different transfer geometries and mechanisms were suggested in the literature and introduced in Ulman's book [26]. These troughs can be used to form a NLO LB film with the ABABAB superlattice. Such a heterogeneous structure is important for photoelectric conversion using an electron acceptor (A), a sensitizer (S), and an electron donor (D) amphiphile, as described later.

2.2.2.3 Molecules for LB Films

Materials that are soluble in water are hydrophilic, while those that dissolve in nonpolar solvents are called hydrophobic. The amphiphile is a molecule that is insoluble in water, with one end that is hydrophilic, and therefore is preferentially immersed in the water, and the other that is hydrophobic, and preferentially resides in the air (or in the nonpolar solvent). A classical example of an amphiphile is stearic acid, where the long hydrocarbon "tail" is hydrophobic, and the carboxylic acid head group is hydrophilic. The reader can find a discussion of the properties of monolayer films of a variety of amphiphiles in Gaines' book [24]. Today many scientists use organic synthesis to construct amphiphiles for different purposes. Such molecular engineering is very important, especially in the design of amphiphiles with functional groups such as chromophores, donors, acceptors, etc. For more details, see the relevant text books [25–27].

2.2.2.4 Characterization of LB Films

In the study of LB films, we are interested in both their surface and bulk properties. Ellipsometry is used to measure the thickness and uniformity of the films. Contact angles with different liquids are measured to evaluate wetting properties, surface-free energy, and uniformity, and to get information on surface order.

Fourier transform infrared (FTIR) spectroscopy, in both grazing-angle and attenuated total reflection modes, is used to learn about the direction of transition dipoles, and to evaluate dichroic ratios, molecular orientation, packing, and coverage. We also used surface potential measurements to get information on the coherence of a film at the air–water interface and on metal surfaces, electron spectroscopy for chemical analysis (ESCA) to study surface composition and monolayer structure, and surface imaging techniques such as fluorescence microscopy (FM) and Brewster angle microscopy (BAM) [28] to learn about the surface morphology. For more details of the techniques, the reader is referred to the relevant text books [24–27].

The present section describes only scanning probe type methods which can be used to learn about the detailed structure of LB films with much higher lateral resolutions. The methods can also be used to fabricate and characterize the molecular devices [29].

With conventional microspectrophotometries and local analysis methods, such as Auger electron spectroscopy (AES) and secondary ion mass spectrometry (SIMS), the lateral resolutions are always limited by diffraction of the corresponding waves in the far field. For example, the resolutions of IR microspectrophotometry, AES, and SIMS are 5, 0.02 and 0.1 µm, respectively. Therefore, new types of spectroscopies, i.e., nanospectroscopies, are required for "nanoanalyses".

The conventional local analyses are classified by the combination of probing and monitoring particles. The microspectrophotometries use photons as the probing and monitoring particles, while AES and SIMS utilize electron–electron and ion–ion particles, respectively. Similar classification will be applicable for scanning probe type spectroscopies already available. The classification will also be useful to develop new types of nanospectroscopies, for which we coin the word SXX'M. Here, X represents a probing particle from the tip or a probing interaction by the tip, and X' represents a monitoring particle from the sample surface or a monitoring physical quantity as a result of the tip–sample interaction. In comparison to scanning probe microscopies (SPM or SXM), SXX'M forces us to pay considerable attention to "cause" and "effect". For example, scanning tunneling spectroscopy (STS) is based on a voltage–tunneling current relation, while scanning near-field optical microscopy (SNOM) including Raman, IR, and fluorescence spectroscopies based on a relationship between probing–monitoring photons. We are developing a variety of SXX'Ms. For example, in scanning tunneling ion mass spectrometry (STIMS) [29], the current from a scanning tunneling microscope (STM) tip is the stimulus and the ions ejected are the response, while in friction force microscopy (FFM) the monitoring heat or temperature is of interest under the stimulus of friction [29]. Needless to say, however, a variety of near-field optical spectroscopies are among the most important because they provide interesting information about the molecule which cannot be obtained by the other SXX'Ms. Scanning near-field optical fluorescence spectroscopy (SNOFS), in particular, is the most sensitive and promising [30].

Fluorocarbon (FC) amphiphiles are known to be phase-separated from hydro-carbon (HC) amphiphiles in the bilayer membrane [29]. The increase in the concentration of FC amphiphiles at the surface of mixed LB films of HC and FC amphiphiles was also studied by the angular dependence of ESCA measurements [29]. The next section describes the characterization of the phase-separated mixed monolayer by some of the scanning probe microscopies.

2.2.2.5 Characterization of the Phase-separated Domains HC–FC Mixed Monolayers by SPM

AFM and FFM. In collaboration with Güntherodt's group we have tried to observe AFM and FFM images of phase separation in mixed monolayers [31] and bilayers [31] of HC and FC amphiphiles polyion complexed with poly 4-methyl-vinylpyridinium cations [32]. Figure 2.2-8 shows AFM and FFM images of the mixed monolayer of arachidic acid ($C_{19}H_{39}COOH$) and PFECA ($C_9F_{19}C_2H_4OC_2H_4COOH$) (1:1) polyion complexed with polyvinylammonium cations deposited on oxidized Si (100). The phase separation is clearly seen in the AFM and FFM images. The different resistivity against wear by the AFM tip between the HC and FC surfaces reveals that the "island" surfaces consist of HC chains, whereas the "sea" surface consists of FC chains. From this investigation, the friction of the FC surface was found to be higher than that of the HC surface. This FFM result implies that chemical differentiation of the outermost layers can be made by friction measurements. A clear contrast in friction between a silicon substrate surface and the outermost surfaces of multibilayers (single and double) of cadmium arachidate was also reported from the Basel group [33]. The size and

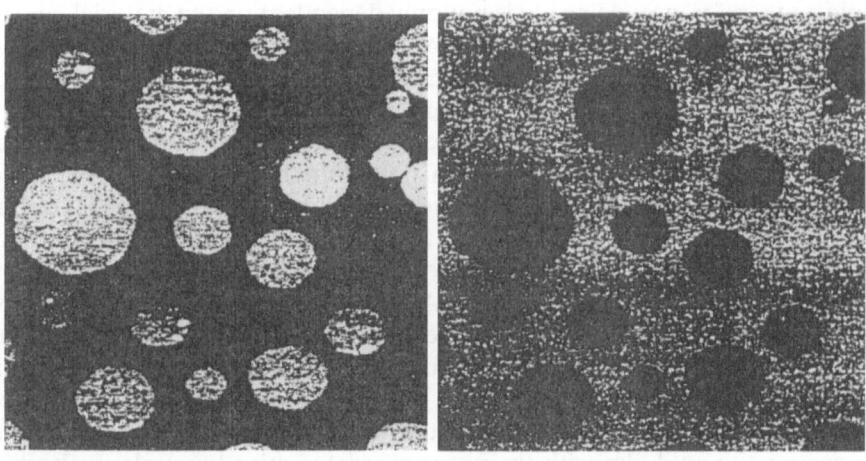

a   b

Fɪɢ. 2.2-8. AFM and FFM of a mixed monolayer of arachidic acid and PFECA (1:1) polyion complexed with polyvinylammonium cations deposited on Si (100) ($4 \times 4\,\mu m^2$). **a**, Topography with a step height of 2.0–2.4 nm; **b**, friction force map

shape of the islands depend on the chain lengths of the amphiphiles [31]. The size can also be changed readily from microns to submicrons by changing the kinds of cationic polymers. The chemical assignment of the surface between the islands and the sea was also confirmed by comparison between the AFM and fluorescence microscopic images and by the use of scanning surface potential microscopy (SSPM), as described below.

Scanning Maxwell Stress Microscopy. The surface dipole moments of the FC and HC chains are known to be significantly different from each other [34]. From macroscopic measurements, it is well known that surface potentials, or contact potential differences (CPD), are highly material-dependent and are related to work functions [35] and surface dipole moments [35]. This prompted us to measure the local surface potential distribution on the nano-scale to confirm the chemical differentiation of the phase-separated surfaces by a different method. High-resolution potentiometry by electrostatic force microscopy [29] or high lateral resolution CPD measurements, i.e., Kelvin probe force microscopy [29], had already been devised for simultaneous topographic and CPD measurements. The method was further modified and called scanning Maxwell stress microscopy (SMM) [36] or SSPM [37].

For the principle of this new type of Kelvin probe force microscope in the light of the conventional measurement of Volta potential differences, the reader is referred to the reference [37]. The concept of the Volta potential and surface potential used in this standard method is quite useful to discuss CPDs measured by the present SSPM. CPDs can be related to the work functions of the substrate metals and to the surface Volta potential difference observed upon introduction of the monolayers onto the substrate. The application of SSPM for characterization of the mixed monolayer will be discussed below. The CPD [35] between a sample and a reference electrode is not only dependent on the material, i.e., the work function, but also on the condition of the surface, such as contamination and monolayer deposition. The CPD can be measured, for example, by the vibrating plate method, called the Kelvin method [24, 27].

A mixed monolayer of perfluorodecanoic acid (Asahi Glass) and arachidic acid (Tokyo Kasei) polyion complexed with poly(4-methylvinylpyridinium) iodide [32] was transferred onto a vapor-deposited gold film on a glass plate [37] or onto an oxidized Si (100) wafer surface [31]. The mixed film transferred onto a vapor-deposited gold film had been studied previously by AFM and the SSPM [37]. Due to the roughness of the gold film, the AFM topographic image of the phase-separated domains, which was clearly seen on the flat Si (100) surface, disappeared in the background of the very rough gold surface. The substrate was the gold film, and thus the CPD distribution was homogeneous over the film before the LB film transfer. The spatial distribution of the CPDs, however, was created after coating the gold film with the phase-separated mixed monolayer, because the surface Volta potential difference would be different between the islands and the sea regions in the mixed film. Although the resolution was ca. 1 μm due to the roughness of the gold film, a clear contrast was observed between the HC islands and the FC sea regions [37]. The surface potential difference between these two regions was attributed to the difference in the surface dipole moments due to the terminal CH_3- and CF_3-groups of the HC and FC chains, respectively [34].

Figure 2.2-9 shows a $5 \times 5\,\mu m^2$ AFM topographic image of a mixed monolayer on oxidized Si (100), and is very similar to that of the mixed monolayer of $C_9F_{19}C_2H_4OC_2H_4COOH$ and $C_{19}H_{39}COOH$ (1:1) observed previously [31]. The flatness of the Si (100) surface allowed observation of the AFM topogaphic image of the mixed film, which was difficult on the gold films. Figure 2.2-9b shows an SSPM image recorded on the same surface as the AFM image in Fig. 2.2-9a. The SSPM image shows a clear contrast in the surface Volta potential difference between the islands and the sea. It is clear from the comparison of the images of Fig. 2.2-9 that almost all islands seen in the AFM topographic image can be seen in the SSPM image, although the lateral resolution of the SSPM image of ~100 nm is not as good as that of the AFM image. However, this resolution is much better than that of ~1 μm observed on the gold vapor-deposited film covered with the mixed monolayer. The improved resolution can be attributed to the increase in the flatness of the conducting surface from the vapor-deposited gold film to the Si (100) surface.

As described above, the surface Volta potential difference, ΔV, between the islands and the sea was initially interpreted [37] as arising from the difference in the surface dipole moments due to the terminal CH_3- and CF_3-groups in the monolayer, modelled in the "side-by-side" structure shown in Fig. 2.2-10. In other words, in higher ΔV and thus the brighter image in the FC sea observed in

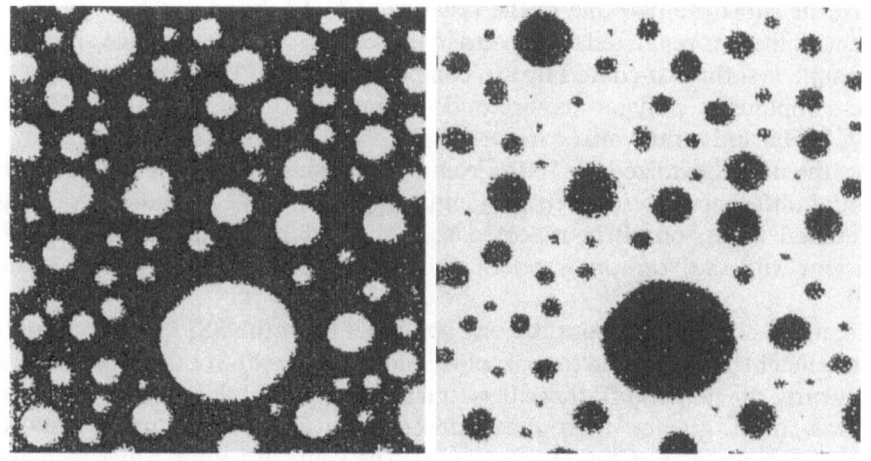

a   b

FIG. 2.2-9. Typical images of a phase-separated mixed monolayer of $C_{19}H_{39}COOH$ and $C_9F_{19}COOH$ (1:1) ion complexed with poly 4-methylvinyl pyridinium cation deposited on oxidized Si (100). **a** AFM topographic image. **b** SSPM image under constant S

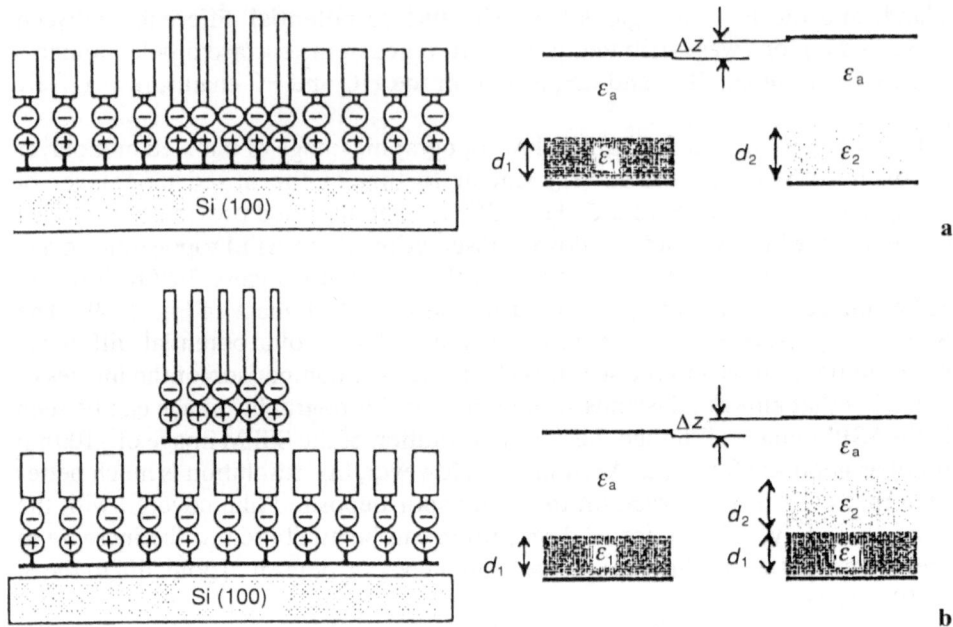

FIG. 2.2-10. Schematic view of the phase separation modes of a mixed monolayer of HC and FC amphiphiles polyion complexed with a cationic polymer and their equivalent parallel plate capacitors. **a** The "side-by-side" structure. **b** The "on-top" structure

Fig. 2.2-9b seemed to be attributable to the large surface dipole moment directed toward the substrate [34] due to the polarized $C^{\delta+}$–$F^{\delta-}$ bond in CF_3-group compared with the less polarized C–H bond in the CH_3-group. The voltage applied to the sample (vs. the Au-coated tip) to compensate the CPDs were measured for single-component polyion complexed monolayers of $C_{19}H_{39}COOH$ and $C_9F_{19}COOH$, and for the mixed monolayer on the oxidized Si (100) together with that of the naked oxidized Si (100). From these measurement, the surface Volta potential differences were calculated and compared with each other. Although the detailed discussion will be reported elsewhere, all these data were found to be consistent with the "on-top" model rather than the side-by-side model in Fig. 2.2-10.

The above conclusion about the on-top model from the SSPM was consistent with the height profile of the topographic image of the AFM of the mixed film. In other words, the height difference between the islands and the sea was ca. 1.9 nm, and was much greater than the difference in molecular lengths between $C_{19}H_{39}COOH$ and $C_9F_{19}COOH$ (ca. 1.1 nm). The value of 1.9 nm, which is smaller than the film thickness of cadmium arachidate (2.8 nm), can be interpreted by the large tilting angle in the polyion complexed film expected from the larger molecular area [32]. A similar deviation in the height difference from the molecular lengths had already been observed for the previous mixed monolayer between $C_{19}H_{39}COOH$ and $C_9F_{19}C_2H_4OC_2H_4COOH$ (1:1) [31]. More interestingly, the monolayer film was easily shorn by the AFM tip inside the islands, and the bottom of the resulting hole was at the same level as the sea and gave the same friction as that observed in the sea. However, from a comparison between the π–A isotherms of the single component, i.e., $C_{19}H_{39}COOH$ and $C_9F_{19}COOH$, and the mixture (1:1), the side-by-side model is more likely. This contradiction implies the possibility of an on-top structure formation during the monolayer deposition, and thus further investigations in relation to the squeeze-out mechanism are now in progress.

The dielectric constant of the HC monolayer was also obtained from SSPM on the assumption of the on-top model in Fig. 2.2-10b [29, 37].

Fluorescence Microscopy and SIMS. The chemical assignment of the islands and the sea had already been confirmed by the comparison between the AFM and fluorescence images of the same mixed monolayers of a pyrene derivative of a fatty acid and PFECA with various mixing ratios [29].

In this section, simultaneous observation of AFM, FFM, and fluorescence microscopy will be described for the mixed monolayer of stearic acid, PFECA, and cationic cyanine dye with two long alkyl chains (1:1:1/40). Addition of a small amount of the dye had little effect on the phase separation behavior of the HC and FC mixture. It is reasonable to assume that the long alkylated cyanine dye is much more soluble in the HC than in the FC phase. In practice, a much stronger emission of the cyanine dye was observed in the islands than in the sea in the phase-separated mixed monolayer [29]. This observation, together with that of the previous pyrene system [29], agrees with the chemical assignment of the islands, which consist of HC amphiphiles. Scanning near-field optical fluorescence microscopy (SNOFM) and SNOFS [30] of the mixed monolayer is now under investigation.

It was also found that the addition of a comparable amount of a cationic amphiphile, such as octadecyltrimethylammonium chloride (ODTMAC), increased the size of the islands dramatically [29]. This increase in the size of the islands enabled us to perform scanning SIMS measurements on the mixed monolayer. Figure 2.2-11 shows an AFM topographic and a scanning SIMS image for fluorine distribution in the mixed monolayer of stearic acid, ODTMAC, and PFECA (1:1:1) formed in the presence of poly vinylamine. As shown in Fig. 2.2-11b, the fluorine mass intensity is much stronger in the sea region, and therefore the assignment of the sea as the FC phase was confirmed [29].

2.2.2.6 Study of Organic Thin Films by Near-Field Optical Microscopies

The resolution of SNOM is determined by the dimensions of the microscopic light source (or detector) and the probe-to-sample separation rather than the diffraction limit [29, 30, 38]. A microlithographic hole, a small aperture in a metal

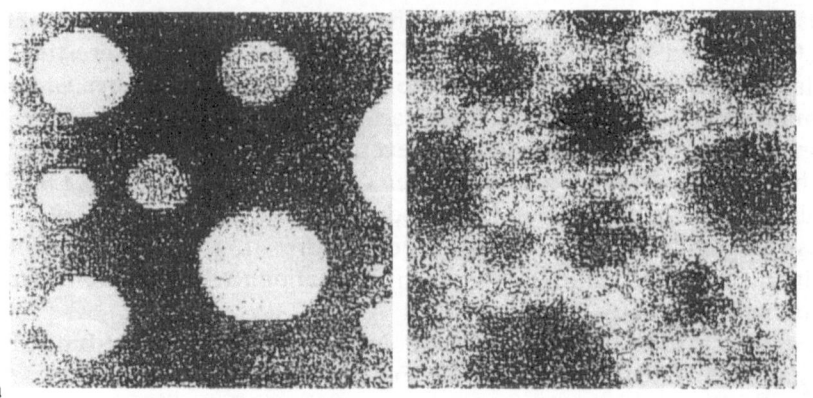

FIG. 2.2-11. AFM and scanning SIMS images of a mixed monolayer of stearic acid, ODTMAC, and PFECA (1:1:1) poly ion complexed with poly vinyl ammonium cations deposited on Si (100) ($10 \times 10\,\mu m^2$). **a** AFM topographic image. **b** Fluorine mass intensity map

film, a micropipette, a fluorescence probe, and an optical fiber tip have all been used as the light source. For separation control, the steep decrease in a tunneling current, a shear force, or the intensity of an evanescent wave with an increase in the separation has been utilized. In our microscope, we took advantage of non-contact scanning force microscopy with a vibrating cantilever [30], not only to control the separation for SNOM without mechanical damage to the sample, but also to give a simultaneous AFM topographic image. By precise control of the position of the tip within 100 nm of the sample surface, we demonstrated that this combined non-contact AFM and SNOM (SNOAM) method can be used as fluorescence microscopy (SNOFM) and nanoscopic fluorescence spectroscopy (SNOFS) with a resolution of ~100 nm [30].

Although the study of SNOM has a long history [38], optical images have begun to be taken only within the last decade. One of the most advanced SNOMs uses the shear force to control the separation as well as an aluminum-coated tapered optical fiber [39]. The other is the combined SNOM and AFM in the contact mode based on a silicon nitride cantilever with a pyramidal tip [40]. In the latter, the tip is used simultaneously with the SNOM optical probe and the AFM tip.

In our SNOAM [30], an optical fiber tip, sharpened and bent by heating, was mounted on a stainless steel cantilever or used as the AFM cantilever itself. The separation between the tip and the sample was controlled by an AFM operation with a laser beam deflection in which the cantilever was vibrated at the resonance frequency. The decrease in amplitude of the vibrating cantilever by van der Waals force was used to control the separation. Figure 2.2-12 shows a schematic diagram of our present noncontact AFM–SNOM.

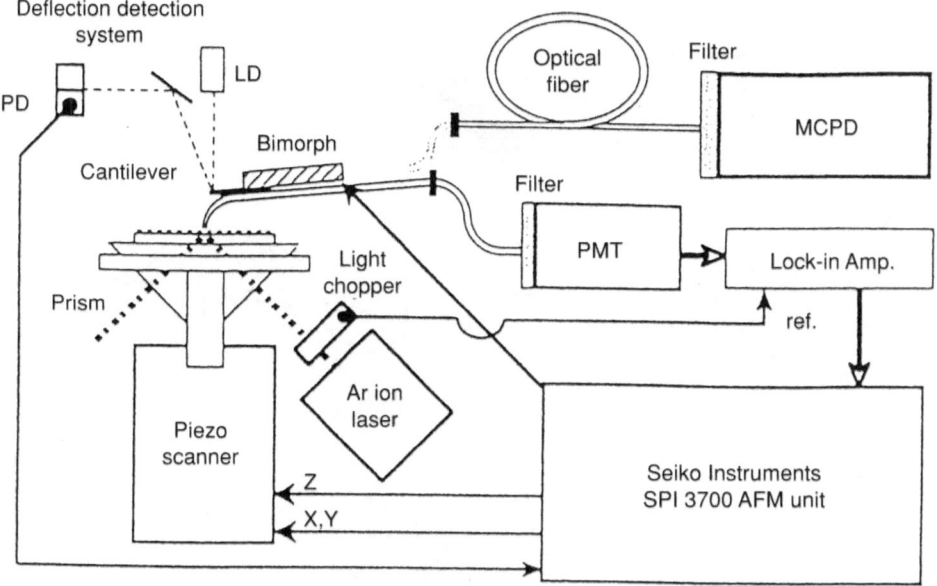

FIG. 2.2-12. Schematic diagram of the noncontact AFM–SNOM (SNOAM) for SNOFM and SNOFS

To obtain experimental results for a test sample for scanning optical absorption measurements of the Ar ion laser beam with the noncontact AFM we used a chromium checker-board pattern on a quartz substrate gifted by Dai Nippon Printing Co., Saitama. For the fluorescence microscopy and spectroscopy, the chromium pattern was spin-coated with a thin polyvinyl alcohol film containing fluorescein [29]. Recently, the method has been applied to observe the phase separation of an HC–FC mixed monolayer.

We also performed SNOM recording on chemically modified surfaces and Langmuir–Blodgett (LB) films. Photocleavage of the chemically modified Si (100) surface and photoisomerization from the *trans*- to the *cis*-form of azobenzene moieties of an amphiphilic azobenzene derivative in the LB films were used for the chemical reaction. Photochemically recorded surfaces were studied with AFM and FFM [41].

2.2.2.7 Applications of SPM to Molecular Devices

In addition to their capacity for local analyses, for example, the SSPM can be used to read information written into localized domains. Figure 2.2-13 illustrates detection by the tip of the SSPM (a noncontact mode) of the charge-separated state of an A–S–D triad surrounded by antenna molecules [42] in the form of a

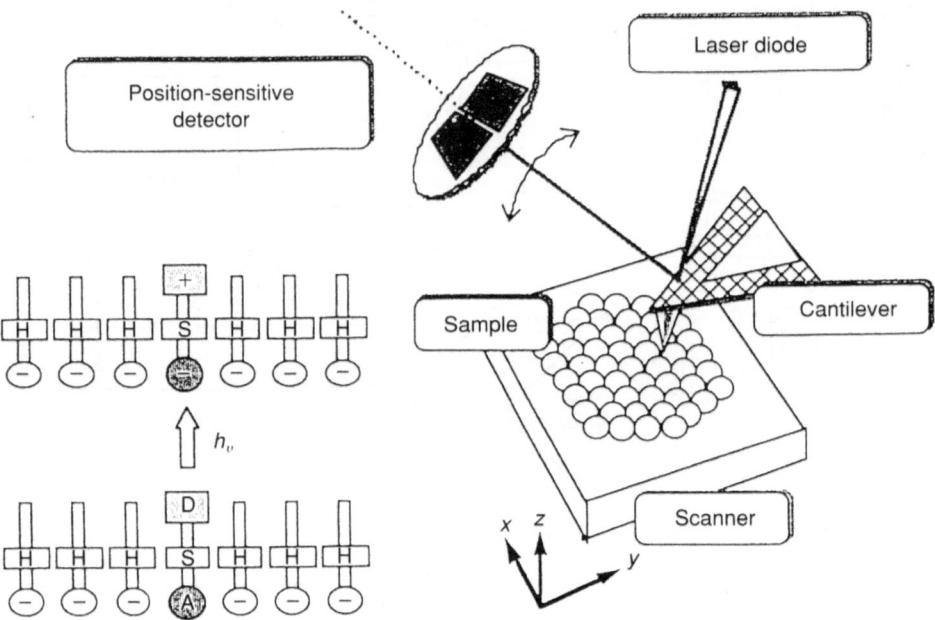

FIG. 2.2-13. Detection of a photoinduced change in surface potentials of an island of a mixed monolayer of an A–S–D triad and antenna H molecules surrounded by an FC sea

single isolated island after photoirradiation [29]. In this case a unit of information is stored in each island. The SSPM can also be used to study the photoresponse of the natural reaction centers (RCs) [29] in the form of monolayer assemblies. In the latter case, each RC can play a role of a unit of information. Recently, we succeeded in obtaining a molecularly resolved FFM image of chromatophores [29]. A localized charge was injected into the surface of an insulator and was detected in the noncontact mode [43].

The high lateral resolution in the illumination can be done by SNOM irradiation with the sharpened tip of an optical fiber. Figure 2.2-14 illustrates that the photoresponse of a single-oriented triad molecular photodiode can be observed in contact mode (via tunneling current) with a SNOM tip having a protruded STM tip. The tunneling through a single molecule is attempted by the use of the STM tip [44]. Here, it should be noted that the electron transfer reaction proceeding in the individual molecule of the RC or the triad looks like one molecular process, but a large number of vibrational modes of the polypeptide chains or of the LB assemblies are involved through the solvent organization energy [29]. In other words, a single molecular device such as the RC and the triad molecule surrounded by a large number of antenna molecules should clearly exhibit its molecular photodiode function without suffering from serious noise by thermal fluctuation. Thus, if the process is repeated many times during the observation, time-averaging of the process of individual molecules may afford a meaningful

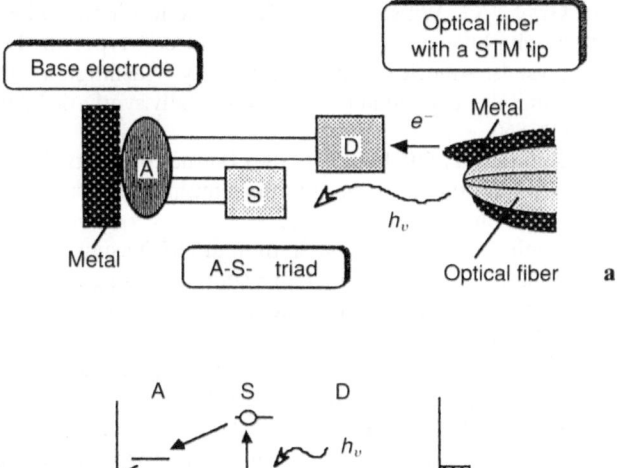

FIG. 2.2-14. Photocurrent measurements of single-oriented triad molecules illuminated with a SNOM tip having a protruded STM tip: **a**, structure; **b**, energy diagram

result. How many events in a molecule, or how many molecules in an event, are necessary as a definite signal source is another interesting subject for future study.

References

1. Fendler JH (1982) Membrane mimetic chemistry. Wiley-Interscience, New York, Chap. 6
2. Bangham AD, Horne RW (1964) Negative staining of phospholipids and their structural modification by surface-active agents as observed in the electron microscope. J Mol Biol 8:660–668
3. Gebicki JM, Hicks M (1973) Ufasomes are stable particles surrounded by unsaturated fatty acid membranes. Nature 3243:232–234
4. Hargreaves WR, Deamer DW (1978) Liposomes from ionic, single-chain amphiphiles. Biochemistry 17:3759–3768
5. Kunitake T (1992) Synthetic bilayer membranes: molecular design and self organization. Angew Chem, Int Ed Engl 31:709–726
6. Fuhrhop JH, Köning J (1994) Membranes and molecular assemblies: the synkinetic approach. Royal Society of Chemistry, Cambridge, Chap. 2
7. Brockerhoff H (1977) In: van Tamelen EE (ed) Bioorganic chemistry vol 3. Academic, New York
8. Tanford C (1973) Micelle shape. In: The hydrophobic effects. Formation of micelles and biological membranes. Wiley-Interscience, New York, pp 71–80
9. Israelachvili JN, Mitchell DJ, Ninham BW (1976) Theory of self-assembly of hydrocarbon amphiphiles into micelles and bilayers. J Chem Soc, Faraday Trans 2:1525–1568

10. Gray WG, Winsor PA (1976) Genetic relationships between non-amphiphilic and amphiphilic mesophases of the "fused" type. In: Friberg S (ed.) Lyotropic liquid crystals. American Chemical Society, Washington, D.C., pp 1–12
11. Kunitake T, Okahata Y (1977) A totally synthetic bilayer membrane. J Am Chem Soc 99:3860
12. Kunitake T, Okahata Y (1980) Formation of stable bilayer assemblies in dilute aqueous solution from ammonium amphiphiles with the diphenylazomethine segment. J Am Chem Soc 102:549–553
13. Kunitake T, Okahata Y, Shimomura M, Yasunami S, Takarabe K (1981) Formation of stable bilayer assemblies in water from single-chain amphiphiles. Relationship between the amphiphile structure and the aggregate morphology. J Am Chem Soc 103:5401–5413
14. Cho I, Park JG (1987) Giant helical superstructures formed by cationic cholesterol-containing polymers. Chem Lett 977–978
15. Menger FM, Yamasaki Y (1993) Hyperextended amphiphiles. Bilayer formation from single-tailed compounds. J Am Chem Soc 115:3840–3841
16. Kunitake T, Kimizuka N, Higashi N, Nakashima N (1984) Bilayer membranes of triple-chain ammonium amphiphiles. J Am Chem Soc 106:1978–1983
17. Kimizuka N, Ohira H, Tanaka M, Kunitake T (1990) Bilayer membranes of four-chained ammonium amphiphiles. Chem Lett 29–32
18. Okahata Y, Kunitake T (1979) Formation of stable monolayer membranes and related structures in dilute aqueous solution from two-headed ammonium amphiphiles. J Am Chem Soc 101:5231–5234
19. Kimizuka N, Kawasaki T, Kunitake T (1993) Self-organization of bilayer membranes from amphiphilic networks of complementary hydrogen bonds. J Am Chem Soc 115:4387–4388
20. Kim JM, Kunitake T (1989) Stabilization of a phosphate molecular bilayer in organic media by complexation with Ca^{2+} ion. Chem Lett 959–962
21. Kunieda H, Nakamura K, Evans DF (1991) Formation of reversed vesicles. J Am Chem Soc 113:1051–1053
22. Ishikawa Y, Kuwahara H, Kunitake T (1994) Self-assembly of bilayers from double-chain fluorocarbon amphiphiles in aprotic organic solvents: thermodynamic origin and generalization of the bilayer assembly. J Am Chem Soc 116:5579–5591
23. Kuwahara H, Hamada M, Ishikawa Y, Kunitake T (1993) Self-organization of bilayer assemblies in a fluorocarbon medium. J Am Chem Soc 115:3002–3003
24. Gaines Jr GL (1966) Insoluble monolayers at liquid gas interfaces. Wiley, New York
25. Kuhn H, Möbius D, Bucher H (1973) Spectroscopy of monolayer assemblies. In: Weissberger A, Rositer BW (eds) Techniques of chemistry vol 1. Wiley, New York
26. Ulman A (1991) An introduction to ultra thin films from Langmuir–Blodgett to self-assembly. Academic Press, New York
27. Roberts G (ed) (1990) Langmuir–Blodgett films. Plenum, New York
28. Hönig D, Möbius D (1991) Direct visualization of monolayers at the air–water interface by Brewster angle microscopy. J Phys Chem 95:4590–4592
29. Fujihira M (1995) Study of thin organic films by various scanning force microscopes. In: Güntherodt H-J, Anselmetti D, Meyer E (eds) Forces in scanning probe methods. Kluwer, Dordrecht, pp 567–591
30. Fujihira M, Monobe H, Muramatsu H, Ataka T (1995) Measurements of lateral distribution of fluorescence intensities and fluorescence spectra of microareas by a

combined SNOM and AFM. Ultramicroscopy 57:118–123; Muramatsu H, Chiba N, Ataka T, Monobe H, Fujihira M (1995) Scanning near-field optic/atomic-force microscopy. Ultramicroscopy 57:141–146; Fujihira M (1996) Fluorescence microscopy and spectroscopy by scanning near-field optical/atomic force microscopy (SNOM–AFM). In: Nieto-Vesperinas M, Garcia N (eds) Optics at the nanometer scale. Kluwer, Dordrecht, pp 205–221

31. Meyer E, Overney R, Lüthi R, Brodbeck D, Howald L, Frommer J, Güntheodt H-J, Wolter O, Fujihira M, Takano H, Gotoh Y (1992) Friction force microscopy of mixed Langmuir–Blodgett films. Thin Solid Films 220:132–137; Overney RM, Meyer E, Frommer J, Brodbeck D, Lüthi R, Howald L, Güntherodt H-J, Fujihira M, Takano H, Gotoh Y (1992) Friction measurements on phase-separated thin films with a modified atomic force microscope. Nature 359:133–135; Fujihira M (1997) Friction force microscopy of organic thin films and crystals. In: Bhushan B (ed) Micro/nanotribology and its applications. Kluwer, Dordrecht, pp 239–260.

32. Fujihira M, Gotoh Y (1992) Polyion complexed Langmuir–Blodgett films. In: Göpel W, Ziegler C (eds) Nanostructures base on molecular materials. VCH Publishers, Weinheim, pp 177–193

33. Meyer E, Overney R, Brodbeck D, Howald L, Lüthi R, Frommer J, Güntherodt H-J (1992) Friction and wear of Langmuir–Blodgett films observed by friction force microscopy. Phys Rev Lett 69:1777–1780

34. Vogel V, Möbius D (1988) Local surface potentials and electric dipole moments of lipid monolayers: contributions of the water/lipid and the lipid/air interfaces. J Colloid Interface Sci 126:408–420

35. Parsons R (1954) In: Bockris JO'M (ed) Modern aspects of electrochemistry. Butterworths, London, vol 1, pp 103–179

36. Yokoyama H, Saito K, Inoue T (1992) Scanning Maxwell stress microscopy. Mol Electron Bioelectron 3:79–88

37. Fujihira M, Kawate H, Yasutake M (1992) Scanning surface potential microscopy for local surface analysis. Chem Lett 1992:2223–2226; Fujihira M, Kawate H (1994) Scanning surface potential microscope for characterization of Langmuir–Blodgett films. Thin Solid Films 242:163–169; Fujihira M, Kawate H (1994) Structural study of Langmuir–Blodgett films by scanning surface potential microscopy. J Vac Sci Technol B 12:1604–1608

38. Pohl DW, Courjon D (eds) (1993) Near field optics. NATO ASI Ser E, vol 242. Kluwer, Dordrecht

39. Bezig E, Trautman JK (1992) Near-field optics: microscopy, spectroscopy, and surface modification beyond the diffraction limit. Science 257:189–195

40. van Hulst NF, Moers MHP, Noordman OFJ, Faulkner T, Segerink FB, van der Werf KO, de Grooth BG, Bölger B (1992) Operation of a scanning near-field optical microscope in reflection in combination with a scanning force microscope. Proc SPIE 1639:36–43

41. Fujihira M, Monobe H, Muramatsu H, Ataka T (1995) Near-field optical microscopic recording on Langmuir–Blodgett films and chemically modified surfaces. Ultramicroscopy 57:176–179.

42. Fujihira M (1995) Photoinduced electron transfer and energy transfer in Langmuir–Blodgett films. In: Birge RR (ed) Molecular and biomolecular electronics. Advances in Chemistry Ser 240, American Chemical Society, Washington, pp 373–394; Fujihira M, Sakomura M, Aoki D, Koike A (1996) Scanning probe microscopies for molecular photodiodes. Thin Solid Films 273:168–176; Fujihira M (1995) Photoinduced electron

transfer in monolayer assemblies and its application to artificial photosynthesis and molecular devices. In: Ulman A (ed) Thin films, vol 20. Academic Press, San Diego, pp 239–277

43. Stern JE, Terris BD, Mamin HJ, Rugar D (1988) Deposition and imaging of localized charge on insulator surfaces using a force microscope. Appl Phys Lett 53:2717–2719

44. Pomeranz M, Aviram A, McCorkle RA, Li L, Schrott AG (1992) Rectification of STM current to graphite covered with phthalocyanine molecules. Science 255:1115–1118

2.3 Specially Prepared Molecular Systems

TAKEO SHIMIDZU (2.3.1)
AKIRA FUJISHIMA and YOSHIO NOSAKA (2.3.2)
MASASHI KUNITAKE and KINGO ITAYA (2.3.3)

2.3.1 Ultrathin Films

A film is regarded as a two-dimensional material which separates two outside phases with a large contact area and transduces energy and/or information through carriers such as electrons, holes, photons, ions, molecules, etc. Recent developments in electronics technology and biological science have awakened interest in so-called ultrathin films. However, the concept and definition of ultrathin films cover many different phases. The various kinds of organic molecules, all of which have their own specific structures and functions, indicate the potential possibility of many ultrathin films, each with a unique function. To realize this possibility, the specific structures, intrinsic properties, and unique functions, with respect to the ultimate function of the ultrathin films, should be considered systematically. There have been many examples of technological applications of ultrathin films, and several phases in the science of ultrathin films of organic molecules, but there are many essential problems yet to be solved.

In this section, these problems are discussed, and several examples of attempts to make functional ultrathin films are given.

2.3.1.1 Ultrathin Films of Organic Molecules

The recent remarkable progress of technologies to fabricate and evaluate ultrathin films in the fields of electronics, optics, photonics, etc., based on inorganic and metallic compounds, gives reason to hope that new types of ultrathin film consisting of organic molecules will become a reality. Unlike inorganic and metallic materials, organic molecules, with their intrinsic functions, could be applied as nanometer-sized functional elements in the ultrathin structure. There are many parameters to be considered in the design of novel structure-specific functions.

The thickness of the film is an essential factor. This governs the kinetics of the transport phenomena through the film, at which the interfacial effect appears. The optimum thickness depends on the kind of functions, e.g., conductivity (electron, hole), superconductivity, magnetism, dielectric property, optical property, selective transport, etc. Electric and photonic functions in ultrathin films are to control electron, hole, and photon transport, not only in a thin but homogeneous medium such as conventional thick films, but also in a band-modulating medium such as a "man-made superlattice" [1], which is the ultimate functional material. When the thickness is reduced to the dimension of the mean free path of an electron in the material concerned, a quantum size effect appears as a result of confinement of the carriers. The stability, degradation, wettability, thermal property (glass transition), etc. of the film are also influenced by its thickness.

The properties of ultrathin films depend on the structures of the molecules and their assemblies. There are many factors to be controlled in ultrathin films of various organic molecules with various functionalities.

Molecular structure is one of the most important factors governing the morphological structure of ultrathin films, and this is directly reflected in the film's properties. The morphology of the molecular assembly is also important. Molecular orientation, structural dimensionality, the regularity of the assembly, and the phase all affect the functions of ultrathin films. There are many possible ways to fabricate functional ultrathin films. For example, the organic molecules themselves have asymmetrical structures, so their ordered assemblies in the film provide a local anisotropic field which perturbs electron transfer and the light path. The local anisotropic field is attributed to the effects of the dipole, Coulombic properties of the molecule. All the parameters come from the structure and dimensionality of the molecule concerned. Macroscopic dielectric polarization also results from these molecular properties. An ultrathin film of oriented molecules is always anisotropic. The fine functions with delicate interactions observed in biological system result from metabolic processes in well-balanced reaction fields.

The properties of ultrathin films of organic molecules are governed by their manner of assembly at molecular or mesoscopic levels as well as bulk level. As an example, a metalloporphyrin thin film prepared by the casting or spin-coating method is an amorphous film, while a chemical vapor deposition method gives crystalline, polycrystalline, or microcrystalline films. It is well accepted that the crystalline boundary acts as a carrier trap to depress the efficiency of a sandwich-type metal/porphyrin/semiconductor solar cell. The quantitative importance of electron and hole transfers is not yet clear. The thermal fluctuations of the molecules as well as their intermolecular interactions should be considered. In addition, the arrangement of the molecules, and their interfacial structure on the inorganic semiconductor or metal substrate are clearly important factors which as yet are not understood. Ultrathin films play a significant role in, for example, the Schottky contact, which controls electron and/or hole transfer by depressing a recombination of paired carriers at the interface. Organic thin films work as an insulator, a semiconductor, and a conductor, depending on their properties [2, 3]. The charged carrier in ultrathin films of a molecular crystal generated by an extrinsic process, and the transport of one through tunneling or by a hopping mechanism, depend on the distance between the molecules.

2.3.1.2 Fabrication

The fabrication of ultrathin films can be by either wet or dry processes. Casting, spin-coating, electrodeposition, electropolymerization, and the Langmuir–Blodgett (LB) method are called wet processes, while vapor deposition, chemical vapor deposition (CVD), plasma polymerization, molecular beam epitaxy (MBE), and metal organic CVD (MOCVD) are dry processes. Such fabrication processes for ultrathin films are summarized in Fig. 2.3-1. Every process has a limiting factor to control the thickness of the film, so there are fewer fabrication processes for ultrathin films with nanometer structures. The processibility and layer controllability are also important factors in fabricating ultrathin films.

In the fabrication of ultrathin films, molecular orientation and the ordering of the molecules are the most important factors, and they govern the functions of ultrathin films and their efficiencies. There are many factors affecting the control of the two-dimensional fabrication of molecular thin films, for example, the shape of the molecule (linear, planar, cubic or bulky) and its intermolecular interaction with surrounding molecules and the medium. If film fabrication is to be attempted, the fabrication conditions, such as the temperature of the substrate, deposition rate, vacuum quality, etc., should be considered. A fundamental consideration with respect to the deposition of molecules in a dry process is that the deposited molecules are not always thermodynamically stable when the deposition is not in the equilibrium condition, and then a rearrangement of the deposited molecules may take place. For example, while the higher temperature of the substrate and a low vacuum condition gave less-ordered stearic acid assembly onto a highly oriented pyrolytic graphite (HOPG) substrate, the higher deposition rate achieved in a high vacuum gave a well-oriented assembly. These phe-

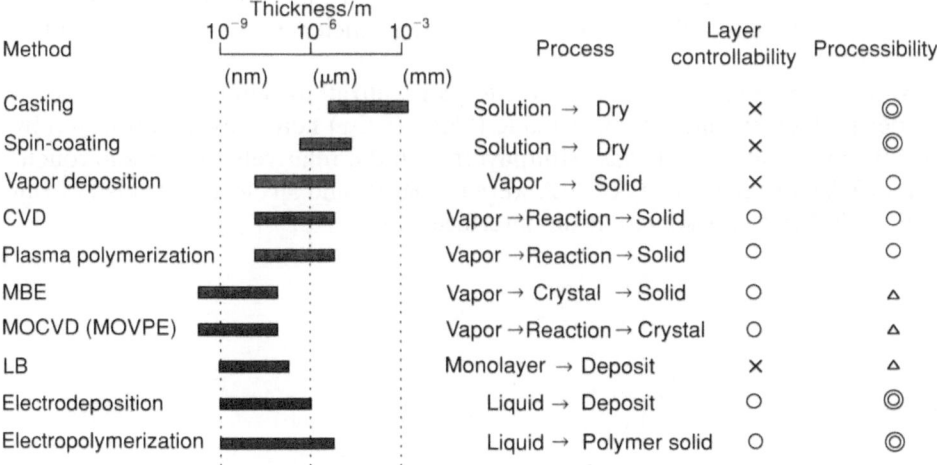

FIG. 2.3-1. Fabrication processes for ultrathin film membranes

nomena tell us that residual gaseous molecules (except the depositing molecules), even in a vacuum, depress the deposition of stearic acid molecules aligned perpendicular to the substrate, and that a reorientation of the deposited molecules may take place. The ultimate thin film is a LB film, whose unit structure is a monolayer. However, the density of the functional moiety, which is usually attached on the hydrophobic part of the amphiphile, is diluted by the other hydrophobic part. However, LB film gives a remarkably high anisotropic ultrathin film.

Molecular manipulation in organic ultrathin films by scanning tunneling microscopy (STM) presents two- and three-dimensional functional structures. The poling procedure forces molecular dipoles to be aligned for nonlinear optical materials.

2.3.1.3 Heterolayer

At present, organic molecular material whose function results from an ultrathin film and the function of that material are attributed to an intrinsic property. However, there are several functional materials which consist of an organic molecular film which is almost an ultrathin film. For example, a field effect transistor (FET) sensor, an organic electroluminescent device (EL), and an electrochromic device (ECD) are based on heterostructures such as a p–n junction. A p–n heterolayer-type organic solar cell, which is more effective than one based on the Schottky junction, has been demonstrated by effectively utilizing the photoresponse of p–n junctions. In a solar cell, Zn phthalocyanine (ZnPc) as the p-type semiconductor layer and 5,10,15,20-tetra(3-pyridyl)porphyrin (T(3-Py)P) as the n-type semiconductor layer were fabricated into a p–n heterolayer junction, defined as $Au/ZnPc(50\,nm)/T(3-Py)P(7\,nm)/Al$ organic cell [2]. The efficiency of this cell decreased with the thickness of the T(3-Py)P layer, which was interpreted as meaning that the thick (T(3-Py)P) layer absorbs the incident light and a negative filter effect decreases the efficiency. This is illustrated in Fig. 2.3.2.

Another example of organic and inorganic ultrathin heterolayers is that of copper phthalocyanine–titanium oxide (CuPc–TiO_x) heterolayers fabricated by reactive evaporation methods. Multilayers with the relatively high-grade roughness of 7 Å and a periodicity of 40 Å were prepared, and an electron transfer from CuPc to TiO_x was observed at the interfaces [3].

FIG. 2.3-2. Charge separation processes induced by light absorption in a Au/ZnPc/TPyP/ Al cell

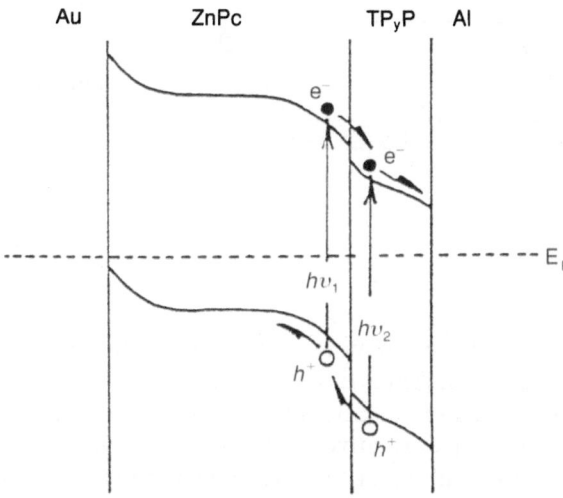

2.3.1.4 Anisotropic Conducting Organic Heteromultilayers

Utilizing individual molecular asymmetry properties and fabricating molecularly aligned ultrathin or layered materials, highly anisotropic and specific materials

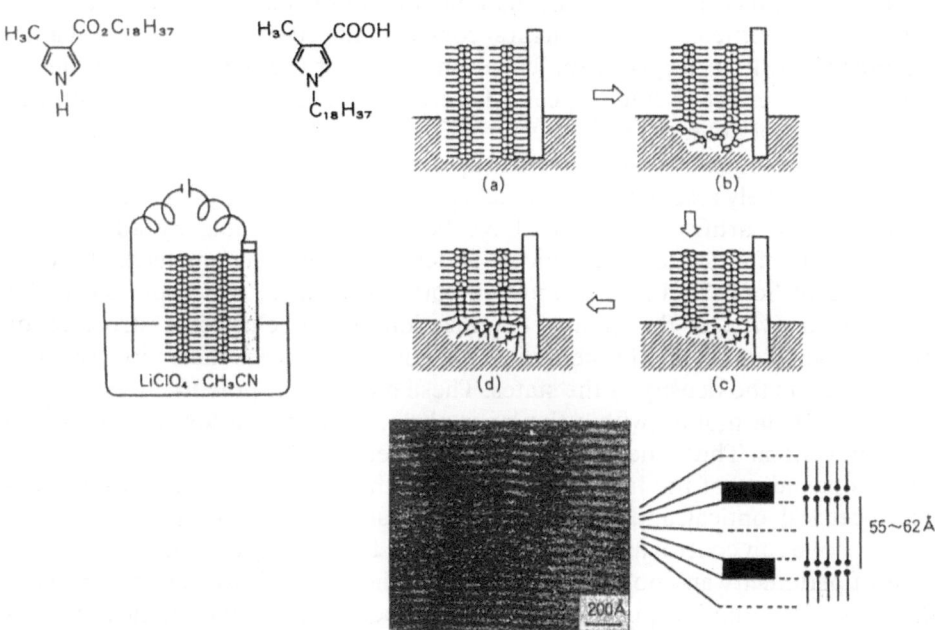

FIG. 2.3-3. Polymerization of pyrrole derivative Langmuir–Blodgett multilayers and transmission electron microscopy picture of the resulting polymer multilayers

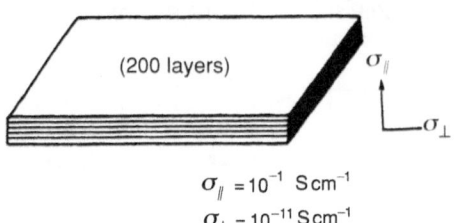

FIG. 2.3-4. Anisotropic conductivity of pyrrole derivative Langmuir–Blodgett multilayers

(200 layers)

$\sigma_{\parallel} = 10^{-1}\ S\,cm^{-1}$
$\sigma_{\perp} = 10^{-11}\ S\,cm^{-1}$

are found even in molecular level ultrathin layers. A highly anisotropic conduct-ing ultrathin film is described below. This film was fabricated by electrochemical polymerization of LB multilayered film consisting of amphiphilic thiophene [4] or pyrrole [5] derivatives. The polymerization and a property of the resulting ultrathin multilayer film are shown in Figs. 2.3.3 and 2.3.4, respectively. The DC electrical conductivities in the lateral and perpendicular directions are 10^{-1} and $10^{-11}\,S\,cm^{-1}$, respectively. The anisotropy reaches a magnitude of 10 orders. The polymerization of each layer is performed as shown in Fig. 2.3.3.

2.3.1.5 Conjugated Polymer Heterolayer Superlattices

Most conjugated polymers are organic semiconductors with mesoscopic struc-tural control. The compositional modulation of copolymer thin film in the depth direction should correspond to the manipulation of the electronic band structure in the film. In particular, with structural control in the 10–100 Å scale, which is the mesoscopic scale in terms of the coherent length of the electron concerned, a quantum-size effect should appear where a carrier is confined in a restricted structure, e.g., a potential well structure. Ever since the paper by Esaki and Tsu in 1970 [1], inorganic semiconductor superlattices and multiple quantum wells have been actively researched in the field of semiconductor physics and materials science. These artificial materials have been made available by advances in ultrathin film fabrication technology under ultrahigh vacuum technology, as described in Sect. 2.3.1.2. The most important advance, which produced this increased activity, is the energy quantization of the electronic structure by mesoscopically modulated potentials. This not only changes the energy scheme, but also alters the density of the states. These mesoscopic potential modulations restrict electron motion within the layer plane, leading to a lower dimensional electron system. Thus a new material can be created. These features give rise to an understanding of the intrinsic functions in these ultrathin materials, such as new electrical, optical, and magnetic properties specific to the designed structure. Because one can control and prescribe all these features by adjusting the param-eters of periodicity and potential difference in the layers, as well as the quality of the band discontinuities, new materials with specific, desirable physical proper-ties can be constructed. This is known as "energy band engineering." Further-more, these structured materials will lead the way to the realization of new devices constructed by "wave function engineering" [6]. In recent years, the high electron mobility transistor, the quantum well laser, and the avalanche photo-diode, for example, have had a profound impact in fields where mesoscopically constructed structures convey many advantages. These excellent devices, the quantum heterostructures, have become known as artificial structured materials or engineered electronic structured materials.

Along the same lines a conjugated polymer heterolayers superlattice has been fabricated by the potential-programmed electropolymerization method [6]. The electropolymerizable monomers have their own redox potentials, so that their rate of polymerization depends on the applied potential in the electropolymerization process. The growing oligomer or polymer in the vicinity

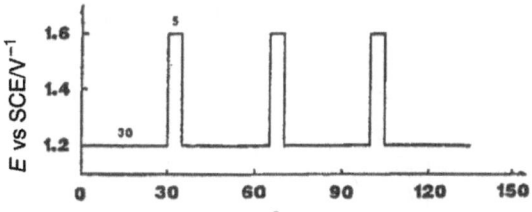

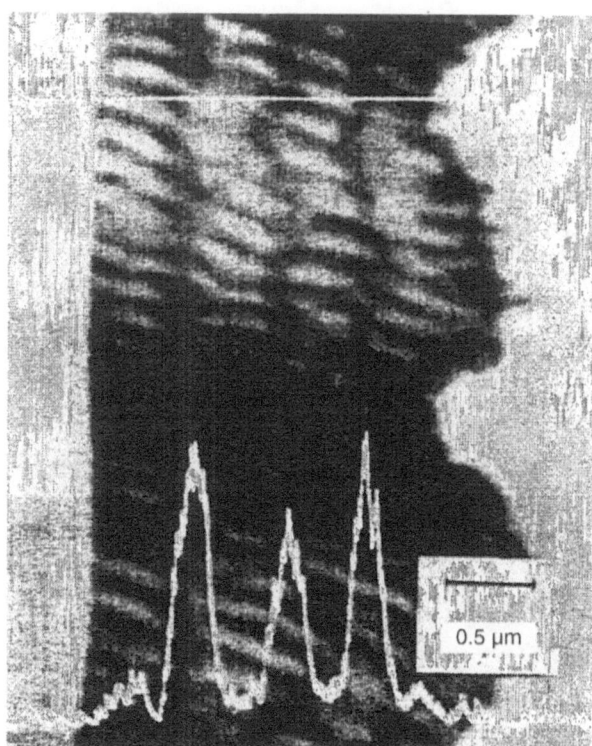

FIG. 2.3-5. *Top*, the applied rectangular potential sweep function. *Bottom*, transmission electron microscopy cross section and electron-probe microanalysis on sulfur of a polypyrrole-poly(3-methyl-thiophene) composite thin film. The analyzing line is the straight white line

of the electrode depends on the polymerization potential, and is then deposited on the electrode if it becomes insoluble in the solvent used. This method can give a structure with any depth profile, or with a layered, sloped, or non-periodic profile. Figures 2.3-5 and 2.3-6 show schematic representations of this method, and an examples of the fabrication of ultrathin heteromultilayers with pyrrole and 3-methylthiophene, respectively. The layer thickness is controlled by polymerization conditions such as polymerization time, etc.

Figure 2.3-7 shows band structures of several homopolymers and the copolymer of bithienyl and pyrrole. Any type of superlattice can be made with a combination of these polymers. Figure 2.3-8 shows the photoluminescence versus the well layer thickness of Type II copolymer heterolayer superlattices consisting of different compositional copolymers of pyrrole and bithiophene. The maxima of the photoluminescences of these multilayer superlattices shifted to higher

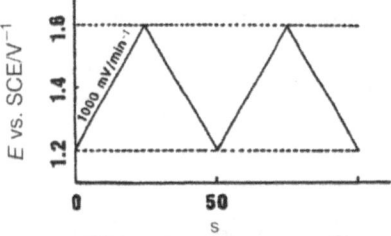

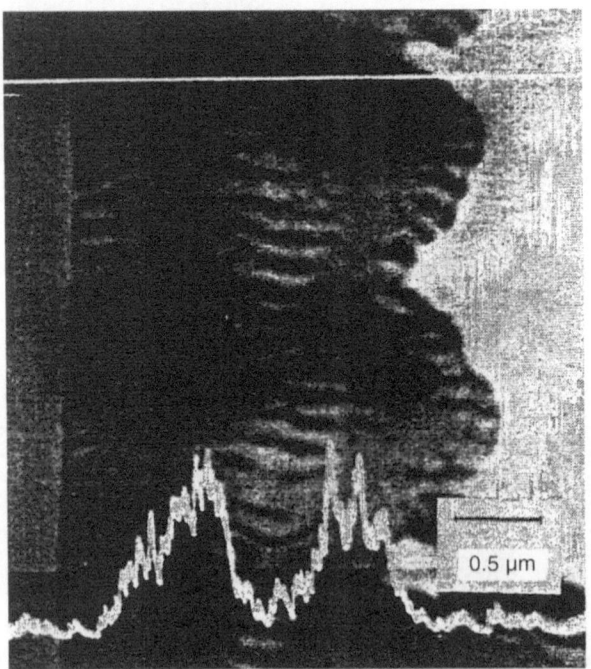

FIG. 2.3-6. *Top*, the applied triangular potential sweep function. *Bottom*, transmission electron microscopy cross section and electron-probe microanalysis on sulfur of a polypyrrole-poly(3-methyl-thiophene) composite thin film. The analyzing line is the straight white line

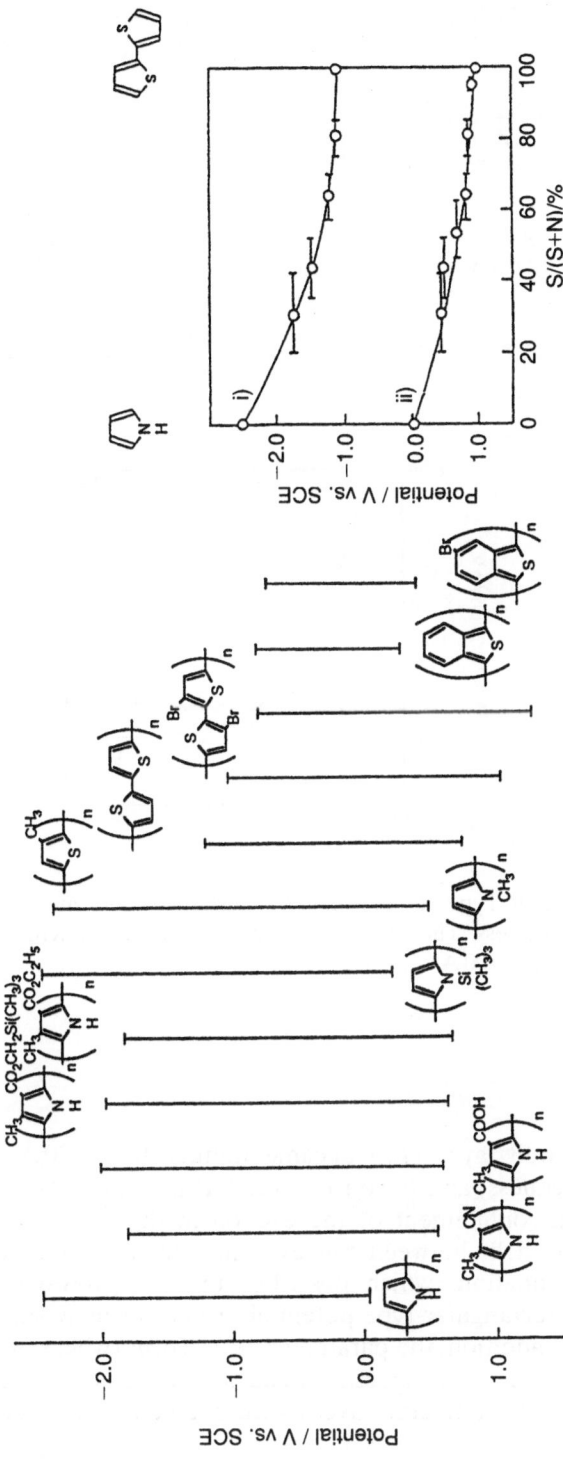

FIG. 2.3-7. Band structures (E_c and E_v) of (*left*) conjugated homopolymers and (*right*) the copolymer

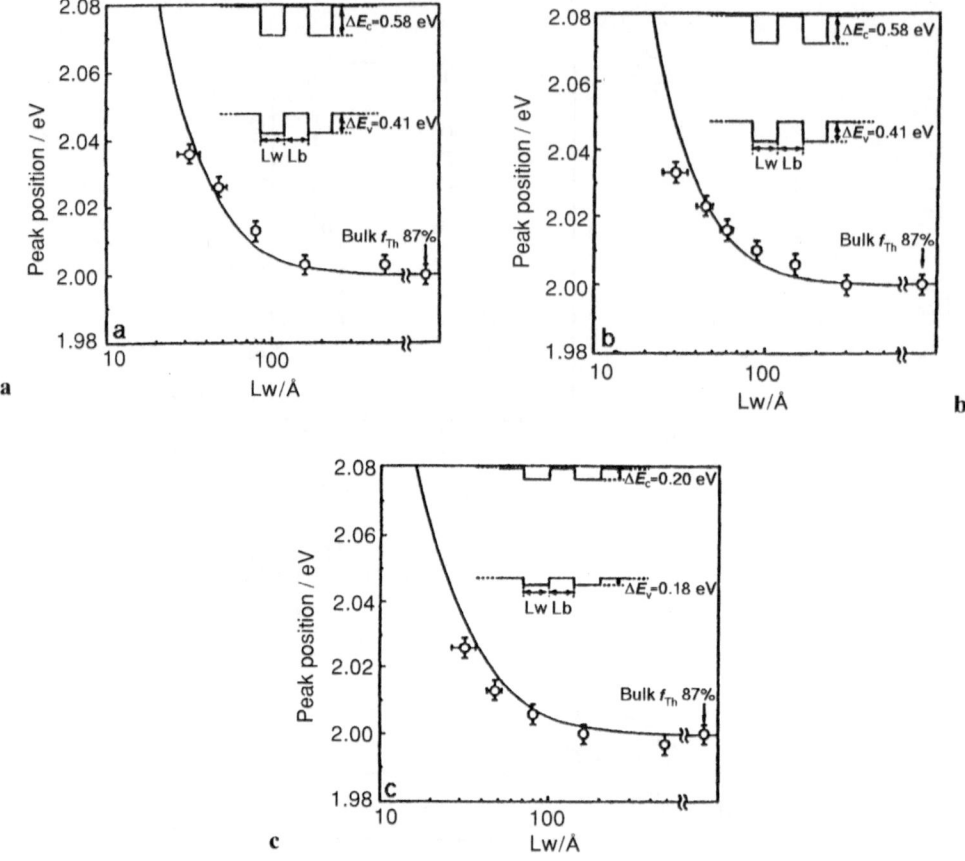

FIG. 2.3-8. Structure of Type II heterolayer superlattices and emission peak shift as a function of layer thickness. The solid line is estimated from the Kronig–Penney model. **a** $Lw/Lb = 0.6$. **b,c** $Lb = 100\,\text{Å}$ constant

energies as the well layer (L_w) became thinner than $120\,\text{Å}$, keeping L_w/L_b (barrier layer thickness) and L_b constant ($100\,\text{Å}$) in both cases. Such high energy shifts result from confinement of the exciton in the fabricated quantum well structure. The good fit between the experimental data and results from the Kronig–Penney equation, which describes the energy–wave number vector relationship in a rectangular-type potential profile, strongly supports the quantum size effect. In addition, the parameter $m^* = 0.6\,m_e$ (where m_e is free electron mass) was obtained. This spectral behavior leads to the conclusion that conjugated polymer multiheterolayers with a quantum size effect have been formed.

2.3.2 Ultrafine Particles

Photoprocesses in solid materials are described in a different manner from those in dye molecules. For molecules, electronic energy states are defined in terms of the particular distribution of electrons around the atoms which form the molecule. On the other hand, electronic levels for solid materials are described by the periodicity of the atomic sequence in the crystallites. Ultrafine particles, whose size is of the order of nanometers, are intermediate between molecules and solids in the representation of electronic states. However, their chemical and physical properties may change gradually.

Solid semiconductor materials are commonly used for photodevices and this has a significant size-effect on their electronic states. This effect originates from the fact that the de Broglie wavelength of the carriers in semiconductors becomes

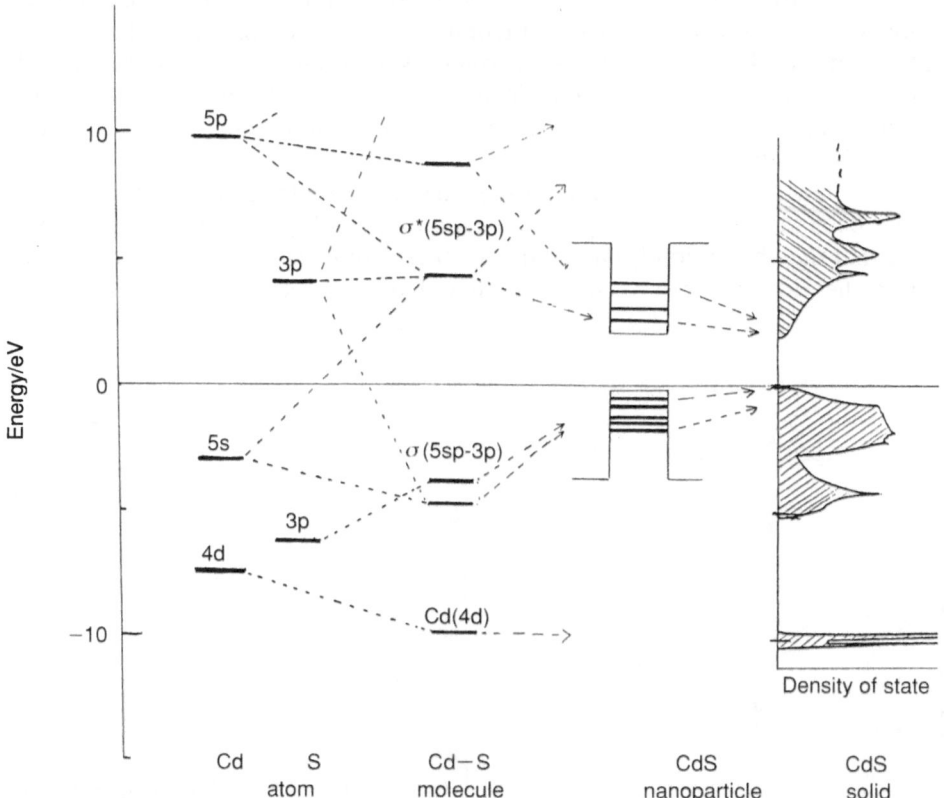

FIG. 2.3-9. Schematic correlation of electronic energy levels. Energy levels for the Cd and S atoms and the Cd–S molecule were calculated with the Gaussian program with STO-3G basis. The density of states in solid CdS was taken from Zunger and Freeman (Phys Rev B (1978) 17:4850)

of the same order as the crystallite size. The energy levels of electrons and holes in semiconductor nanoparticles are easily understandable on the basis of the particle-in-a-box model in quantum mechanics. Since the subject of nanometer-sized particles is interdisciplinary and stems from various scientific roots, these materials are designated by many different terms: ultrafine particles, ultrasmall particles, nanocrystallites, microclusters, quantized particles, Q-particles, quantum dots, etc.

Cadmium sulfide, which has been used both as a yellow pigment and in photo-cells, is one of the most intensively investigated nanoparticle semiconductors. Figure 2.3-9 shows changes in the electronic energy levels of isolated Cd and S atoms, a hypothetical diatomic Cd–S molecule, a spherical CdS quantum-size structure, and solid CdS. In this figure, the energy levels below the null line are all occupied by electrons, and light absorption occurs transversely across this line. Thus, the size effect leads to a distinctive absorption spectrum for the nanoparticles.

Figure 2.3-10 shows a plot of peak wavelengths in the absorption spectrum as a function of particle size [7]. Experimental data are shown as the symbols $\bigcirc$, $\bigcirc$, and $\triangle$. The solid line was calculated using a simple finite-depth quantum well model. Although other more sophisticated methods have been applied to calculate electronic states, the importance of coupling with angular momentum has recently been confirmed [8].

The chemical structure of some of the CdS clusters (designated $\bigcirc$ in Fig. 2.3-10) has been identified as a polynuclear cadmium–thiolate complex. It should be noted that the absorption wavelengths for these complexes also lie on the same line as the CdS particles. When thiophenol is used as a capping agent, the

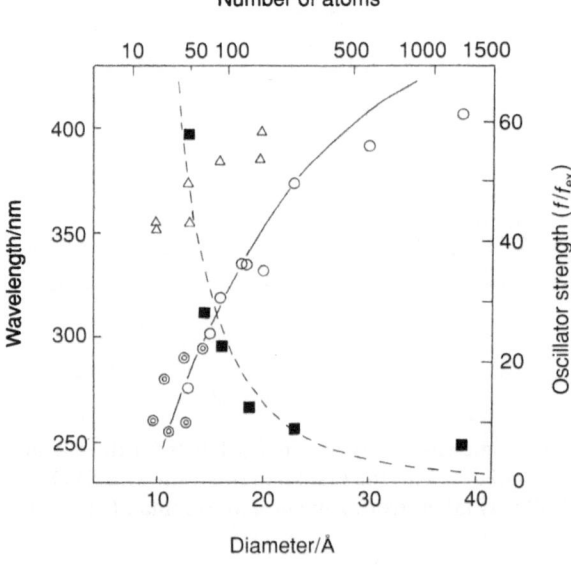

FIG. 2.3-10. Wavelength of the absorption peaks ($\bigcirc$, $\bigcirc$, and $\triangle$) and corresponding oscillator strength ($\blacksquare$) of CdS nanoparticles plotted as a function of the diameter [9]

absorption peak shifts to a longer wavelength (designated $\triangle$). This discrepancy indicates that there is some electronic interaction between the CdS solid and the surface-attached phenyl group, and may provide an example which could help our understanding of the surface electronic states of nanoparticles.

A remarkable observation in the spectrum is the increase in absorbance with decreasing size. Since theoretical calculations predict that the oscillator strength of a particle is independent of its size, the oscillator strength per atom is reciprocally proportional to the cube of the diameter. Weller and co-workers [9] proved this relationship experimentally, as replotted in Fig. 2.3-10. A sudden increase in the oscillator strength normalized to that of the bulk exciton (f_{ex}) is observed at a diameter of 15 Å. This increase gives rise to a large optical third-order nonlinearity $\chi^{(3)}$ [9] of the semiconductor nanocrystallite.

The optical nonlinearity of materials attracts much interest in the development of future optical signal-processing systems. Bistable optical switching with rise and fall times of 25 ps has been demonstrated using a filter glass doped with microcrystalline CdS_xSe_{1-x}, as shown in Fig. 2.3-11 [10]. Optical memory function basically involves the fact that two values of output power can be exhibited at one input power, depending on whether higher input-power was received previously. In order to achieve a high-speed, low-power device, increases in nonlinearity are essential. It has been shown that optical nonlinearity increases with the cooperation of a large number of nanoparticles which are aligned to form a distinctly spaced lattice [11]. A coating made up of CdSe nanoparticles with spacer molecules for the self-organization of three-dimensional quantum dot superlattices has been reported [12].

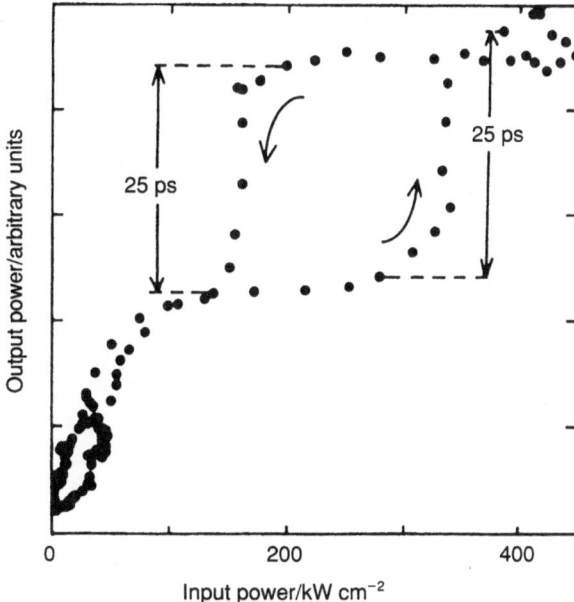

FIG. 2.3-11. Optical bistability with 25-ps switching time [10]

As described above, the most distinctive feature of semiconductor nanoparticles is that the excitation energy changes with size. Consequently, the color of the substance can be varied by changing the size. This property is applicable to multicolored electroluminescent devices. CdSe nanoparticles have been embedded in semiconducting p-paraphenylene vinylene [13] and polyvinylcarbazole [14]. The electroluminescence and photoluminescence spectra are nearly identical, and are tunable from 530 to 650 nm by varying the size of the particles [14].

Although silicon is one of the most widely used materials in electronics, luminescence of crystalline Si is forbidden due to the symmetry of its crystal structure. However, Si nanocrystals show high luminescence at room temperature [15]. The size effect is primarily kinetic, and apparently serves to isolate the electron–hole pairs from each other. There is a close correspondence between porous Si luminescence and Si nanocrystal luminescence [15].

Semiconducting materials have found an application as photocatalysts for chemical reactions. Several advantages of decreasing size have been demonstrated [16]: (i) an increase in redox activity due to shifts in the energy levels; (ii) increases in the rate of surface trapping of photoinduced electron and holes; (iii) large surface areas capable of adsorbing large numbers of reactant molecules. The third advantage has been used in nanocrystalline semiconductor electrodes in which a sponge-like porous semiconducting film is in contact with a conductive glass electrode [17–19]. The electrolyte is in contact with the individual nanoparticles, and electron hole separation takes place efficiently, as illustrated in Fig. 2.3-12 [19]. The application of light to electrical energy conversion in solar cells seems promising.

Materials which are similar to the nanoparticle electrodes, also with nanoscopic dimensions, have been prepared using template methods [20]. For example, electrochemical plating of gold within the pores of an aluminum

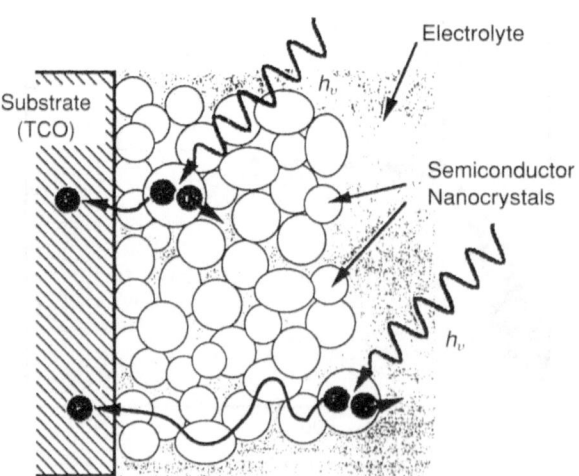

FIG. 2.3-12. Model of the charge carrier separation and charge transport in a nanocrystalline film [19]

anodic oxide template provides a nanometal array. The color resulting from the plasmon resonance can be controlled by the size of the ultrafine metal particles.

Recently, the advantages of preparing nanostructured materials using micelle vesicles, lipid membranes, and other related hosts have been extensively reviewed [21], where epitaxial crystal growth on Langmuir–Blodgett films and other modern wet colloid-chemical techniques are described.

2.3.3 STM of Adsorbed Molecules

Recently, the investigation of adsorbed organic molecules by means of scanning tunneling microscopy (STM) has attracted the attention of researchers working in many different fields. This is primarily owing to the capabilities of STM to visualize topographic and even electronic features of the adsorbed molecules on the atomic scale. Research for chemisorbed molecules on metals has mostly been conducted in ultrahigh vacuum (UHV) conditions, because it is easy to prepare clean, reactive metal surfaces in UHV. In 1988, the first high-resolution STM images of a coadsorbed CO and benzene on Rh(111) were reported [22]. Figure 2.3-13 shows a high-resolution STM image of (3×3) arrays of benzene–CO [22]. The ordered structure of the benzene molecule, i.e., a smooth circular shape with a central hollow spot, can be recognized. In this case, CO molecules hidden between the benzene molecules were not resolved. Ordered and disordered STM images for naphthalene chemisorbed on Pt(111) were also reported [23]. Weiss and Eigler [24] demonstrated the capacity of STM to image an isolated benzene molecule adsorbed on Pt(111) at low temperature (4K). They reported that the shape of benzene in the STM images depended on the adsorption sites on Pt(111). Lippel et al. [25] showed STM images of isolated Cu–phthalocyanine molecules on Cu(100) with atomic-scale features. The image agreed well with the

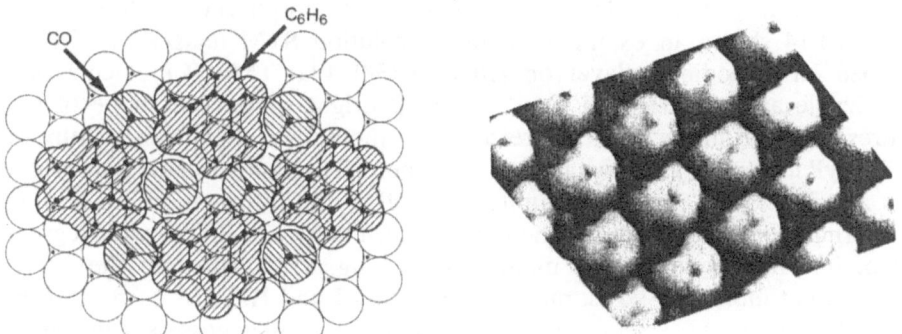

a

b

FIG. 2.3-13. **a** Model structure (3×3) of CO–benzene on Rh(111). **b** An STM image [22]

calculated molecular orbital structure. These results directly demonstrated that STM has potential to reveal electronic structural features in molecules.

However, STM images of isolated molecules were usually low resolution without internal structures because of molecular surface diffusion on the substrate. On the other hand, close-packed molecular adlayers seem to be more suitable systems for investigation by STM. One of the most intensively investigated systems is the ordered layers of liquid crystal compounds. In 1988, the first images with high atomic resolution of 4-n-octyl-4'-cyanobiphenyl (8CB) and 4-(trans-4n-pentylcyclohexyl)-benzonitrile on HOPG were obtained [26]. Many liquid crystal materials have now been investigated on either HOPG or MoS_2 [27]. STM was also employed to examine several other ordered molecular systems: polymers on HOPG prepared by epitaxial polymerization [28], thiol compounds on Au [29], and adsorbed amphipathic molecules on HOPG in air [30]. Organic superconductors such as TTF-TCNQ [31] and β-(BEDT-TTF)I_3 [32] were also successfully imaged by STM with high atomic resolution because of their high conductivity.

STM and related techniques such as atomic force microscopy (AFM) can be used not only in UHV or air, but also in electrolyte solutions. All electrochemical reactions, including many common processes, occur at electrode–electrolyte interfaces [33]. STM has provided, for the first time, a means of observing the atomic structures of electrode surfaces as well as those of adsorbed layers in electrolyte solutions [33]. Many important electrochemical reactions such as metal deposition, reconstruction, etching of semiconductors, and adsorption of organic and inorganic materials have been investigated in solution at atomic resolutions [33, 34].

The formation and characterization of the ordered adlayers of organic molecules at electrode–electrolyte interfaces are very important from a fundamental point of view. Lindsay and co-workers [35, 36] have succeeded in obtaining images of DNA bases such as adenine, guanine and cytosine onto HOPG and Au(111). Srinivasan et al. presented STM images of condensed guanine on HOPG [37].

However, we recently reported successful in situ imaging with extremely high resolution of an ordered adlayer of a water soluble porphyrin in solution [38]. Figure 2.3-14 shows an example of high resolution STM images of TMPyP adsorbed on an iodine adlayer on Au(111) [38]. Flat TMPyP molecules are recognizable as a square with four additional bright spots. The shape of the features observed in the image corresponds to the known chemical structure of the TMPyP molecule, as shown in the same figure. The characteristic four bright spots are due to the pyridinium units. As well as the information concerning the internal molecular structure, the image shown in Fig. 2.3-14 revealed a great deal of information about the molecular packing arrangement in the array. Two different molecular rows, marked by arrows I and II, cross each other at ca. 60°. It can be clearly seen that every TMPyP molecules possesses the same molecular orientation in the row marked I. On the other hand, every second molecule along the row marked II shows the same orientation. The rotation angle

FIG. 2.3-14. A high-
resolution STM image of the
TMPyP adlayer obtained in
0.1 M HClO$_4$ [38, 39]

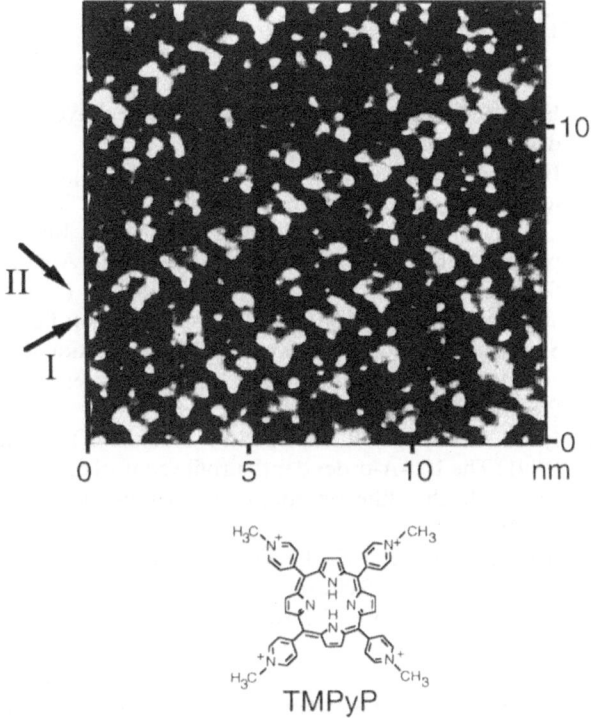

TMPyP

of 45° can be recognized between two neighboring molecules in row II. The high
resolution achieved in solution is comparable to the atomic resolution obtained
in STM studies carried out in UHV [25]. The internal molecular structures,
orientations, and packing arrangements of various organic adlayers have recently
been determined with near-atomic resolution under electrochemical conditions
[39].

The adsorption of organic molecules on electrode surfaces in electrolyte solutions has long been an important subject in electrochemistry for elucidating the
role of molecular properties and the atomic structure of electrode surfaces [40].
Now, we believe that in situ STM allows us to investigate details of molecules,
and of intermolecular and molecule–substrate interactions.

References

1. Esaki L, Tsu R (1970) Superlattice and negative differential conductivity in semiconductors. IBM J Res Dev 14:61–65
2. Harima Y, Yamashita K, Suzuki H (1984) Spectral sensitization in an organic p–n junction photovoltaic cell. Appl Phys Lett 45:1144–1145
3. Takada J, Awaji H, Koshioka M, Nevin WA, Imanishi M, Fukada N (1994) Copper phthalocyanine–titanium oxide multilayers. J Appl Phys 75:4055–4059
4. Sagisaka S, Ando M, Iyoda T, Shimidzu T (1993) Preparation and properties of amphiphilic polythiophene Langmuir–Blodgett films. Thin Solid Films 230:65–69
5. Shimidzu T, Iyoda T, Segawa H (1994) Functionalization of conducting polymer for advanced materials. In: Prasad PN (ed) Frontiers of polymers and advanced materials. Plenum, New York
6. Iyoda T, Toyoda H, Fujitsuka M, Nakahara R, Tsuchiya H, Honda K, Shimidzu T (1991) The 100-Å-order depth profile control of polypyrrole-poly(3-methylthiophene) composite thin film by potential-programmed electropolymerization. J Phys Chem 95:5215–5220
7. Nosaka Y, Shigeno H, Ikeuchi T (1995) Formation of polynuclear cadmium–thiolate complexes and CdS clusters in aqueous solution studied by means of stopped-flow and NMR spectroscopies. J Phys Chem 99:8317–8322
8. Ekimov AI, Hache F, Schanne-Kline MC, Ricard D, Flytzanis C, Kudryavtsev IA, Yazeva TV, Rodina AV, Efros Al L (1993) Absorption and intensity-dependent photoluminesence measurements on CdSe quantum dots: assignment of the first electronic transitions. J Opt Soc Am B 10:100–107
9. Vossmeyer T, Katsikas L, Giersig M, Popovic IG, Diesner K, Chemseddie A, Eychmueller A, Weller H (1994) CdS nanoclusters: synthesis, characterization, size dependent oscillator strength, temperature shift of the excitonic transition energy, and reversible absorbance shift. J Phys Chem 98:7665–7673
10. Yumoto J, Fukushima S, Kubodera K (1987) Observation of optical bistability in CdS_xSe_{1-x}-doped glasses with 25-psec switching time. Opt Lett 22:832–834
11. Takagawara T (1993) Enhancement of excitonic optical nonlinearity in a quantum dot array. Optoelectronics 6:545–555
12. Murray CB, Kagan CR, Bawendi MG (1995) Self-organization of CdSe nanocrystallites into three-dimensional quantum dot superlattices. Science 270:1335–1338
13. Colvin VL, Schlamp MC, Alivisatos AP (1994) Light-emiting diodes made from cadmium selenide nanocrystals and a semiconducting polymer. Nature 370:354–357
14. Dabbousi BO, Bawendi MG, Onitsuka O, Rubner MF (1995) Electroluminescence from CdSe quantum-dot/polymer composites. Appl Phys Lett 66:1316–1318
15. Brus L (1994) Luminescence for silicon materials: chains, sheets, nanocrystals, nanowires, microcrystals, and porous silicon. J Phys Chem 98:3575–3581
16. Nosaka Y (1994) Ultrasmall particles of semiconductor for photocatalysts. J Catal Soc Jpn 36:507–514
17. Kamat PV (1995) Tailoring nanostructured thin films. Chemtech 22–28
18. Weller H, Eychmueller A (1995) Photochemistry of semiconductor nanoparticles in solution and thin-film electrodes. Adv Photochem 20:165–216

19. Hagfeldt A, Graetzel M (1995) Light-induced redox reactions in nanocrystalline systems. Chem Rev 95:49–68
20. Martin CR (1994) Nanomaterials: a membrane-based synthetic approach. Science 266:1961–1966
21. Fendler JH, Meldrum FC (1995) The colloid chemical approach to nanostructured materials. Adv Mater 7:607–632
22. Ohtani H, Wilson RJ, Chiang S, Mate CM (1988) Scanning tunneling microscopy observations of benzene molecules on the Rh(111)–(3 × 3) (C_6H_6 + 2CO) surface. Phys Rev Lett 60:2398–2401
23. Hallmark VM, Chiang S, Brown JK, Woll C (1991) Real-space imaging of the molecular organization of naphthalene on platinum(111). Phys Rev Lett 66:48–51
24. Weiss PS, Eigler DM (1993) Site dependence of the apparent shape of a molecule in scanning tunneling microscopy images: benzene on Pt(111). Phys Rev Lett 71:3139–3142
25. Lippel PH, Wilson RJ, Miller MD, Woll C, Chiang S (1989) High-resolution imaging of copper–phthalocyanine by scanning tunneling microscopy. Phys Rev Lett 62:171–174
26. Foster JS, Frommer JE (1988) Imaging of liquid crystals using a tunnelling microscope. Nature 333:542–545
27. For example, Rabe JP (1992) Molecular at interfaces: in materials and life sciences. Ultramicroscopy 42:41–54
28. For example, Sano M, Sasaki DY, Kunitake T (1992) Polymerization-induced epitaxy: scanning tunneling microscopy of a hydrogen-bonded sheet of polyamide on graphite. Science 258:441–443
29. Alves CA, Smith EL, Porter MD (1992) Atomic scale imaging of alkanethiolate monolayers at gold surfaces with atomic force microscopy. J Am Chem Soc 114:1222–1227
30. Rabe JP, Buchholz S (1991) Commensurability and mobility in two-dimensional molecular patterns on graphite. Science 253:424–427
31. Sleator T, Tycko R (1988) Observation of individual organic molecules at a crystal surface with use of a scanning tunneling microscope. Phys Rev Lett 60:1418–1421
32. Yoshimura M, Shigekawa H, Yamauchi H, Saito G, Saito Y, Kawazu A (1991) Surface structure of the organic conductor beta-(BEDT-TTF)$_2$I$_3$ observed by scanning tunneling microscopy (where BEDT-TTF is bis(ethylenedithio)tetrafulvalene). Phys Rev B 44:1970–1972
33. Bard AJ, Abruna HD, Chidsay CE, Faulkner LR, Feldberg SW, Itaya K, Majda M, Melroy O, Murray RW, Porter MD, Soriaga MP, White HS (1993) The electrode/electrolyte interface—a status report. J Phys Chem 97:7147–7173
34. Wiesendanger R, Güntherodt HJ (eds) (1992) Scanning tunneling microscopy II. Springer, Berlin
35. Tao NJ, Lindsay SM (1992) In situ scanning tunneling microscopy study of iodine and bromine adsorption on gold(111) under potential control. J Phys Chem 96:5213–5217
36. Tao NJ, DeRose JA, Lindsay SM (1993) Self-assembly of molecular superstructures studies by in situ scanning tunneling microscopy: DNA bases on Au(111). J Phys Chem 97:910–919
37. Srinivasan R, Murphy JC, Fainchtein R, Pattabiraman N (1991) Electrochemical STM of condensed guanine on graphite. J Electroanal Chem 312:293–300

38. Kunitake M, Batina N, Itaya K (1995) Self-organized porphyrin array on iodine-modified Au(111) in electrolyte solutions: in situ scanning tunneling microscopy study. Langmuir 11:2337–2340
39. Batina N, Kunitake M, Itaya K (1996) Highly ordered molecular arrays formed on iodine-modified Au(111) in solution: in situ STM imaging. J Electroanal Chem 405:245–250
40. Lipkowski J, Ross PN (eds) (1992) Adsorption of molecules at metal electrodes. VCH Publishers, New York

3. Molecular Systems and Their Applications to Information Transduction

MASAHIRO IRIE

Information transduction using organic molecular systems opens up new aseas in the field of optoelectronic devices. Although at present inorganic materials dominate most electronic and optoelectronic devices, they are being challenged on several fronts by organic materials. In the field of display, for example, organic liquid crystals provide flat-panel displays with light weight and low power consumption, and these displays are indispensable to carry-on type electronic computers and others. It is expected that electrochromic and electroluminescent organic materials will be used in future display devices. Photoconductors using organic dyes are now replacing Se-based inorganic photoconductors in electrophotography. In optical recording, photoreactive materials, such as photochromic and photochemical hole-burning materials, are candidates for future ultra-high-density memory. Organic molecules and molecular organisms are the most suitable materials for detecting bio-originated compounds. This part introduces the working principles of functional molecular systems, and describes the design, synthesis, and evaluation of organic materials for information transduction.

3.1 Electroactive Molecular Systems

Katsumi Yoshino and Tsuyoshi Kawai (3.1.1)
Shogo Saito (3.1.2)

3.1.1 Electrochromic Molecules

Some organic and inorganic compounds have been known to show clear color changes upon electrochemical oxidation or reduction. This chromic phenomenon is called "electrochromism." The first electrochromic materials to be studied were inorganic compounds such as VO_2 and WO_3. However, their lifetime, response time, and chromic color give them some disadvantages. As well as inorganic electrochromic compounds, considerable interest has recently been focused on some organic compounds which have superior properties to the inorganic electrochromic materials. Here, we describe organic electrochromic materials with various different characteristics and which are originated by different electrochromic mechanisms.

3.1.1.1 Electrochromism in Conducting Polymers

Conducting polymers having highly extended conjugated π-electron systems on their main chain show reversible insulator–metal transition upon electrochemical and chemical doping [1, 2]. Since the color of the conducting polymer in the insulating state should be determined by the optical absorption spectrum due to interband transition, while that in the metallic state might be determined from the plasma-reflection of the conducting electrons, clear electrochromism has been expected upon electrochemical insulator–metal transition. This idea has been confirmed experimentally [3, 4]. Subsequently, the chromic phenomena in conducting polymers was explained by consideration of new electronic states such as the polaron and the bipolaron, which appear upon doping.

As shown in Fig. 3.1-1, the fundamental structure of an electrochromic device is relatively simple. Two transparent glass electrodes, which are coated with an indium tin oxide layer, are separated with spacers. The conducting polymer film is formed on one of the electrodes. An electrolyte solution is introduced between the electrodes. Upon application of a positive bias, an anion is inserted into the conducting polymer layer. Conversely, the doped anion is extracted upon

Poly(phenylacetylene)
[PPA]

Poly(thiophene)
[PTh]

Poly(aniline)
[PAn]

Poly(p-phenylene)
[PPP]

Poly(pyrrole)
[PPy]

Poly(3-alkylthiophene)
[PAT]

Poly(isothianaphthene)
[PIN]

Poly(p-phenylene vinylene)
[PPV]

Poly(naphthalene vinylene)
[PNV]

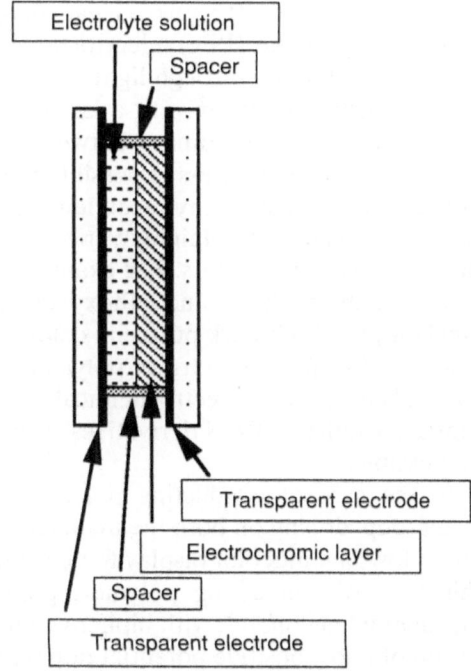

Electrolyte solution

Spacer

Transparent electrode

Electrochromic layer

Spacer

Transparent electrode

FIG. 3.1-1. An electrochromic cell

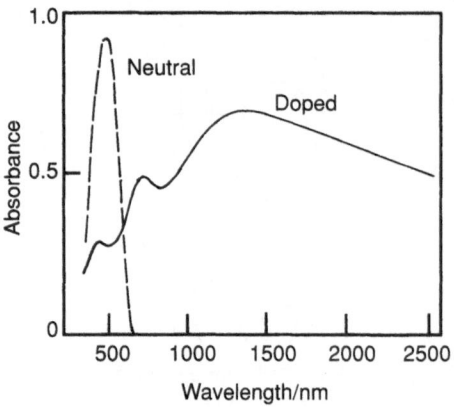

FIG. 3.1-2. Absorption spectra of polythiophene in the neutral and the doped state

application of a negative bias or when these is a short-circuit. Accompanying these doping–undoping reactions are clear changes in color and thus in the absorption spectrum. In some cases, doping of cations (n-type doping) is also possible.

Figure 3.1-2 shows the changes in absorption spectrum of polythiophene (PT) upon electrochemical doping. In this case, $LiBF_4$ and acetonitrile were used as the supporting salt and electrolyte solvent, respectively. BF_4^- anions were doped upon application of positive bias and the color of the polymer turned from red to blue reversibly. The color and absorption spectrum of the conducting polymers in both the doped and undoped states depend strongly on the molecular structure [3, 5]. For example, PT shows a clear change in color between red and blue. In the case of polypyrrole (PPy), electrochemical doping results in a change of color from yellow to blue through light brown, while poly(p-phenylene) (PPP), which is a conducting polymer with a large band gap, and poly(phenylene vinylene) (PPV) show electrochromism between yellow and blue. Even electrochromism between red and colorless is possible in some polyacetylene derivatives such as poly(o-trimethylsilylphenyl acetylene) [6]. Chemical modifications and derivations are effective in obtaining various kinds of electrochromic material showing different chromism colors. For example, PPV is yellow in color in the undoped state but its methoxy and ethoxy derivatives are red and poly(naphthalene vinylene) (PNV) is dark purple in color in its undoped state [5, 7–10]. Figure 3.1-3 shows absorption spectra of polyaniline (PAN). As evident in this figure, the spectral change after electrochemical doping covers the whole visible wavelength range. By utilizing PAN derivatives, multicolor or full-color display devices may be possible.

In the case of conducting polymers having a conjugated moiety as their side group, doping on both the main chain and the side chain are possible, which also allows a multicolor display to be achieved. Figure 3.1-4 shows an example of this type of conducting polymer, poly(bis(4-methoxy-4'-hexyloxy biphenyl)-dipropargyl malonate), with biphenyl units in the side chain [11]. With the application of a low positive potential doping on the main chain occurs, and with the

Fig. 3.1-3. Absorption spectra of polyaniline at various applied electrochemical
potentials versus an Li electrode

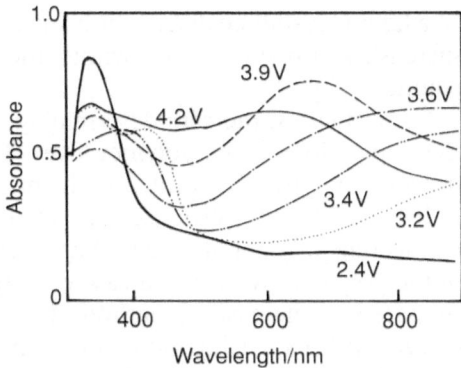

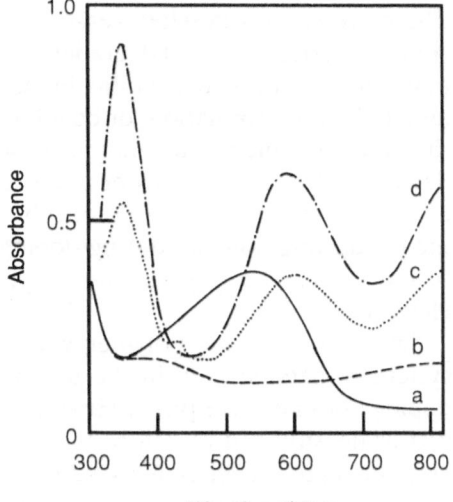

Fig. 3.1-4. Absorption spectra of BF$_4^-$doped
poly(bis(4-methoxy-4'-hexyloxy biphenyl)-
dipropargyl malonate) at various doping
levels. *a* neutral; *b* 14%; *c* 140%; *d* 230%

application of a higher potential doping on the side biphenyl group occurs. In this
conducting polymer, clear electrochromism from red to colorless and then to blue
has been observed with increasing potential.

Electrochemical doping of a conducting polymer with a narrow band gap also
results in suppression of absorbance in the visible region and in the colorless
state, since the density of states of conduction and valence bands decrease upon
doping. For example, poly(isothianaphthene) (PIN) with a band gap energy of
1.5 eV has been reported to be generally colorless in the doped state and to show
clear electrochromism upon both p-type and n-type doping.

Electrochromism of conducting polymers has various advantages compared
with the usual twisted nematic-type liquid crystalline display, e.g., a clear color
change with no polarizer, a memory effect, and no marked restriction in the
visual angle. The response time is several times 10 ms, which is comparable to that

of a liquid crystalline display utilizing a nematic liquid crystal, and decreases with increasing applied voltage within the stable voltage range of the electrolyte solution [12].

In most cases, the rate-determining step of the doping process is diffusion of the dopant ions in the polymer layer. Therefore, concentration, type and size of dopant ion, type of solvent, and the morphology and thickness of the conducting polymer affect the response time of the electrochromism. The structure of the conducting polymer also sometimes restricts the kinetics of the dopant migration in the polymer layer. In the case of poly(3-alkylthiophene) (PAT) film, which is prepared by the casting method of the soluble polymer, the response time decreases with increasing numbers of repetitive doping–undoping cycles [13], and the diffusion coefficient of the dopant ions in the polymer layer increases with repetitive doping–undoping cycles [14].

One of the most serious problems in using a conducting polymer is its lifetime. There are several possible reasons for the degradation of electrochromism in conducting polymers: (1) degradation in the chemical structure of the conducting polymer itself; (2) irreversibility in the dopant migration, such as dopant accumulation, dendrite formation, or counter-insertion of cations upon the undoping of the anion; (3) mechanical degradation of the conducting polymer film and the detachment of that film from the electrode substrate; (4) irreversible chemical and electrochemical reactions of solvent and electrolyte. A change in volume of the conducting polymer during doping–undoping reactions is one of the origins of the mechanical degradation of polymer film and its detachment from the electrode.

Since a conducting polymer with a two-dimensionally extended π-electron system is metallic even in the undoped state, insulator–metal transition upon doping does not take place. However, its electrochromism is still possible since carrier density, and therefore plasma reflection, changes upon doping. Indeed, graphite film, which is a two-dimensionally and quasi-infinitely extended π-electron system, shows changes in color between blue, cuprous and golden.

An all-solid polymeic chromic device can be realized by utilizing a solid polymer electrolyte and conducting polymers as positive and negative electrodes. PAN and WO_3 were chosen as the anode and cathode electrodes, respectively, since both are blue in color in the oxidized and in the reduced states, and the electrochromic contrast is enhanced by the coloration of both electrodes [15].

3.1.1.2 Electrochromism in Organic Molecules

Viologen compounds have been attracting much interest as electrochromic materials [16]. A typical reaction scheme and the cyclic voltammetry of a viologen molecule are shown Fig. 3.1-5. The viologen dication changes upon reduction to the radical cation with an accompanying color change, as shown in Fig. 3.1-6. Methylviologen is soluble in aqueous solutions in both the dication and cation radical state, which restricts practical applications of it for display devices. However, dialkyl viologen, which has an alkyl chain longer than the pentyl group,

FIG. 3.1-5. Cyclic voltam-
mogram of methylviologen
and a schematic illustration of
the electrochromic reaction of
dialkylviologen (*inset*)

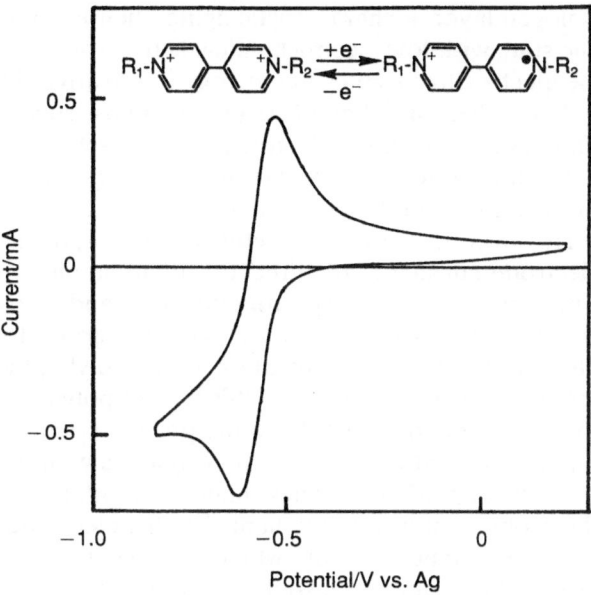

FIG. 3.1-6. Absorption spectra of alkylvi-
ologen in the reduced (*solid line*) and oxidized (*dashed line*) forms

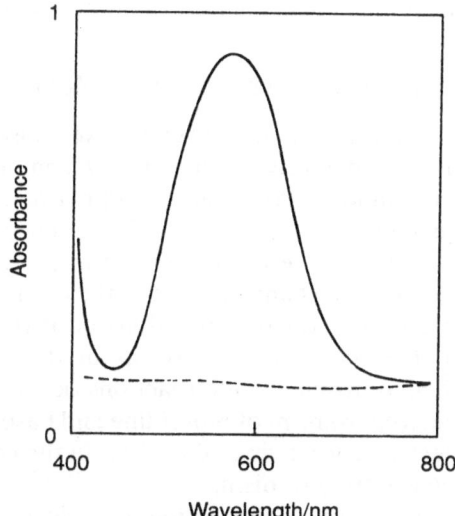

is not soluble in the reduced cation radical state [17]. Consequently, upon electrochemical reduction of dihexylviologen, the cation radical salt precipitates on the
test electrode and forms a film of blue color. The precipitation of the cation is
attributed to a strong intermolecular interaction between cations which form
dimers with no electron-spin-resonance (ESR) signal [17]. In the precipitated

viologen layer, a characteristic aging phenomenon occurs, resulting in a crystal-line state with low electrochemical activity and also relatively low reversibility of the electrochromism of the viologen derivatives [18].

A number of other molecular systems, such as quinone derivatives, show changes in color after electrochemical reactions [19]. Vacuum-evaporated films of tetranitrofluorenone and related molecules also show reversible color changes. These films are relatively stable in both the reduced and oxidized states [20].

However, in most-redox active molecules, the fixation of chromophores to the electrode surface is required to obtain stable and reversible electrochromism with a memory effect. Tetrathiafulvalene and carbazole have been fixed in a non-conjugated polymer matrix as a pendant group to form a stable electrochromic film [21, 22]. As already mentioned, the polyheptadien derivative, which has side groups, is also an example of this type of polymer with relatively high conductiv-ity in the doped state [11]. Some polyion complexes of organic dyes have been reported as stable electrochromic systems with fixed chromophores [23].

Most aromatic molecules might be candidates for electrochromic material if the problem of how to fix them onto the electrode surface could be solved. Even a photochromic molecule which changes its color upon photoirradiation shows electrochromism in the solution phase [24]. Some diarylethene-type photochro-mic dyes change color from red to yellow upon electrochemical oxidation. These dyes can be fixed in a solid polymer electrolyte in an optical pattern and then electrochemical changes can be achieved.

3.1.1.3 Electrochromism in Metal Complexes

A number of metal complexes also show electrochromism. One of the most extensively studied metal complex systems is based on phthalocyanine compounds [25]. Some porphyrin derivatives, phenanthroline derivatives, basophenanthroline derivatives, and bipyridine compounds are also known to show reversible electrochromism [26–28]. In some cases two different electro-chemical mechanisms exist as the origin of the electrochromism. Electrochemical oxidation–reduction reactions occurring in the electronic state which localizes on the ligand moiety result in the electrochromism of phthalocyanine and phenanthroline, while reactions occurring on the central metal ion result in the electrochromism of bipyridine and basophenanthroline complexes. In both cases, electrons occupying 3d orbitals in the central metal ion play an important role in the electrochromism.

Some diphthalocyanine complexes containing lanthanoid metals such as lutetium, ytterbium, and erbium have been studied, since clear electrochromism, and even multicolor electrochromism, can be observed corresponding to several stable oxidation states of different colors [29]. Phthalocyanine complexes of molybdenum, copper, and cobalt have also been studied for electrochromic material [30]. The electrochemical oxidation of the metal naphthalocyanine and of metal diphthalocyanine complexes are based on the change in oxidation state of the ring moiety, although H_2Pc dose not show electrochemical activity.

These facts indicate strong electronic interaction between the central metal and ligand ring moiety [30]. In the case of a Co–naphthalocyanine complex, the central metal acts as the redox site in the electrochemical reduction, showing electrochromism from green in the neutral state to purple in the reduced state, while the ligand acts as the redox center in the oxidation [31], i.e., the HOMO of the complex is localized on the ring moiety. Langmuir–Blodgett films of phthalocyanin derivatives were prepared to avoid thermal decomposition of the molecule upon evaporation [32].

Ruthenium bipyridyl derivatives, iron basophenanthroline complexes, and Prussian blue were also studied as the electrochromic materials [27, 33, 34]. In these cases, it is also important to fix the electrochromic molecules to the electrode surface strongly. In the case of metal bipyridyl complexes, an organic ion exchange polymer was used as a binding matrix to form a poly-ion complex.

As mentioned above, electrochromic devices have various advantages, such as clear changes in color and tone, a favorably wide viewing angle, a relatively fast response which is comparable to a liquid crystalline display, a memory effect, and low driving voltage and electric power. It is also relatively easy to construct a large display in the case of electrochromic devices. Thus, these electrochromic devices are promising for practical applications. If chromism induced by an electric field rather than by electrochemical reactions is possible, it would have a long lifetime and a very fast response time. Some molecules which exhibit quite a large Stark effect, Frany–Keldish effect, or field-induced structural change could be candidates for such new types of chromic device.

3.1.2 Electroluminescent Systems

Organic thin films composed of π-conjugated molecules have several interesting properties such as semiconducting properties, photoluminescence and so on. The π-conjugated molecules are easily excited to their singlet states. The excited molecules are usually created not only by the absorption of UV-light, but also by the recombination of conducting electrons and holes. If one succeeds in utilizing these fundamental processes, one can propose organic electroluminescent devices, organic solar cells and so on.

3.1.2.1 Principle and Structures of Electroluminescent (EL) Devices

The working principle of organic EL devices is similar to that of inorganic light-emitting diodes: creation of the excited molecules by recombinations of injected electrons and holes in an organic fluorescent thin film and spontaneous emissions from the fluorescent film with decays of the excited molecules to their ground states. Therefore, in general, the EL spectrum is very similar to the photoluminescence (PL) spectrum.

There were early difficulties in injecting electrons and holes from outer electrodes into organic emissive films. Several years ago, it was discovered that the insertion of an organic hole transport film, composed of an organic semiconductor, between an outer electrode and an emissive film was extremely effective at improving the hole injective capability [35, 36]. Four types of EL device have been proposed so far:

1. ITO/hole transport layer (HTL)/emissive layer (EML)/metallic electrode—SH-A,
2. ITO/emissive layer (EML)/electron transport layer (ETL)/metallic electrode—SH-B,
3. ITO/HTL/EML/ETL/metallic electrode—DH,
4. ITO/EML/metallic electrode—SL,

where ITO is a transparent electrode composed of indium–tin-oxide and SH, DH, and SL mean single hetero-junction, double hetero-junctions, and single layer, respectively.

In order to get strong EL from an EML, it is extremely important to select the most suitable structure among the four devices. One should understand the carrier transporting properties in the EML first. If an EML has a unipolar carrier transporting tendency, one should select an SH-device. Conversely, an EML with a bipolar transporting tendency should be used in a DH- or SL-device. Generally, the multilayer EL devices which are recognized as partial-charge types have higher efficiencies for emission than SL-types.

3.1.2.2 Materials for EL Devices

A variety of organic dyes and pigments which contain a π-conjugated chromophore in their molecular structures are utilized as emissive and carrier transport materials. Typical examples for an EML material are 8-hydroxyquinoline-Al complex (AlQ$_3$), naphthostyryl derivatives (NSD), and bis-styrylanthracene derivatives (BSA). Triphenylamine derivatives (TAD) are well known as HTL materials. Oxadiazole derivatives (PBD) and AlQ$_3$ are typical ETL materials, and their molecular structures are shown below. Recently, several investigators have tried to use π-conjugated polymers such as poly(p-phenylene vinylene) and transparent polymer films doped with fluorescent dyes as EML materials.

As the metallic electrode (cathode), one should use a metal of low work function, which generally means an alloy such as Al, MgAg, as Ca AlLi.

3.1.2.3 Rational Selections of Organic Materials for Multilayer EL Devices

An approach for increasing the EL efficiency is to achieve confinement of the injected carriers and generated excitons within the EML.

To understand the carrier injection characteristics across a hetero-junction between two semiconducting layers, it is very useful to analyze the carrier transport by using an electronic energy band diagram. In spite of the amorphous nature of organic thin films, these have been attempts to discuss carrier injection and transport in terms of the ionization potential I_p and electron affinity E_a of the organic thin films [37]. The luminance–current relation as an EL characteristic has been measured with BSA as the EML for the following three types of device: (i) ITO/TAD/BSA/MgAg (SH-A); (ii) ITO/BSA/PBD/MgAg (SH-B); (iii) ITO/TAD/BSA/PBD/MgAg (DH).

The SH-A device exhibited much stronger EL emission than the SH-B and DH devices by factor of 10^2. This result can be explained from the values of I_p and E_a listed in Table 3.1-1. The BSA layer is more effective for electron injection from a MgAg electrode than PBD. The TAD layer plays an important role in accelerating hole injection from the ITO electrode into the BSA layer, and is also very effective in blocking electron flow from the BSA to the ITO electrode. This result suggests that a BSA layer prepared for EML also behaves as an excellent ETL.

It can be concluded that the requirements for effective carrier confinement are

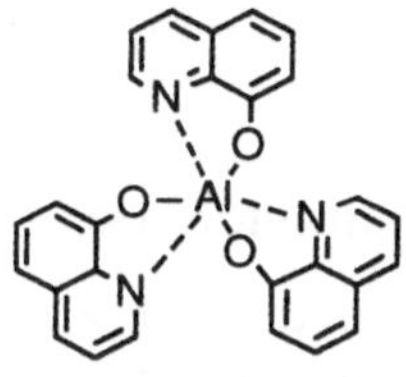

AlQ$_3$

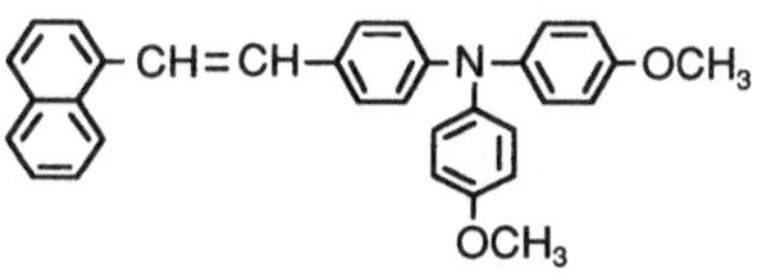

NSD

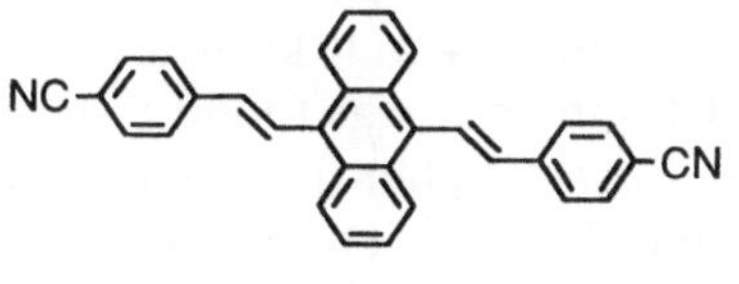

BSA

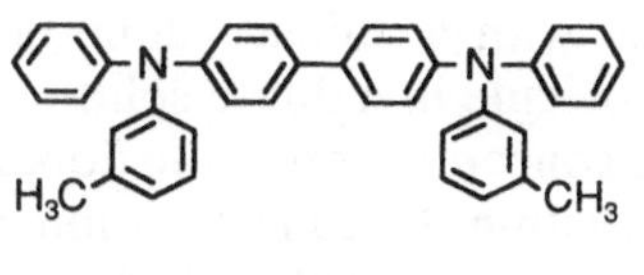

TAD

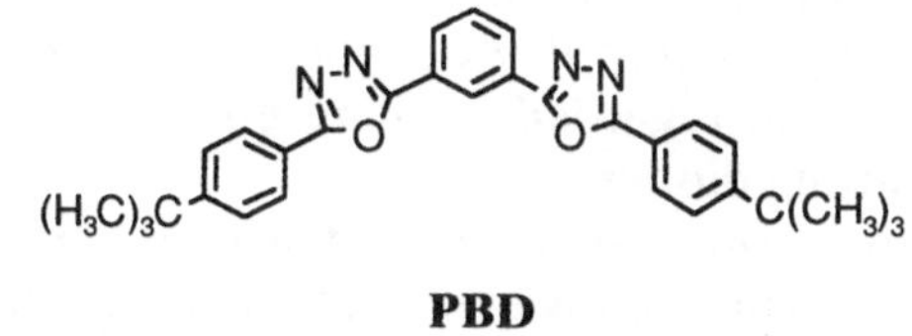

PBD

TABLE 3.1-1. Ionization potentials, electron affinities and excitation energies of materials for EL devices

		I_p (EV)	E_a (eV)	E^* (eV)
EML	BSA	6.0	3.5	2.21
ETL	PBD	5.9	2.6	3.10
HTL	TAD	5.5	2.4	2.88
Cathode	MgAg	3.7		
Anode	ITO	4.9		

I_p was determined from measurements of photoelectron emission, and E_a was assumed to the difference between I_p and the optical absorption edge. E^* was determined from the wavelength of the PL maximum.

$$I_p(\text{ITO}) < I_p(\text{HTL}) \simeq I_p(\text{EML}) \ll I_p(\text{ETL})$$
$$E_a(\text{HTL}) \ll E_a(\text{EML}) \simeq E_a(\text{ETL}) < E_a(\text{metal}) \tag{3.1-1}$$

To achieve a high emission efficiency, it is also important to confine the excitons generated by the recombinations of electrons and holes within EML. The requirement for avoiding the transfer of excitation energy from the excitons to the charge transport layers is

$$E^*(\text{HTL}) > E^*(\text{EML}) < E^*(\text{ETL}) \tag{3.1-2}$$

where E^* is the excitation energy, which is estimated from the frequency of the fluorescence maximum.

3.1.2.4 Effects of Dopant Molecules on EL Characteristics

One of the most important requirements in fluorescent dyes for EML is that they should emit strong fluorescence in their solid thin films. In most cases, however, a concentrated quenching effect cannot entirely be avoided. Doping with a small amount of another fluorescent dye into the EML is well worth studying. The doping effect was studied in multilayer EL devices with an EML composed of AlQ$_3$ doped with a quinacrydone derivative, which is well known as a highly emissive molecule in a dilute solution. The EL intensity was found to be dependent on the concentration of the dopant, and became higher than that of undoped AlQ$_3$ EML within the concentration range up to 2% [38]. The EL was confirmed from the dopant molecules. A dopant concentration of 0.47% gave the highest result. The luminance value was 68000 cd/m^{-2} at a current density of 1 A cm^{-2}. This result corresponds with an outer emission efficiency of about 5 lm W^{-1}. Very recently, many groups have achieved the surprisingly high efficiency of about 10 lm W^{-1}. These devices are regarded as more efficient light-emitting devices than inorganic light-emitting diodes.

A specific zone in the EML can be doped with a small number of other fluorescent molecules which have a lower excitation energy than that of the EML molecules. If the excitons in the doped specific zone are created by the carrier recombination, one can detect the EL spectrum from the dopants excited by the energy transfer from the excitons of the host molecules. Thus, one can determine the distribution of emission sites in the EML by using the doping method in which an extremely thin doped zone is formed at a specific depth from the interface between the EML and the carrier transport layer. Examples [39] of the distribution of emission sites determined by this method are shown in Fig. 3.1-7. In two-layer devices, the emission sites extend from the interface between the two organic layers for a few tens of nanometres. In the single-layer device, the emission sites are distributed over a much wider region. To understand the distribution of emission sites, several factors should be taken into account.

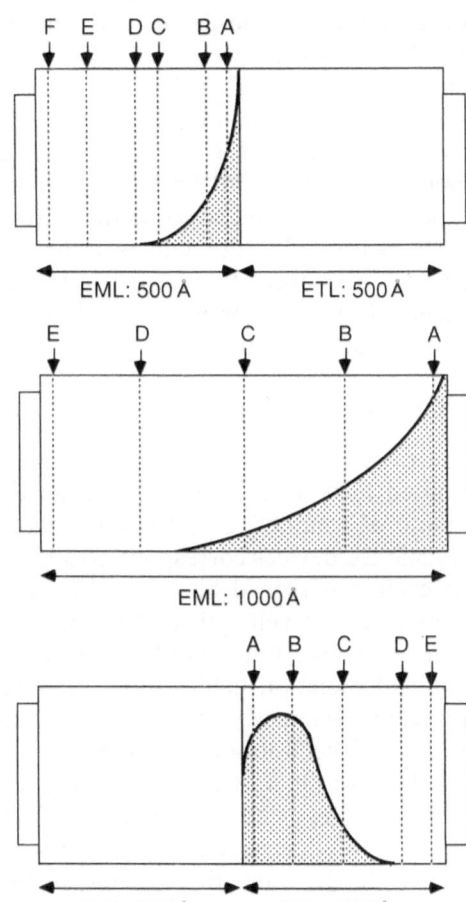

FIG. 3.1-7. Distributions of emission sites in electroluminescent devices [39]. A–F indicate the doped positions

1. Carrier injection into the EML and the carriers transport properties in the EML.
2. Blocking of the carrier flow into the outer electrode without recombination.
3. Non-radiative decay of the excitons due to the energy transfer into the outer electrode.
4. Evanescent wave propagation into the ITO electrode with a high refractive index.

3.1.2.5 The Smallest Thickness of EML—a Device of Molecular Size

It is very interesting to examine the smallest thickness of EML in which confinements of the injected carriers and the created excitons can be attained. A DH type of EL device with a cyanine dye (OCD) thin film as the EML was prepared by a combination of vacuum sublimation and the Langmuir–Blodgett (LB) technique [40]: ITO/TAD (HTL)/OCD (EML)/PBD (ETL)/MgAg. Only the EML layer was prepared by the LB technique. In these EL devices, the requirements for confinement of the injected carriers and created excitons within the EML were fulfilled. In the device with an OCD bimolecular layer, no emission from TAD, PBD, or the exciplex between TAD and PBD, but only the emission from the OCD layer, was observed. This result shows that recombination of carriers and creation of excitons occurred only within the OCD bimolecular layer. In the device with an OCD monomolecular layer, not only the emission from the monolayer but also an emission from the exciplex between TAD and PBD were observed. The result suggests that direct contacts between TAD and PBD were partially formed at the defects in the OCD monolayer during the vacuum sublimation procedure. If a defect-free OCD monolayer could be prepared, it is expected that the confinement of carriers and excitons within the OCD monolayer could be attained.

3.1.2.6 Size Effects in Organic EL Devices

The total thickness of organic layers is generally less than the wavelength of emitted light. Remembering that an extremely thin EML is sandwiched by a couple of conductive and highly reflective electrodes, one can recognize that organic EL devices correspond to a kind of optical microcavity. When a molecular exciton is placed between two parallel reflective mirrors whose spacing is close to the wavelength of light, the emission from the exciton can no longer be irreversible. One should consider the influence of emitted light within a confined radiation field around the exciton. In this microcavity, enhancement and suppression of the spontaneous emission from the exciton are expected to occur. The florescent lifetime from the EML in the DH device shown in Fig. 3.1-8 was measured as a function of the spacing d between the EML and a mirror (MgAg electrode) [41]. The lifetime changed remarkably with a variation of d: the lifetime at one specific d was shorter and that at another specific d became longer than the lifetime of the fluorescence from EML without the metallic mirror. To understand the variation of the lifetime with d, one should

FIG. 3.1-8. Multilayer electroluminescent device for measuring the emission intensity and
fluorescent lifetime as a function of the thickness of the electron transport layer [41]

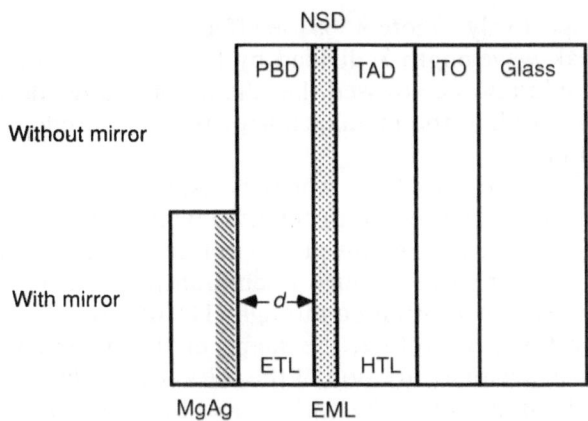

FIG. 3.1-9. Microcavity effects
on the exciton decay rates and
the outer emission intensity. b_r,
b_{nr}^{therm}, and b_{nr}^{ET} are the rates of
radiative decay, non-radiative
thermal decay, and non-radiative decay due to energy transfer into the mirror, respectively

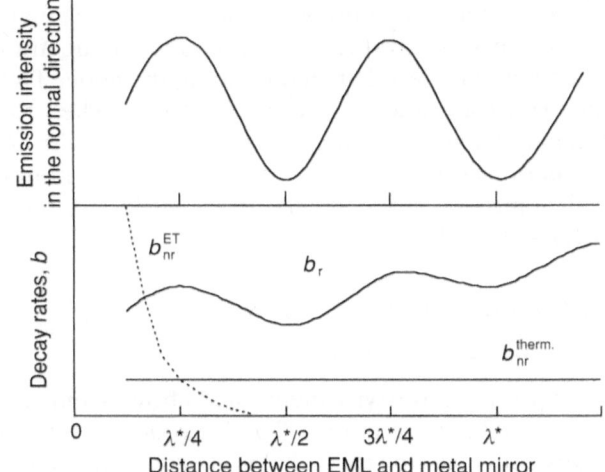

consider that the decay of the created excitons is composed of three different
processes:

1. radiative decay which yields fluorescence;
2. non-radiative thermal decay;
3. non-radiative decay due to the energy transfer into the metallic mirror.

Then the total decay rate, which corresponds with the inverse of the fluorescent
lifetime, is given by the sum of the three decay rates. By applying Chance's
theory [42] to the above model, we can interpret the observed periodical variation of the lifetime with d. A schematic representation of the variations in decay
rates is shown in Fig. 3.1-9. It should be pointed out that the minima and maxima
in the lifetime vs. d curve appear at a spacing of $(2n - 1)\lambda^*/4$ and $2n\lambda^*/4$,

respectively, where λ^* is the effective wavelength, and is given by the emission-peak wavelength λ divided by the refractive index of the organic medium. It should also be stressed that the non-radiative decay process due to the energy transfer into the metallic mirror becomes predominant in the range of d below 50 nm.

The second effect which is expected to occur in the micro-cavity is an interference effect of the emitted light. Only a portion of the emitted light from the molecular excitons placed in front of the two mirrors is transmitted directly into outer space through a semitransparent ITO. The remaining portion of the emission is transmitted through ITO after repeated reflections, or along the glass substrate. Therefore, the shape of the outer emission pattern is expected to change with a variation of the thickness or dimension of the DH device.

To understand the variation of the luminance with d, it is helpful to use a simple model such as a Fabry–Perot cavity structure, in which the EML is represented by vibrating dipoles without thickness. The amplitude of the light transmitted directly into outer space can be calculated by using the transmittance of the glass substrate. The amplitude of the transmitted light after multiple reflections can be calculated by using the above factor and the reflectance of the glass substrate and the MgAg electrode. Then the variations in the observed luminance and the outer emission pattern with d can be interpreted through calculations using the above model. Again it should be stressed that the maxima of the emission perpendicular to the surface of the EL device appear at spacings of $(2n - 1)\lambda^*/4$.

3.1.2.7 Strongly Directed Emission from Controlled-Spontaneous-Emission EL Devices

The optical microcavity mentioned above seems to be of low quality because of the low reflectance of the ITO electrode. By replacing the ITO electrode with a highly reflective mirror, a Fabry–Perot microcavity can be introduced into conventional thin-film EL devices. Another key to developing a Fabry–Perot cavity is to discover a specific dye which has a very sharp photoluminescence spectrum. This problem has been solved by adopting an Eu-complex ((4,7-diphenyl-1,10-phenanthroline)-Eu complex) which forms stable homogeneous films with vacuum sublimation as an EML [43].

The device structure proposed [43] is Dielectric reflector/ITO/TAD (HTL)/Eu-complex (EML)/AlQ$_3$ (ETL)/MgAg. The dielectric reflector is a quarter-wave stack made of four pairs of SiO$_2$/TiO$_2$ layers. Five devices with different total organic thicknesses were developed. All the devices exhibited intense red emission when driven with d.c. voltage, and a sharply directed emission was clearly observed. For instance, in the device where the total thickness of the organic layers was 142 nm, the emission was only observable at a specific viewing angle of about 40° off normal to the device surface, and no emission was detected with the naked eye at different viewing angles. This behavior indicates that the emission is strongly directional along a cone surface. Figure 3.1-10 shows the emission patterns of the five devices [43]. The maximum emission angle mea-

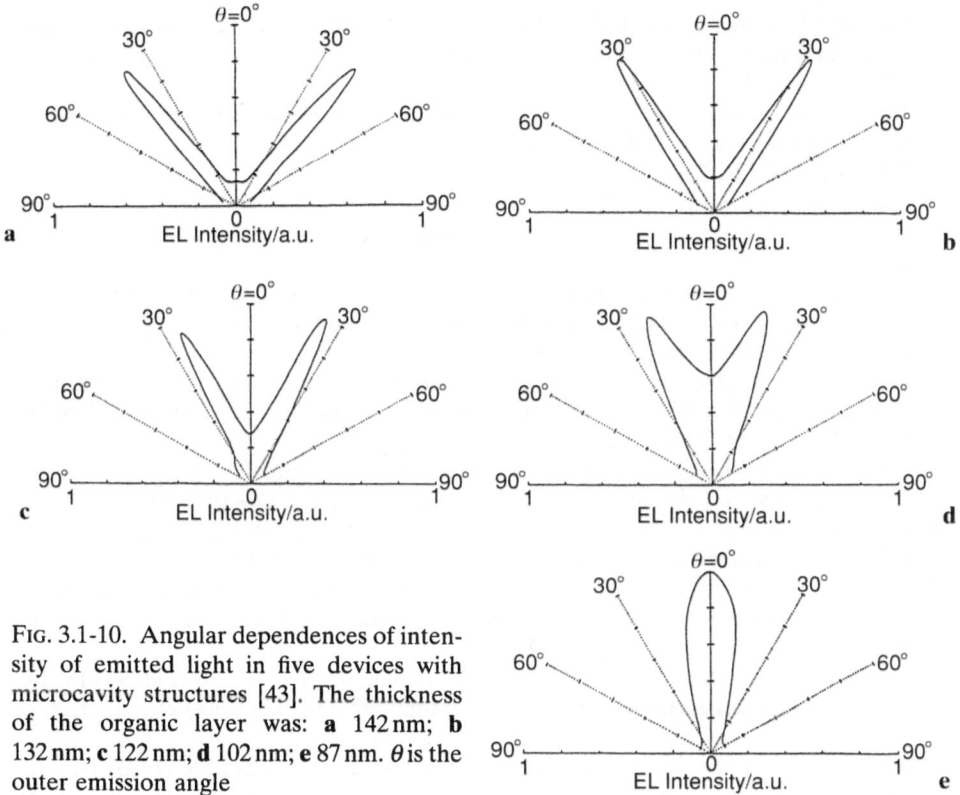

FIG. 3.1-10. Angular dependences of intensity of emitted light in five devices with microcavity structures [43]. The thickness of the organic layer was: **a** 142 nm; **b** 132 nm; **c** 122 nm; **d** 102 nm; **e** 87 nm. θ is the outer emission angle

sured from a direction normal to the emission surface decreases with an decrease in the thickness of the organic layers.

The EL emission spectrum from a Fabry–Perot cavity is clearly different from that from a conventional EL device without a microcavity structure. The spectrum from a conventional device without a cavity contains several small peaks characteristic of emissions from an Eu ion in free space. These small peaks disappear completely in a device with a microcavity. This is further evidence that a micro-cavity structure is effective in attaining single resonance in a cavity.

There are many possible future applications of organic EL devices. Conventional devices are promising for high-quality full-color flat panel displays, and optical microcavities are promising for various applications such as light sources for signal processing, holography, optical interconnections, and possibly for novel laser devices with a low threshold.

References

1. Yoshino K (1988) Fundamentals and application of conducting polymer (in Japanese). IPC, Tokyo

2. Chiang CK, Fincher Jr CR, Park YW, Heeger AJ, Shirakawa H, Luis EJ, Gau SC, MacDiarmid AG (1977) Electrical conductivity in doped polyacetylene. Phys Rev Lett 39:1098–1100
3. Yoshino K, Kaneto K, Inuishi Y (1983) Proposal of electro-optical switching and memory devices utilizing doping and undoping processes of conducting polymer. Jpn J Appl Phys 22:L157–L158
4. Kaneto K, Yoshino K, Inuishi Y (1983) Characteristics of electrooptic devices using conducting polymers, polythiophene and polypyrrole. Jpn J Appl Phys 22:L412–L414
5. Yoshino K, Takiguchi T, Hayashi S, Park DH, Sugimoto R (1986) Electrical and optical properties of poly(p-phenylene vinylene) and effects of electrochemical doping. Jpn J Appl Phys 25:881–886
6. Yoshino K, Takahashi H, Morita S, Kawai T, Sugimoto R (1994) Electrochemical characteristics of stable polyacetylene derivative: poly(o-trimethylsilylphenyl acetylene). Jpn J Appl Phys 33:L254–L257
7. Murase I, Ohnishi T, Taniguchi T, Hirooka M (1984) Highly conducting poly(p-phenylenevinylene) prepared from a sulfonium salt. Polym Commun 25:327–329
8. Murase I, Ohnishi T, Taniguchi T, Hirooka M (1985) Alkoxy-substituent effects of poly(p-phenylenevinylene) conductivity. Polym Commun 26:362–363
9. Onoda M, Manda Y, Iwasa T, Nakayama H, Amakawa K, Yoshino K (1990) Electrical, optical, and magnetic properties of poly(2,5-diethoxy-p-phenylene vinylene). Phys Rev B42:11826–11832
10. Onoda M, Morita S, Iwasa T, Nakayama H, Amakawa K, Yoshino K (1991) In situ absorption spectra measurements in poly(1,4-naphthalene vinylene) during electrochemical doping. J Phys D: Appl Phys 24:1152–1157
11. Yoshino K, Yin XH, Morita S, Kawai T, Ozaki M, Jin SH, Choi SK (1993) Unique electrical and optical properties of conducting polymeric liquid crystal. Jpn J Appl Phys 32:L1673–L1676
12. Kaneto K, Agawa H, Yoshino K (1987) Cycle life, stability, and characteristics of color switching cells utilizing polythiophene films. J Appl Phys 61:1197–1201
13. Yoshino K, Kuwabara T, Kawai T (1990) Learning effect in the doping process of conducting polymer. Jpn J Appl Phys 29:L995–L997
14. Kawai T, Kuwabara T, Wang S, Yoshino K (1990) Electrochemical characteristics of chemically prepared poly(3-alkylthiophene). J Electrochem Soc 137:3793–3797
15. Jelle BP, Hagen G (1993) Transmission spectra of an electrochromic window based on polyaniline, Prussian blue and tungsten oxide. J Electrochem Soc 140:3560–3564
16. Schoot SC, Ponjee JJ, van Dam HT, van Doorn RA, Bolwijn PT (1973) New electrochromic memory display. Appl Phys Lett 23:64–65
17. van Dam HT, Ponjee JJ (1974) Electrochemically generated colored films of insoluble viologen radical compounds. J Electrochem Soc 121:1555–1558
18. Bewick A, Cunningham DW, Lowe AC (1987) Electrochemical and spectroscopic characterisation of structural reorganisation in N,N′-dialkylpyridinium cation radical deposits. Macromol Chem, Macromol Symp 8:355–370
19. van Uitert LG, Zydzik GJ, Singh S, Camlibel I (1980) Anthraquinoide red display cells. Appl Phys Lett 36:109–111
20. Yasuda A, Seto J (1988) Electrochromic properties of vacuum evaporated organic thin films. Part 1. J Electroanal Chem 247:193–202
21. Kaufman FB, Engler EM (1979) Solid state spectroelectrochemistry of crosslinked donor bound polymer films. J Am Chem Soc 101:547–549

22. Lacaze PC, Dubois JM, Desbene-Monvernay A, Desbene PL, Basselier JJ, Richard D (1983) Polymer-modified electrodes as electrochromic material. Part III. J Electroanal Chem 147:107–110

23. Akahoshi H, Toshima S, Itaya K (1981) Electrochemical and spectroelectrochemical properties of polyviologene complex and electrodes. J Phys Chem 85:818–822

24. Kawai T, Koshido T, Yoshino K (1993) Electrochemical properties of cis-1,2-dicyano-1,2-bis(2,4,5-trimethyl-3-thienyl) ethene and its application. Chem Express 8:553–556

25. Nicholson MM, Pizzarello FA (1979) Charge transport in oxidation product of lutetium diphthalocyanine. J Electrochem Soc 126:1490–1495

26. Yamana M, Kashiwagaki N, Yamamoto M, Nakano T (1987) A storage type electrochromic display utilizing poly-Co-QTPP films. Jpn J Appl Phys 26:L1113–L1116

27. Itaya K, Akahoshi H, Toshima S (1982) Polymer modified electrodes. II. J Electrochem Soc 129:762–767

28. Elliott CM, Redepenning JG (1986) Stability and response studies of multicolor electrochromic polymer modified electrodes prepared from tris(5,5'-dicarboxyester-2,2'-bipyridine)ruthenium(II). J Electroanal Chem 197:219–223

29. Nicholson MM, Pizzarello FA (1980) Galvanostatic transients in lutetium diphthalocyanine films. J Electrochem Soc 127:821–827

30. Collins GCS, Shiffrin DJ (1982) The electrochromic properties of lutetium and other phthalocyanines. J Electroanal Chem 139:335–369

31. Yanagi H, Toriida M (1994) Electrochromic oxidation and reduction of cobalt and zinc naphthalocyanine thin films. J Electrochem Soc 141:64–70

32. Yamamoto H, Sugiyama T, Tanaka M (1985) Electrochromism of metal-free phthalocyanine Langmuir–Blodgett films. Jpn J Appl Phys 24:L305–L308

33. Oyama N, Anson FC (1979) Polymeric ligands as anchoring groups for the attach-ment of metal complexes to graphite electrode surfaces. J Am Chem Soc 101:3450–3456

34. Itaya K, Akahori H, Toshima S (1982) Electrochemistry of Prussian blue modified electrodes. J Electrochem Soc 129:1498–1500

35. Hayashi S, Etoh H, Saito S (1986) Electroluminescence of perylene films with a conducting polymer as an anode. Jpn J Appl Phys 25:L773–L775

36. Tang CW, Vanslyke SA (1987) Organic electroluminescent diode. Appl Phys Lett 51:913–915

37. Saito S, Aminaka E, Tsutsui T, Era M (1994) J Luminescence 60/61:902–905

38. Wakimoto T, Murayama R, Nakada H, Imai K (1991) Organic electroluminescent diodes with doped quinacridone derivatives. Poly Preprints Jpn 40:3600–3602

39. Adachi C, Tsutsui T, Saito S (1991) Electroluminescent mechanism of organic multi-layer thin film devices. Optoelectronics 6:25–36

40. Era M, Adachi C, Tsutsui T, Saito S (1992) Organic electroluminescent device with cyanine dye LB film as an emitter. Thin Solid Films 210/211:468–470

41. Takada N, Tsutsui T, Saito S (1993) Controlled spontaneous emission in organic electroluminescent devices. Optoelectronics 8:403–412

42. Chance RR, Prock A, Silbey R (1976) Comments on the classical theory of energy transfer. 2. J Chem Phys 65:2527–2531

43. Takada N, Tsutsui T, Saito S (1994) Strongly directed emission from controlled-spontaneous-emission electroluminescent diodes with europium complex as an emitter. Jpn J Appl Phys 33:L863–L866

3.2 Photoactive Molecular Systems

MASAHIRO IRIE (3.2.1)
KAZUYUKI HORIE (3.2.2)
HACHIRO NAKANISHI (3.2.3)
IWAO YAMAZAKI (3.2.4)

3.2.1 Photochromic Molecules

Photochromism is defined as the reversible transformation of a molecule between two forms whose absorption spectra are distinguishably different, the change being induced in at least one direction by photoirradiation. This is illustrated schematically in Fig. 3.2-1. Both forms, A and B, are in the ground state. The initial A form converts to the B form, and the back-reaction from B to A can occur either thermally or photochemically.

$$A \underset{hv', \Delta}{\overset{hv}{\rightleftharpoons}} B \qquad (3.2\text{-}1)$$

Although most photochromic reactions are based on unimolecular reactions, in some photochromic systems bimolecular reactions are also involved.

$$A + C \underset{hv', \Delta}{\overset{hv}{\rightleftharpoons}} B \qquad (3.2\text{-}2)$$

$$A \underset{hv', \Delta}{\overset{hv}{\rightleftharpoons}} B + C \qquad (3.2\text{-}3)$$

These three reactions (3.2.1–3.2.3) are called photochromism.

Photochromism was reported for the first time in 1867 by M. Fritsche, who observed that orange-reddish tetracene converts to a colorless form with light in the presence of air, and the colorless form (an endoperoxide of tetracene) returns to colored tetracene upon heating. Since then, a great number of photochromic molecules and molecular systems have been reported. However, interest in photochromism was rather limited before the discovery of photochromic spiropyrans in 1952 [1]. The new photochromic compounds attracted considerable interest because of their potential use for non-silver photographic films and variable-density shutters. Although much effort was put into improving the

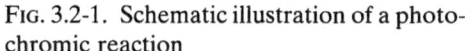

Fig. 3.2-1. Schematic illustration of a photo-chromic reaction

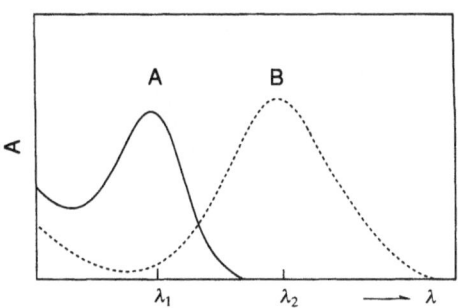

properties of spiropyrans and other compounds for practical use, active research almost ceased at the end of the 1960s because of a failure to synthesize fatigue-resistant and thermally irreversible photochromic systems.

Remarkable progress in optoelectronic technology revived interest in photo-chromism at the beginning of the 1980s. The instant image-forming property without processing has led to consideration of their use in photonic devices such as optical memory media or displays. Research was resumed to develop photo-chromic molecules or molecular systems which meet the requirements for photonic devices. These requirements are as follows:

1. thermal stability of both isomers;
2. low fatigue (can be cycled many times without loss of performance);
3. sensitivity at longer wavelengths and a rapid response;
4. nondestructive readout capability.

These new efforts resulted in considerable progress in the understanding of the mechanism of photochromism, both theoretically and experimentally. This section describes recent progress of basic photochromic properties and possible new uses for them.

Photochromic reactions induce changes not only in optical characteristics, but also in other properties such as polarity or geometrical structure. These changes can be applied to control various physical and chemical properties of polymers in which the photochromic molecules are incorporated [2]. Photoresponsive polymers will be considered in Sect. 4.3.

3.2.1.1 Basic Physico-Chemical Properties

Absorption Spectra. Several classes of photochromic compound are shown in Fig. 3.2-2. Among these classes spiropyrans are the subject of particularly active research and development. The spiropyrans synthesized before 1970 were summarized by Bertelson [3]. A typical spiropyran, NSP, changes color by UV irradiation from colorless to blue, and the open-ring form PMC has an absorption around 580 nm. The main research interest in the 1980s was how to give color variation to the compounds. A spectral shift to longer wavelengths was achieved by replacing the oxygen atom with a sulfur atom, as shown below [4].

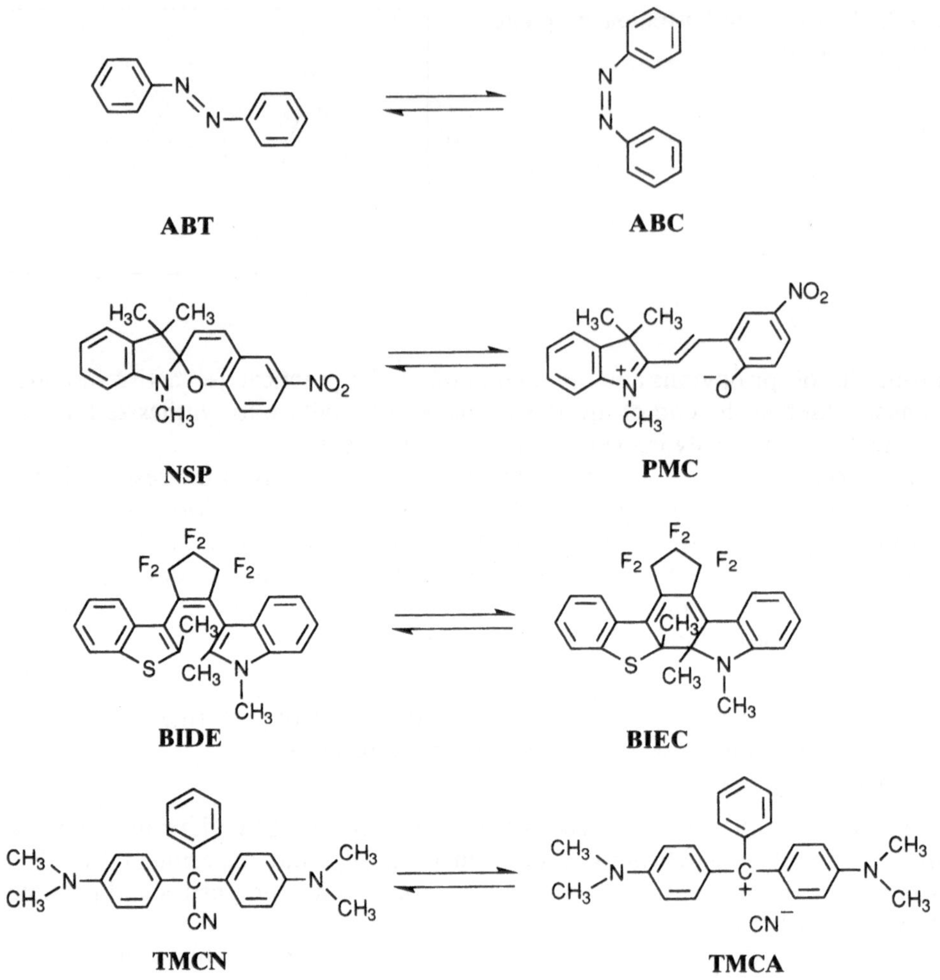

FIG. 3.2-2. Photochromic reactions

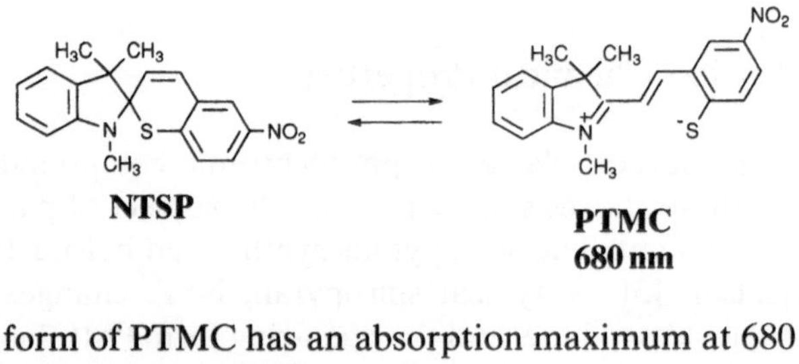

The open-ring form of PTMC has an absorption maximum at 680 nm and can be bleached with a diode laser of 830 nm.

Spironaphthoxazines recently attracted much attention because of their fatigue-resistant property. The coloration/decoloration cycles can be repeated more than 5000 times in isopropylalcohol [5]. The color of the open-ring form of unsubstituted naphthoxazine SINC is blue.

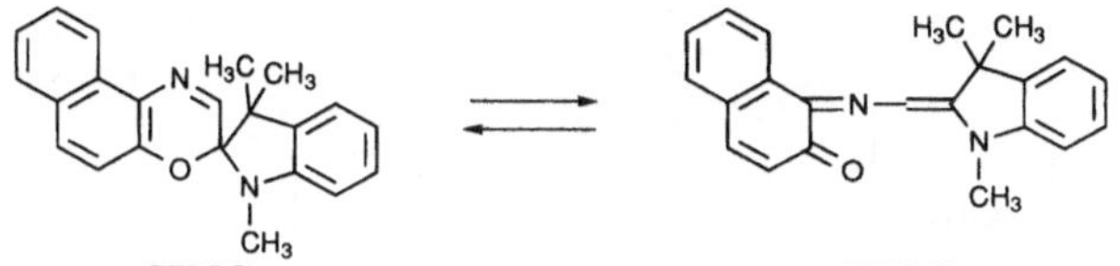

SINO   **SINC**

With the aim of obtaining red–yellow coloring compounds, various kinds of naphthoxazine derivatives have been synthesized. One possible approach is to shorten the π-conjugation as SNOP compounds [5]. However, the absorption maximum is longer than 560 nm.

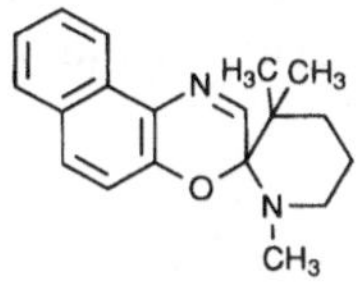

SNOP
562 nm
(in methanol)

Another approach is to introduce an electron-donating group into the naphthalene ring [6]. The open-ring forms of spironaphthoxazines shown below have absorptions shorter than 560 nm. Now we have substituted spiropyrans with absorption maxima from 470 to 660 nm.

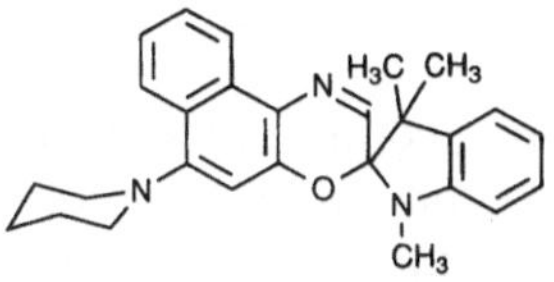

SINP   **SINT**   **SINZ**
560 nm   **530 nm**   **478 nm**
(in Toluene)   **(in Toluene)**   **(in Toluene)**

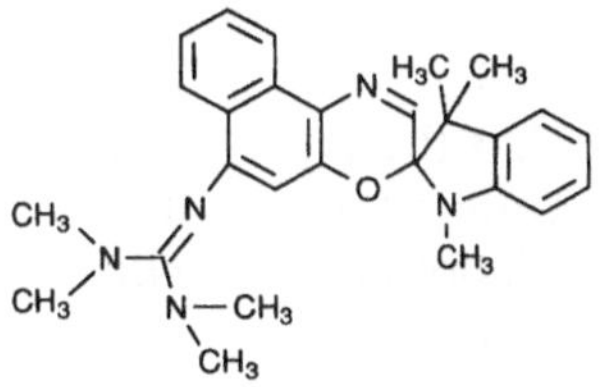

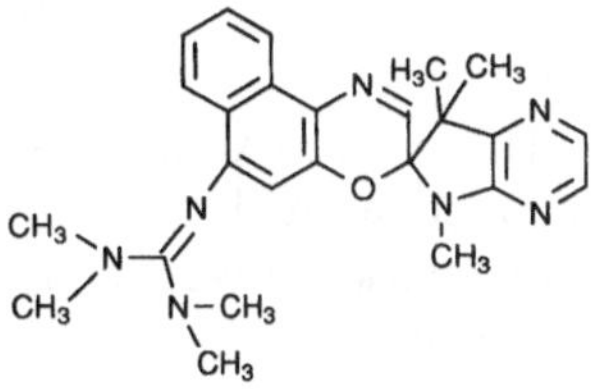

Another class of photochromic compounds with color variation is diarylethenes with heterocyclic rings, as compound BIDE in Fig. 3.2-2 [7–10]. During photoisomerization, diarylethene changes the π-conjugation length. In the open-ring form the absorption spectrum is the superposition of the two isolated aryl groups, while in the closed-ring form the π-electron delocalizes between the two aryl groups and the absorption spectrum has characteristics of a long polyene structure. The length of the π-conjugation controls the absorption wavelengths of the closed-ring forms. Typical diarylethenes, which turn to yellow, red, and blue after photoirradiation, are shown below.

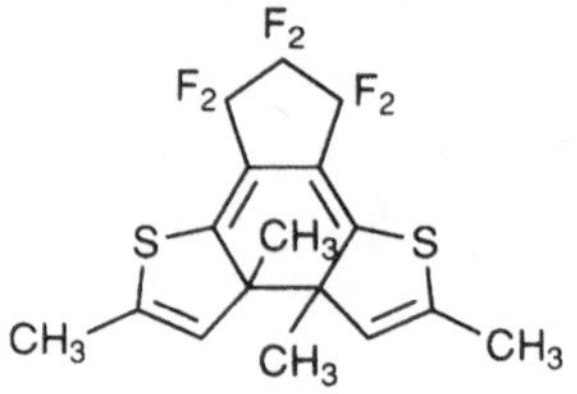

RTCE
yellow
(425 nm)

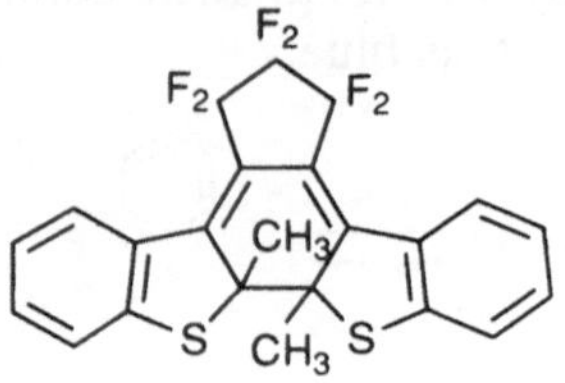

BBCE
red
(526 nm)

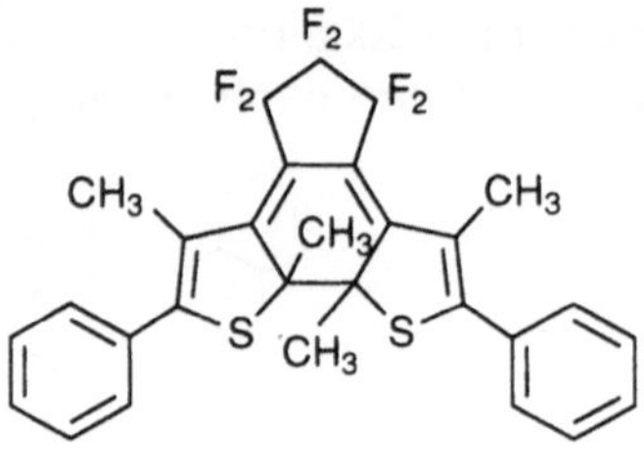

PTCE
blue
(562 nm)

Introduction of electron-donor and -acceptor groups into the two aryl groups further shifts the absorption maxima of the closed-ring forms to longer wavelengths [10, 11]. The closed-ring forms of diarylethenes shown below have absorptions longer than 650 nm.

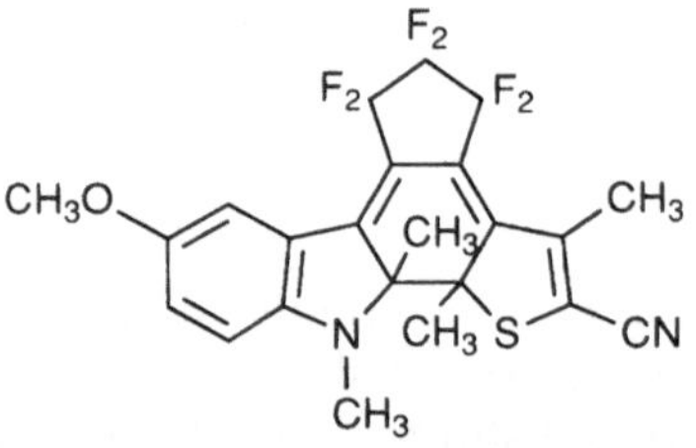

ITDE
665 nm

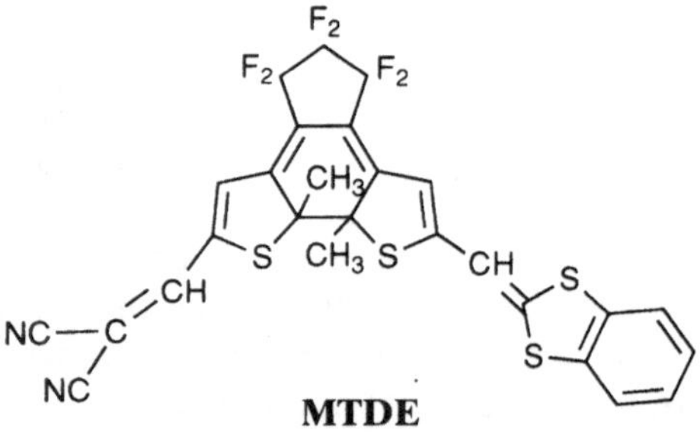

MTDE
828 nm

The above examples show that the color of photochromic compounds can be changed at will by incorporating various substituents to the same molecular framework. Now it is possible to obtain photochromic compounds with any color variation.

Refractive Index. Another interesting property of photochromic compounds is a photoinduced refractive index change. Although much attention has been paid to developing optical devices capable of signal processing such as switching, deflection, and modulation, almost all devices are based on electro-, acousto-, or magneto-optical-induced refractive index changes. The device that can control light directly by using a photooptical effect (photoinduced refractive index change) is of interest, because it can operate without mechanical motion and electric noise [12, 13]. A material which shows such a photooptical effect is a photochromic compound. Refractive index changes can also be used for a holographic display or memory.

The acridizinium dimer ADD shown below shows a large diffractive index change by photodissociation [14].

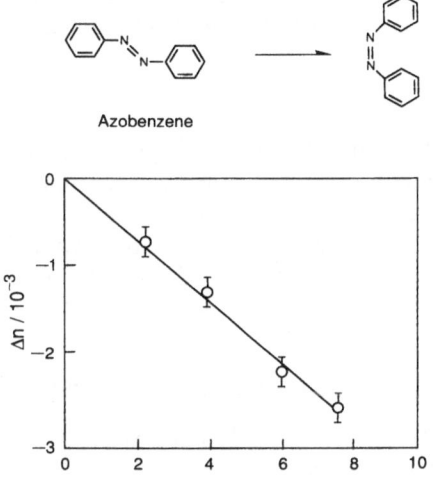

FIG. 3.2-3. Photoinduced refractive index change (Δ_n) at 633 nm for poly(methyl methacrylate) containing azobenzene

It has been found that the molar refraction for many organic compounds can be estimated as the sum of the atomic and bond refractions of the molecule. The dimer to broken-dimer conversion involves breaking four carbon–carbon single bonds and forming two carbon–carbon double bonds. From the table of bond refractions, the molar refraction change of the above reaction, ΔR, can be calculated to be 3.32 cm^3 mol^{-1}. The experimental ΔR was determined to be 29.1 cm^3 mol^{-1} [14] at 488 nm. The large difference between the value calculated from the bond refractions and the experimental value is attributable to the dispersion effect. During photodissociation, the absorption spectrum changes. The spectral shift can account for the larger refractive index change.

For practical applications, photochromic compounds are dispersed in polymer matrices. It is interesting to know how much the refractive index of the polymer film containing photochromic compounds changes after photoirradiation. Figure 3.2-3 shows an example of the refractive index change of a poly(methyl methacrylate) containing azobenzene [13]. Although the calculated ΔR of azobenzene based on the bond refractivity is zero, the experimental index decreases

TABLE 3.2-1. Changes in molar refractions of photochromic compounds with photoisomerization

Compound	ΔR cm^3 mol^{-1}	
	Calculated[a]	Experimental[b]
Azobenzene (ABT)	0	−2.4 (633 nm)
Spirobenzopyran (NSP)	1.7	85.0 (633 nm)
Diarylethene (BBDE)[c]	−1.7	18.5 (633 nm)
Acridizinium dimer (ADD)	3.3	29.1 (488 nm)

[a] Calculated based on bond refractions.
[b] Experimentally determined from the changes in refractive index.
[c] 1,2-Bis(benzo[b]thiophen-3-yl) perfluorocyclopentene.

along with the *trans* to *cis* photoisomerization. The decrease is solely due to the dispersion effect. Table 3.2-1 summarizes the calculated and experimental molar refractions.

Absorption spectra and refractive indexes are two basic properties which photochromic compounds can change with light. The following section presents recently developed new functional properties.

3.2.1.2 New Functional Properties

Thermal Stability. The most important property which photochromic compounds should have for photonic devices is thermal stability of both isomers. In general, one of the isomers of photochromic compounds is thermally unstable and returns to the initial form in the dark. Such thermal-reversible photochromic compounds cannot be used for optical memory media or switching devices. Although several attempts have been made to stabilize the photogenerated isomer by dispersing the compound in a polymer matrix and decreasing the matrix temperature below glass transition temperature T_g, the thermal reverse reaction cannot be prevented by the matrix rigidity. Recently, a guiding principle for the synthesis of thermally irreversible hexatriene–cyclohexadiene-type photochromic compounds has been established [15].

Typical examples of the hexatriene molecular framework are diarylethenes, which have benzene or heterocyclic rings, and fulgide derivatives. Semiempirical MNDO calculations were carried out for compounds STB and DFE [15]. Figure 3.2-4 shows the state correlation diagrams for the conrotatory cyclization reactions from STB to DHP and from DFE to DHF. According to the Woodward–Hoffmann rule based on the π-orbital symmetries, photocyclization reactions are brought about via the conrotatory mode.

First, that we should consider is the stability of the photogenerated closed-ring forms. As seen from Fig. 3.2-4, the stability depends on the energy barrier of the cycloreversion reaction in the ground state. The energy barrier correlates with the ground state energy difference between the open-ring and closed-ring forms,

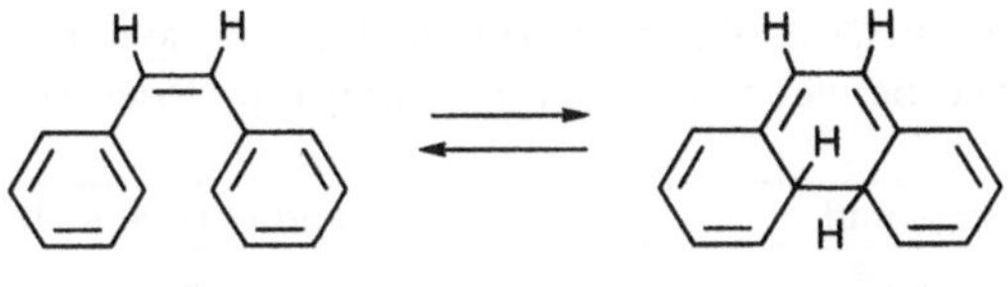

STB   DHP

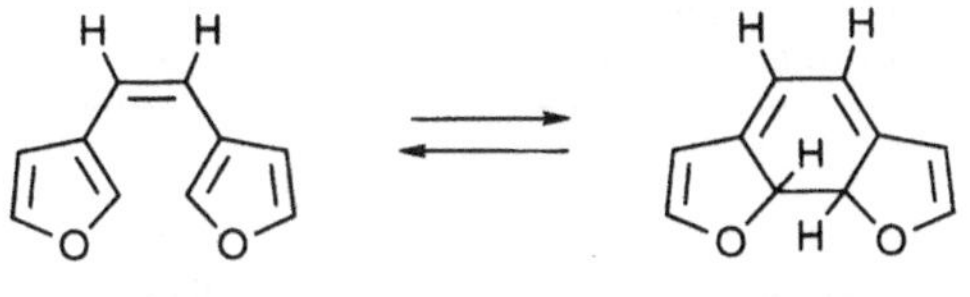

DFE   DHF

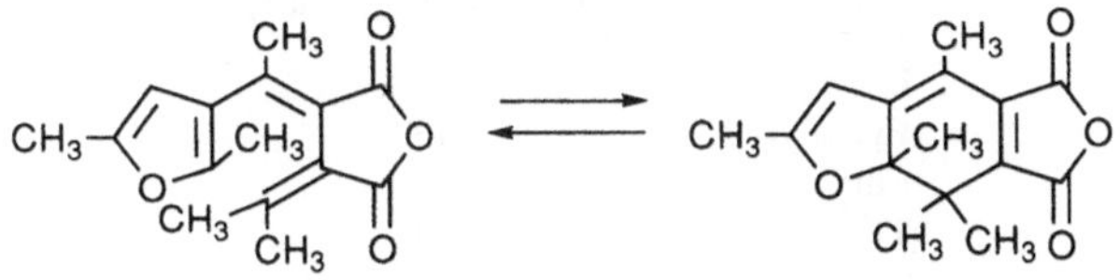

FFO   FFC

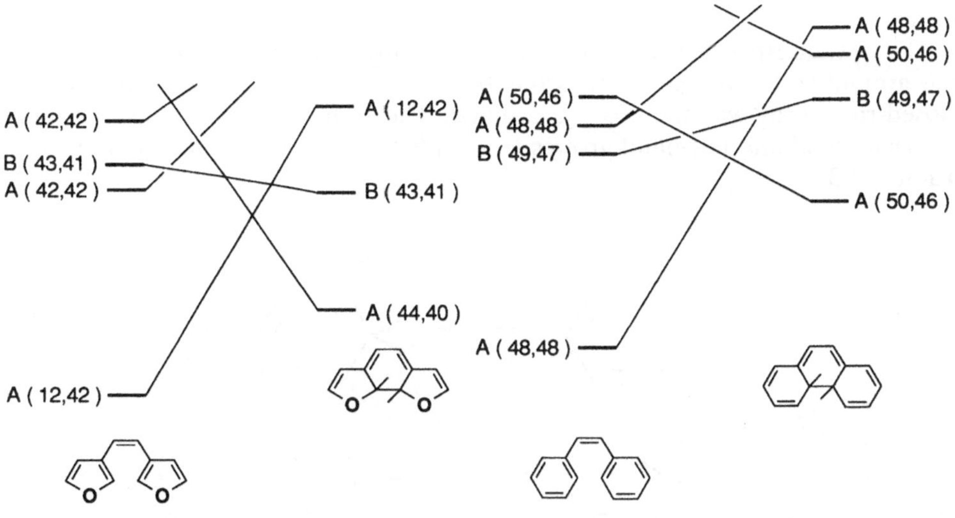

FIG. 3.2-4. State correlation diagrams in conrotatory mode for the reactions of 1,2-difurylethene and 1,2-diphenylethene

calculated values of which are shown in Table 3.2-2. When the energy difference is large, as in the case of 1,2-diphenylethene, the energy barrier becomes small. Conversely, the barrier becomes large when the difference is small, as shown for 1,2-furylethene. In this case, it is unlikely that the reaction would occur. The

TABLE 3.2-2. Relative ground-state energy difference between the open-ring and the closed-ring
forms

Compound	Conrotatory (kcal mol^{-1})
1,2-Diphenylethene	27.3
1,2-Di(3-pyrrolyl)ethene	15.5
1,2-Di(3-furyl)ethene	9.2
1,2-Di(3-thienyl)ethene	−3.3

TABLE 3.2-3. Aromatic stabilization energy difference

Group	Energy (kcal mol^{-1})
Phenyl	27.7
Pyrrolyl	13.8
Furyl	9.1
Thienyl	4.7

ground state energy difference controls the thermal stability of the closed-ring forms.

The next question is what kind of molecular property causes the difference in the ground state energies. The aromaticity change from the open-ring to the closed-ring form was therefore examined. The difference in energy between the right-hand and left-hand groups (shown below) was calculated, as shown in Table 3.2-3.

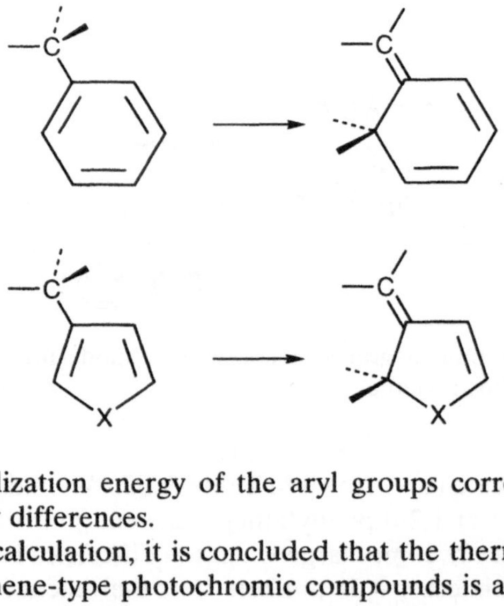

The aromatic stabilization energy of the aryl groups correlates well with the ground state energy differences.

From the above calculation, it is concluded that the thermal stability of both isomers of diarylethene-type photochromic compounds is attained by introduc-

TABLE 3.2-4. Thermal stability of closed-ring forms of diarylethenes

Stable		Intermediate	Unstable
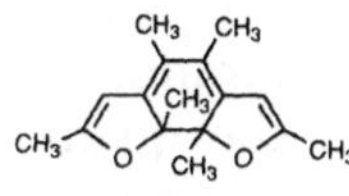 DE1	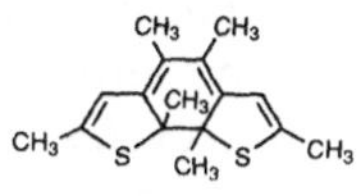 DE2	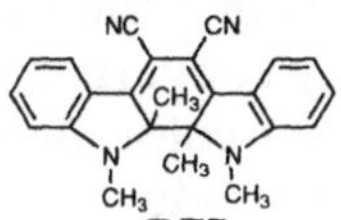 DE7	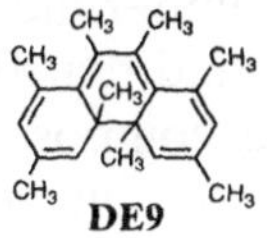 DE9
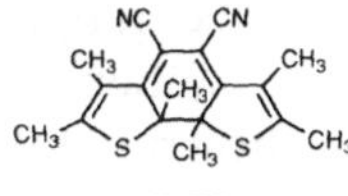 DE3	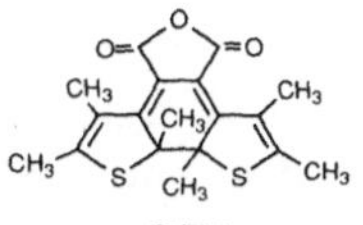 DE4	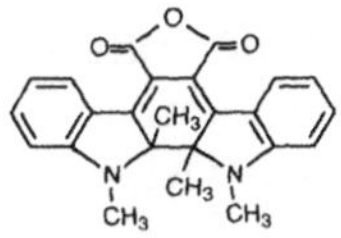 DE8	
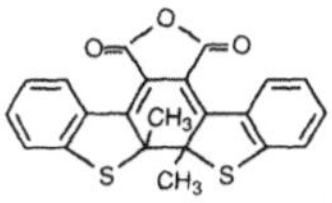 DE5	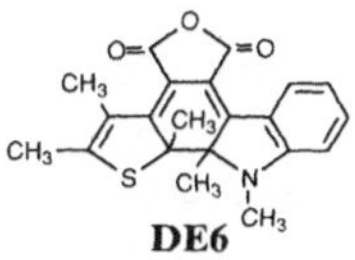 DE6		

ing aryl groups such as furan or thiophene rings, which have low aromatic stabilization energy.

The theoretical prediction was confirmed by the synthesis of diarylethenes with various types of aryl group, as shown in Table 3.2-4. The stability depends on the type of aryl groups. When the aryl groups were furan or thiophene rings, DE1–DE6, which have low aromatic stabilization energy, the closed ring forms were stable for more than 12 h at 80°C. On the other hand, photogenerated closed-ring forms of diarylethenes with phenyl or indole rings, DE7–DE9, were thermally unstable. The photogenerated yellow color of DE9 disappeared with a half-lifetime of 1.5 min at 20°C. The closed-ring form returned quickly to the open-ring form.

Furylfulgide FFO also undergoes a thermally irreversible photochromic reaction. The thermal irreversibility can be explained by the above principle. This compound has a vinyl group, and therefore the energy difference between the open- and closed-ring forms is considered to be small. This explains the thermal stability of the photogenerated closed-ring form. Various derivatives of fulgides have been synthesized to improve the photochemical ring-opening/cyclization quantum yields [16, 17].

Fatigue-Resistant Character. Photochromic reactions are always attended by a rearrangement of chemical bonds. During the bond rearrangement undesirable side reactions take place to some extent. This limits the cycles of photochromic reactions. The difficulty in obtaining fatigue-resistant photochromic compounds can be easily understood from a simple calculation. We assumed the following reaction scheme, in which a side reaction to produce B′ is involved in the forward process.

$$B' \xleftarrow{\Phi_S} A \rightleftharpoons B \tag{3.2-4}$$

Even under the conditions that the side-reaction quantum yield, Φ_s, is as low as 0.001 and that B perfectly converts to A, 63% of the initial concentration of A will decompose after 1000 coloration/decoloration cycles. Thus the quantum yield for conversion to byproducts should be less than 0.0001 in order for the cycles to be repeated more than 10000 times.

With the aim of clarifying photochemically robust structures, the fatigue-resistant properties of spirobenzopyran (NSP), furylfulgide (FFO), azobenzene (ABT), and diarylethenes DTMA, BBMA, EBMA, and BBDE (shown below) were measured [5, 8, 18].

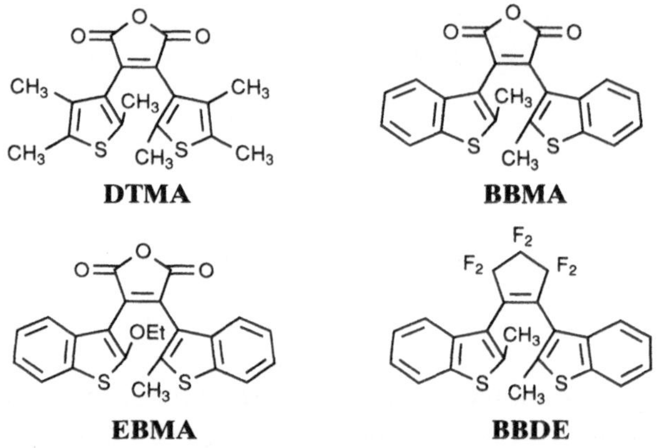

DTMA

BBMA

EBMA

BBDE

The measurement was carried out in the following way. Photochromic compounds were diluted in benzene as much as 10^{-4} M and the benzene solution was irradiated with UV light until 90% of the photostationary state was attained. Then the photogenerated color was completely bleached either thermally (spirobenzopyran) or photochemically (azobenzene, furylfulgides, diarylethenes). The coloration/decoloration cycles of spirobenzopyran, furyl-fulgide, and dithienylmaleic anhydride (DTMA) were limited to fewer than 100. The number of cycles of azobenzene was around 500. However, it is possible to have 5000 cycles for diarylethenes BBMA, EBMA, and BBDE. When the thiophene rings of diarylethenes were replaced with benzothiophene rings, the possible number of cycles increased considerably. Diarylper-fluorocyclopentene derivatives, such as 1,2-bis(2-methylbenzothiophen-3-yl)perfluorocyclopentene BBDE, showed excellent fatigue-resistant properties even in the presence of air.

Gated Photochromic Reactivity. Photochromic reactions, in general, proceed in proportion to the number of photons absorbed by the molecules. Such a linear-response characteristic cannot be used as the basis of photonic devices. Memories or images are destroyed during storage under room light or after many readout operations. One possible way to avoid such inconvenience is to introduce gated photochromic reactivity to the molecule. Gated reactivity means the property that irradiation at any wavelength causes no molecular change, while a photoreaction occurs when another external stimulation, such as a chemical, heat, or light of another wavelength, is present. Several photochromic compounds have been constructed which show such a nonlinear response.

A diarylethene with heterocyclic rings has two conformations with the two rings in mirror and C_2 symmetries, and its cyclization reaction can proceed only from the conformation with the rings in C_2 symmetry. This means that photocyclization is prohibited if the heterocyclic rings are fixed in the mirror symmetry, or parallel orientation, while the reaction is allowed when the conformation converts to the C_2 symmetry, or antiparallel orientation, as shown in Fig. 3.2-5.

Although photoreaction of CBDE was observed in ethanol, it was completely prohibited in cyclohexane [19, 20]. The addition of a very small amount of ethanol to the cyclohexane solution allowed photochemical reactivity. An NMR study clearly indicated that the molecule was in a parallel conformation in cyclohexane, and it converts to an antiparallel conformation upon the addition of small amount of ethanol. It cyclohexane, the intrahydrogen bonds fasten the

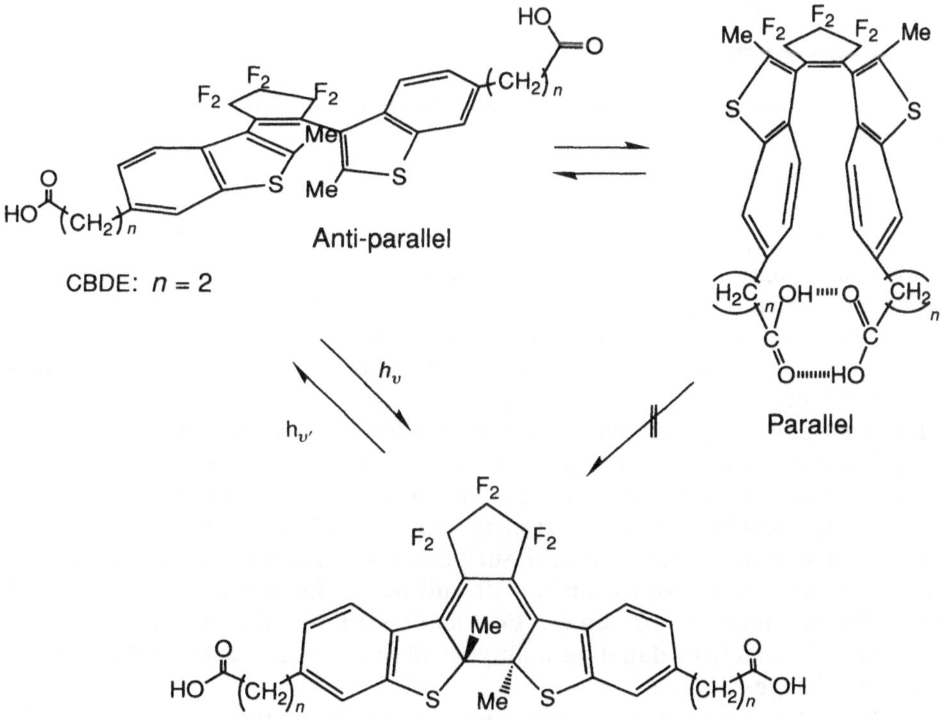

FIG. 3.2-5. Gated photochromic reactions of diarylethenes with intralocking arms

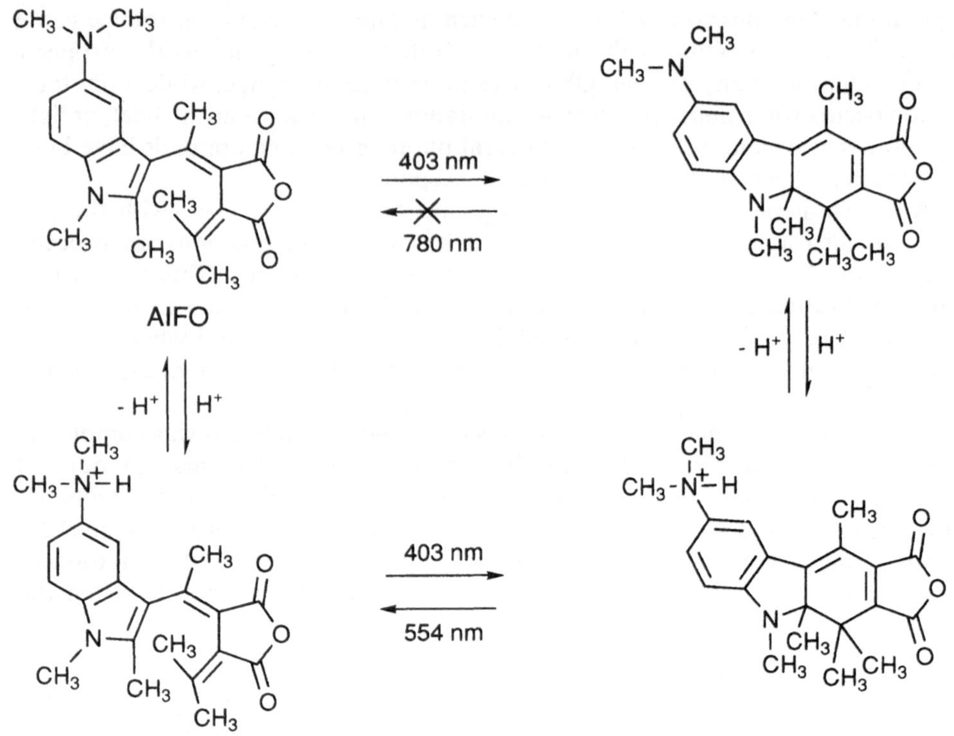

FIG. 3.2-6. Acid-gated photochromic reactions of an indolylfulgide with a dimethylamino substituent

molecule into a parallel conformation and make it photochemically inactive. Conversely, ethanol acts as a switch to unlock the system.

The hydrogen bonds can also be broken upon heating. In decalin, photocyclization did not occur below 60°C, while it was clearly observed at temperatures higher than 100°C. The molecule has both chemical- and thermal-gated reactivity.

Indolylfulgide (AIFO), having a dimethylamino substituent at the 5 position of the indole ring, undergoes an acid-gated photochromic reaction, as shown in Fig. 3.2-6 [21]. In the absence of acid the open-ring form converts to the closed-ring form, but the reverse process is strongly suppressed. The quantum yield is less than 10^{-4}. Upon the addition of acid, such as trichloroacetic acid, the absorption maximum showed a hypsochromic shift and the cycloreversion quantum yield increased as much as 10^3 times. Protonation to the dimethylamino group decreased the electron donating ability of the substituent and resulted in the reactivity change.

Photon-gated photochromic reactivity was found in the compound shown below [22].

NPY   **NKE**   **BCH**

When the light intensity is weak ($0.039\,mW\,cm^{-2}$), photodecoloration of yellow naphthopyran NPY was barely observed even after 7 h irradiation, while as much as 20% of the yellow color disappeared after 5 min irradiation with high-intensity light ($3.9\,mW\,cm^{-2}$). The decoloration rate changed in proportion to the square of the light intensity. These results indicate that two photons are required for isomerization. A photon produces a reaction intermediate, possibly an open-ring keto form NKE, as shown above. The intermediate is thermally unstable and returns to the initial naphthopyran NPY immediately after the light is switched off. Therefore, any net reaction was not observed after irradiation with weak light. When the light intensity is high, the keto form has a chance to capture a second photon and converts to the bicyclohexene isomer BCH. The bicyclohexene is colorless and thermally stable. The isomer again returns to the original naphthopyran after irradiation with 334 nm light. The compound has a nonlinear response to the light intensity.

Photoswitching Functions. A switching function, which allows for the reversible modulation of a given physical property by photoirradiation, is of basic importance for the development of molecular devices. Such a function can be introduced to a molecule by incorporating a photochromic unit into the molecule. A switching device which can control electron flow in a molecular system is shown Fig. 3.2-7 [11]. In the open-ring form, two pyridinium rings are separated from each other and there is no appreciable interaction between them. This is the "off" state. However, in the photogenerated closed-ring form, the π-conjugation delocalizes in between the two pyridinium rings and the absorption spectrum shifts to a longer wavelength (from 352 to 662 nm). This is the "on" state. Cyclic voltammetry indicated that whereas no electrochemical process occurs for the open-ring form in the +0.6 to −0.6 V region, a clearly reversible and monoelectronic reduction wave was observed for the closed-ring form at a potential $E_{1/2} = -230\,mV$ vs. a standard calomel electrode. The compound represents a prototype of a switchable molecular wire in which electron flow may occur in the closed state, on/off switching being triggered by irradiation with light of well-separated wavelengths.

Energy flow in a molecular system can also be switched by incorporating a photochromic unit in the system. A supermolecule, D–F–A, was synthesized to

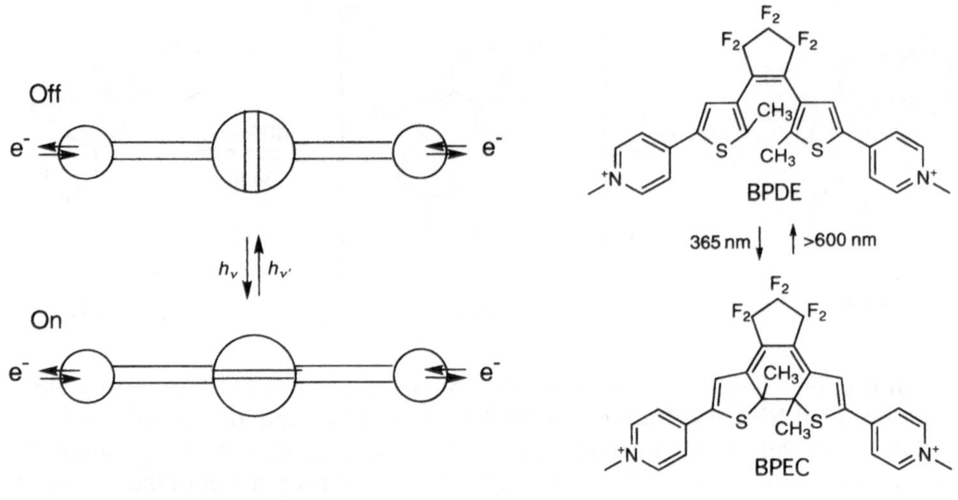

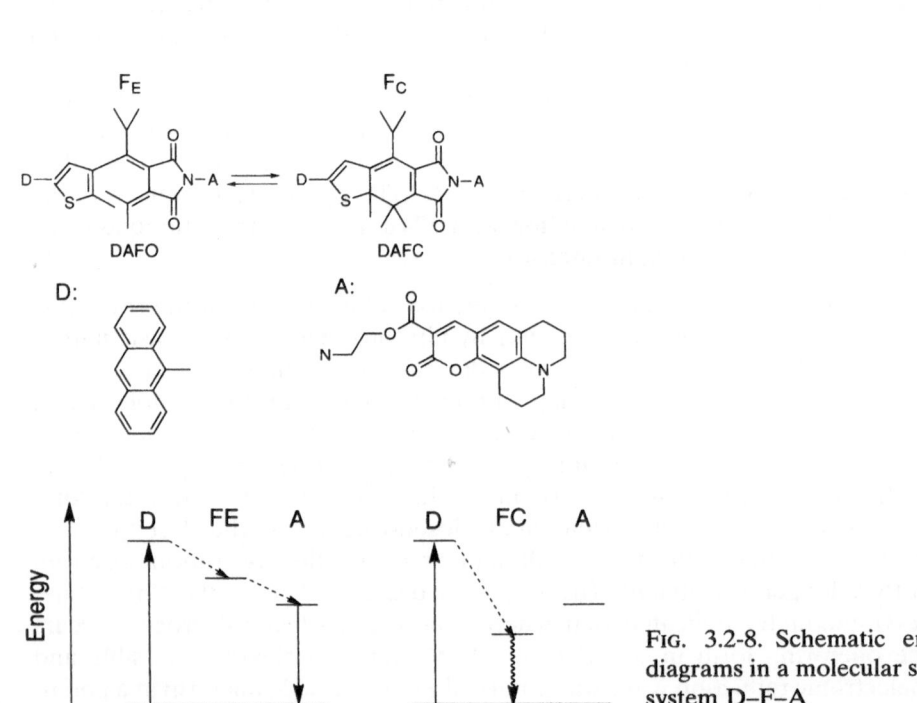

FIG. 3.2-7. Schematic illustration of photoswitching of the electron flow in a molecule

FIG. 3.2-8. Schematic energy diagrams in a molecular switch system D–F–A

allow or prevent energy transfer from D (donor) to A (acceptor) by switching F (fulgimide) with light of an appropriate wavelength from one isomer to the other [23]. The structure and energy diagram of the molecule is shown in Fig. 3.2-8. When the thienylfulgimide is in the open-ring form, the photoexcited single excited energy of D can transfer to A through F. However, the energy is trapped by F when the fulgimide is in the closed-ring form. The idea was confirmed by the fluorescence and fluorescence excitation spectra of the compound DAFO. The fluorescence of DAFO was of the coumarin type, and the fluorescence excitation spectrum demonstrated that excitation of anthryl bands leads to coumarin emission. However, when the fulgimide converts to the closed-ring form, the coumarin fluorescence almost disappeared. The result is interpreted as a result of trapping, i.e., the fulgimide part is a trap for energy absorbed in the donor, and the energy is radiationlessly deactivated by the fulgimide in the closed-ring form.

Chirality offers some interesting possibilities in photochromic chemistry. Photoreversible systems based on interconversion of enantiomers (or pseudoenantiomers) of inherently dissymmetric molecules were synthesized and their unique optical properties were explored [24]. The inherently dissymmetric compound TPTC combines a helical structure with bistability, as a stilbene-type photoisomerization can occur.

TPTC   **TPTT**
M (cis)   **M (trans)**

Irradiation with 300 nm yielded a mixture of 64% M-*cis* and 36% P-*trans*. However, irradiation at 250 nm gave a photostationary state containing 68% M-*cis* and 32% P-*trans*. The two forms have opposite helicity and their ORD and CD spectra are different. Therefore, alternate irradiation with 300 nm and 250 nm light resulted in a modulation of the ORD and CD signals. This is an example of the photoswitching of chiroptical properties by using a photochromic reaction.

3.2.2 Photochemical Hole Burning

Photochemical hole burning (PHB) is a phenomenon in which very narrow, stable photochemical holes are burnt at very low temperatures into the absorption bands of dye molecules molecularly dispersed in an amorphous solid by narrow-band excitation with a laser beam [25, 26]. Proton tautomerization of free-base porphyrins and phthalocyanines, and hydrogen bond rearrangement of quinizarin are typical photochemical reactions which provide PHB spectra.

The principle of PHB is shown in Fig. 3.2-9. Absorption lines of each dye molecule at very low temperatures become very sharp. In real matrices, however, they lie in a slightly different microenvironment, leading to a distribution of the resonant frequencies of the dye molecules, giving an inhomogeneous broad absorption band. Laser irradiation onto this inhomogeneous absorption band induces site-selective excitation and reaction of dye molecules which resonate to the laser frequency, and hence a hole is formed onto the absorption spectrum.

PHB is attracting considerable interest as a means for realizing ultra-high density optical storage of around 10^{11} bit/cm^2, by introducing frequency (wavelength) domain as a new dimension of memory system. PHB can be combined with a photon-echo or Stark effect, providing time-domain or electric-field-domain optical storage and processing devices [25].

PHB is also becoming increasingly important as a new technique for high-resolution solid-state spectroscopy at low temperatures. By using the site-selectivity of PHB phenomenon, the electron–phonon interaction of the systems, the low energy excitation mode, and the structural relaxation of matrix polymers, some very fast relaxation processes which form excited states such as energy transfer and electron transfer can be investigated [25, 26]. Recently, the spectrum

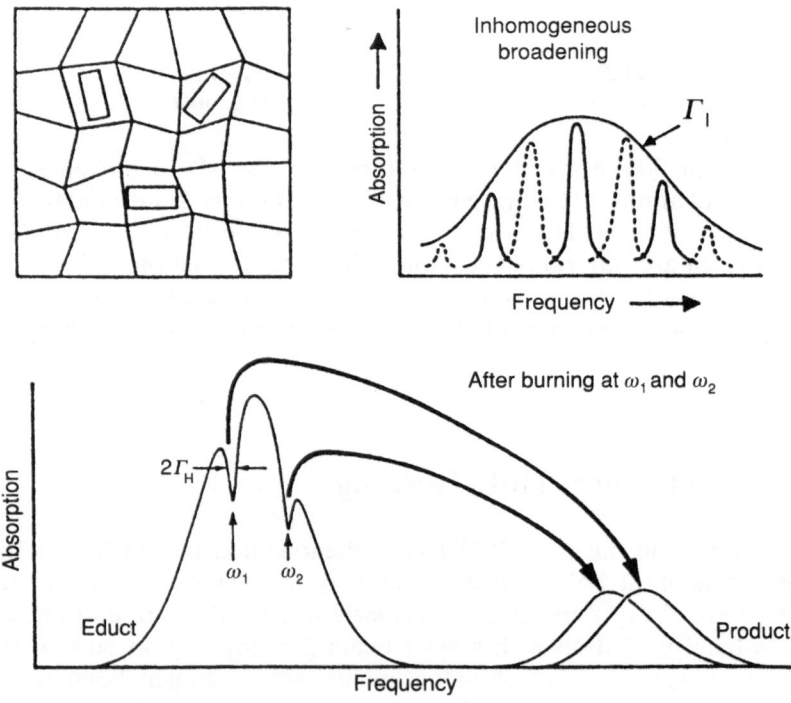

FIG. 3.2-9. Principle of photochemical hole burning

and dynamics of a single molecule in several molecular crystals and polymers have been observed as the extension of PHB studies [27].

For the application of PHB to frequency–domain high-density optical storage, several recessary conditions, such as high-temperature hole burning and storage, high-speed recording and readout, and non-destructive readout, should be met. Some of the recent developments in these topics are discussed below.

3.2.2.1 High-Temperature PHB Materials

For the realization of a PHB memory, hole formation and hole storage at high-temperatures are of essential importance. The study of high-temperature PHB is also important to gain an insight into the microenvironments of dyes at various temperatures.

Since its discovery in 1974, PHB has been considered to be a phenomenon which occurred only below liquid-helium temperature (4 K). Hole storage at room temperature, i.e., hole observation at 4 K for a hole burnt at 4 K after an excursion to room temperature, was reported in 1985 for the $Sm^{2+}/BaClF$ system [28], and hole formation at liquid-nitrogen temperature (80 K) was first observed in 1988 for free-base porphyrins in poly(vinyl alcohol) [29] and phenoxy resin [30]. After the report of hole storage at room temperature for a dye in a molecular crystal system [31], a spectral hole was first burnt and observed at room temperature for a $Sm^{2+}/SrCl_{0.5}Br_{0.5}$ ionic crystal with a broadened inhomogeneous band [32, 33]. Recently, room temperature hole burning has been reported for Sm^{2+} in amorphous glasses [34, 35] and for neutron-irradiated diamond [36].

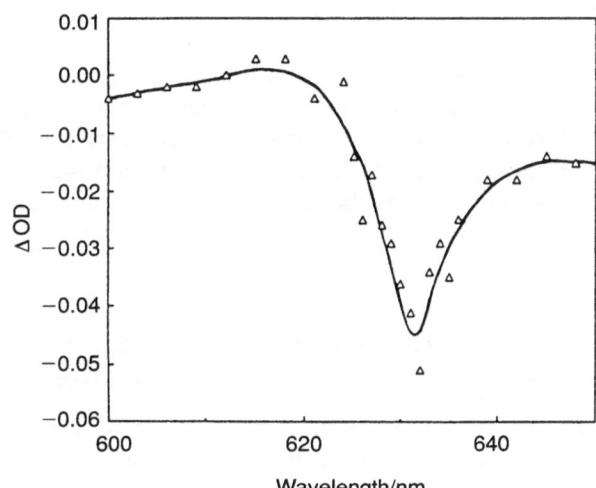

FIG. 3.2-10. Hole profile for a ZnTTBP/TiO$_2$ system at 140 K, with 20 min irradiation at 3.8 mW/cm^{-1} with a dye laser

Although no report has yet appeared for room temperature PHB of an organic dye system, several attempts are being made to restrict the structural relaxation of the matrices and confining dyes in restricted domains. Octaethyl-porphyrin absorbed on a γ-Al_2O_3 surface [37] and zinc phthalocyane in an alumophosphate molecular sieve [38] show hole formation at 90 K with a small temperature dependence of the Debye–Waller factor. Metal-free mioglobin also shows hole formation above 100 K.

Organic dyes have the advantage of having a larger absorption cross section and higher quantum efficiency of hole formation compared with inorganic rare earth ions. Inorganic matrices generally show a larger Debye–Waller factor, and a small temperature dependence compared with organic polymer matrices. Therefore an organic–inorganic hybrid system is a possible candidate for high-temperature PHB. Figure 3.2-10 shows a hole profile at 140 K of zinc tetratolyltetrabenzoporphin (ZnTTBP) doped in TiO_2 by a sol-gel method [39]. The mechanism of hole formation would be one-photon ionization of the dye and electron transfer to the conduction band of the n-type semiconductor matrix. To date, 140 K is the highest temperature of hole formation and observation for organic dye systems.

3.2.2.2 Laser-Induced Hole Filling

To realize high-speed readout with 30 ns per bit in PHB data storage, the formation of deep holes with more than 10% depth is inevitable [40]. However, multiple deep-hole formation with a narrow hole distance in the frequency domain brings about a decrease in the area of previously burnt holes. This phenomenon is called laser-induced hole filling (LIHF).

The mechanism of LIHF for a free-base porphyrin/polymer system was studied by measuring the wavelength-dependence of the extent of LIHF during laser irradiation of a porphyrin (TPPS)/poly(vinyl alcohol) (PVA) sample with multiple holes formed in advance (Fig. 3.2-11a) [41]. The result, in Fig. 3.2-11b, shows the characteristic wavelength-dependence of LIHF. Holes at the longer-wavelength side of the newly burnt hole are filled more thoroughly than holes at the shorter-wavelength side, which can be attributed to the contribution of molecules excited non-site-selectively through a phonon side band (E_s), and/or vibronic sublevels (V_1) in the excited state [42]. However, holes at the shorter-wavelength side of the newly burnt hole are also filled to some extent by the non-site-selective excitation through thermal excitation of ground-state molecules [43].

3.2.2.3 Hole Burning in the S_2–S_0 Absorption Band and the Possibility of Hole Burning in Various Organic Dyes

For organic dye systems, it was generally thought that site-selective persistent spectral holes can be burnt only for the lowest electronic spectral band, i.e., to

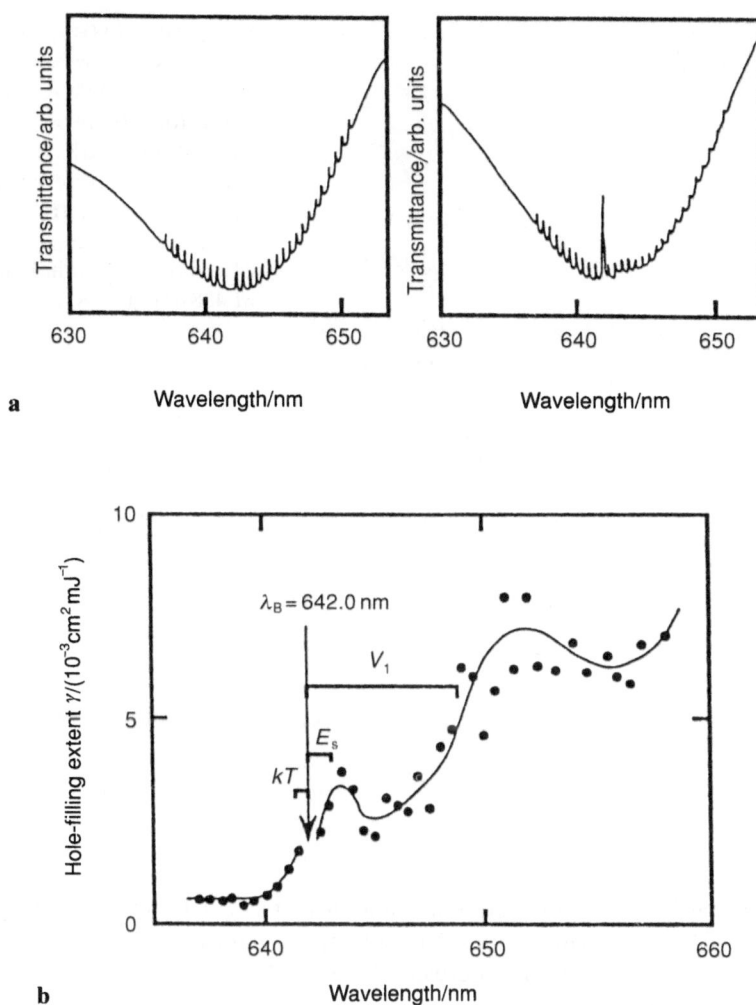

FIG. 3.2-11. **a** Experiments of laser-induced hole filling for previously burnt multiple holes in TPPS/PVA at 20 K. *Left*, before burning for hole filling; *right*, after burning for hole filling. **b** Wavelength-dependence of the extent of LIHF for TPPS/PVA at 20 K [17]

electronic transitions to the first excited singlet of the molecule, or to vibrational transitions in the infrared region. However, the first results on the formation of spectral holes in the Soret (S_2–S_0) band of ZnTTBP in poly(methyl methacrylate) with chloroform (Fig. 3.2-12) have been provided by using a photon-gating technique [44]. Zero-phonon holes with a homogeneous width of $5.1\,cm^{-1}$ at 20 K can be formed over the whole region of the Soret band through electron transfer from a higher excited-triplet state to chloroform by two-photon excitation. Zero-

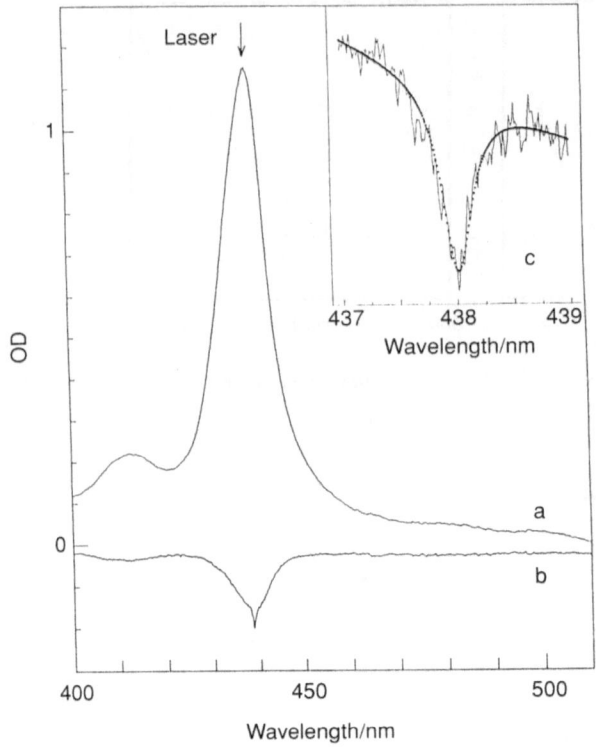

FIG. 3.2-12. PHB for the Soret absorption band of ZnTTBP. in PMMA with chloroform by two-color irradiation at 20 K. *a*, absorption spectra before burning; *b*, difference spectrum after burning at 438 nm; *c*, details of a zero-phonon hole burnt at 438 nm at 20 K

phonon hole width yeilds an S_2 state lifetime of 1.7 ps at 20 K. The decrease in the whole absorption band due to non-site-selective excitation can also be seen in Fig. 3.2-12. A strong correlation between S_2–S_0 and S_2–S_1 transitions is observed through satellite hole analysis.

To date, the number of molecules reported to show hole formation is restricted compared with a wide variety of chemical structures of organic molecules. This is because it is generally believed that for hole formation in PHB, the existence of a zero-phonon line in guest molecules should be accompanied by the occurrence of some photochemical reaction of the guest molecules. A well-known exception to this is the so-called non-photochemical hole burning (NPHB) [45]. Recently, triplet–triplet energy transfer of a guest molecule to a host photoreactive matrix has been reported to be a new family [46] of PHB systems.

Two prerequisites for the formation of persistent holes in organic dyes would be: (i) the existence of a zero-phonon line for a dye molecule, and (ii) a change in energy level of the dye molecule interacting with the surrounding matrix after photoexcitation. The possibility of site-selective excitation, i.e., the existence of a zero-phonon line for certain dye molecules, can be evaluated by comparatively small Stokes shifts (less than $1000 \, \text{cm}^{-1}$) of the dye molecules [47]. Desirable systems for observing zero-phonon holes for a wide variety of dye molecules which have small Stokes shifts but do not undergo chemical reaction by themselves are summarized as follows: (i) the use of poly(vinyl alcohol) as a matrix with hydrogen-bonding rearrangement for non-photochemical hole burning (NPHB); (ii) the use of an energy transfer (sensitization) technique [46] to photoreactive matrices; (iii) the use of electron transfer or photoionization to halogenated matrices or semiconductors.

3.2.3 Nonlinear Optics

Optical nonlinearities are related to the polarization induced by high-order terms of optical electric fields. When a molecule is put under light with electric field E, the molecular polarization p is generally given by

$$p = p_g + \alpha E + \beta E \cdot E + \gamma E \cdot E \cdot E + \cdots \qquad (3.2\text{-}5)$$

where p_g is the molecular polarization in ground state, α is the polarizability of the molecule, and β and γ are the second-order (or first) and third-order (or second) hyperpolarizabilities of the molecule, respectively. Similarly, polarization P for molecular aggregates under electric field E is written as

$$P = P_g + \chi^{(1)} E + \chi^{(2)} E \cdot E + \chi^{(3)} E \cdot E \cdot E + \cdots \qquad (3.2\text{-}6)$$

where P_g is ground-state polarization, $\chi^{(1)}$ is linear optical susceptibility, and $\chi^{(2)}$ and $\chi^{(3)}$ are the second- and third-order nonlinear optical susceptibilities, respectively. The nth-order hyperpolarizability and nonlinear optical susceptibility are $(n+1)$th-rank tensors. Although nonlinear terms of second- and higher orders of E can be ignored under conventional light, those terms should not be neglected under strong electric fields using laser beams. A very important relation between susceptibilities and structures is that even-order effects vanish in centrosymmetric media. Since the coefficients become small with an increase in the order of E, more than fourth-order terms are hard to use for practical applications even using lasers. Typical nonlinear optical phenomena are frequency conversion, phase modulation, and changes in absorptivity and refractivity. The most familiar ones are second- and third-harmonic generations (SHG and THG), which are the nonlinear optical processes which give output frequencies of 2ω and 3ω, respectively, from a fundamental beam of ω.

From the later half of 1970s, π-conjugated organic compounds have attracted attention for their strong nonlinear optical properties, because π-electron systems essentially show high optical nonlinearity and ultrafast response times. Much instructive literature on organic nonlinear optical materials has been published [48–50].

3.2.3.1 Molecular System Design for Second-Order Nonlinear Optics

Large second-order optical nonlinearity is realized by two steps, i.e., finding the chromophore molecules with large β and aligning them in a noncentrosymmetric molecular system. Since the molecular structure of a π-conjugated system substituted by a donor and acceptor at each end, i.e., a D–π–A structure, has been considered to show large β, many D–π–A compounds have been synthesized and examined. The most typical example, and the simplest, is 4-nitroaniline. Actually, 4-nitroaniline has a large β value. However, as a crystal, it shows no second-order optical nonlinearities because of its centrosymmetric molecular arrangement. In order to break the centrosymmetry of the molecular arrangement, several methodologies have been investigated. Fig. 3.2-13 summarizes the general methods used to create a noncentrosymmetric molecular system from a selected D–π–A compound.

Various methods have been reported to make molecules crystallize in noncentrosymmetric structures such as those in Fig. 3.2-13a and b, e.g., the introduction of a steric effect, intermolecular hydrogen bonding and chirality, and complex formations. A simple substitution, e.g., at the ortho- or meta-position of a benzene ring, sometimes works effectively, as in 2-methyl-4-nitroaniline. When a pair of D–π–A molecules are connected with a methylene group, the molecular shape becomes Λ-type and the molecule can stack easily in a polar orientation, e.g., bis(N-(4-nitrophenyl)amino)methane. Intermolecular hydrogen bonds tend to give a polar array of chromophores, e.g., 2-(N,N-dimethylamino)-5-nitroacetoanilide. Since chiral molecules essentially crystallize in a chiral crystal structure, the introduction of chiral groups onto the chromophore is one of the most reliable ways to obtain a noncentrosymmetric crystal structure, e.g., (S)-N-(4-nitrophenyl)prolinol. The formation of an inclusion complex with the host having a chiral structure has been reported, e.g., 4-nitroaniline incorporated into cyclodextrin. For an ionic chromophore, a noncentrosymmetric structure can be obtained by a complex formation with a proper counter ion.

The Langmuir–Blodgett (LB) technique is also one of the most reliable ways to produce noncentrosymmetric structures. In order to add amphiphilic properties to the chromophores for LB film formation, one or more long alkyl group is generally attached. If necessary, a hydrophilic group is also provided. The X- and Z-type films (Fig. 3.2-13c) are prepared by head-to-tail type deposition, although they are generally unstable, and rapidly change into more stable centrosymmetric Y-type films. Thus, the alternating deposition of two components was proposed to attain polar and stable hetero Y-type films (Fig. 3.2-13d). Even in Y-type films composed of one component, second-order optical nonlinearities are observed when the film has a herringbone Y-type structure (Fig. 3.2-13e). The polarization of the film in Fig. 3.2-13c and d is in a direction perpendicular to the substrate, and that of the film in Fig. 3.2-13e is in the dipping direction.

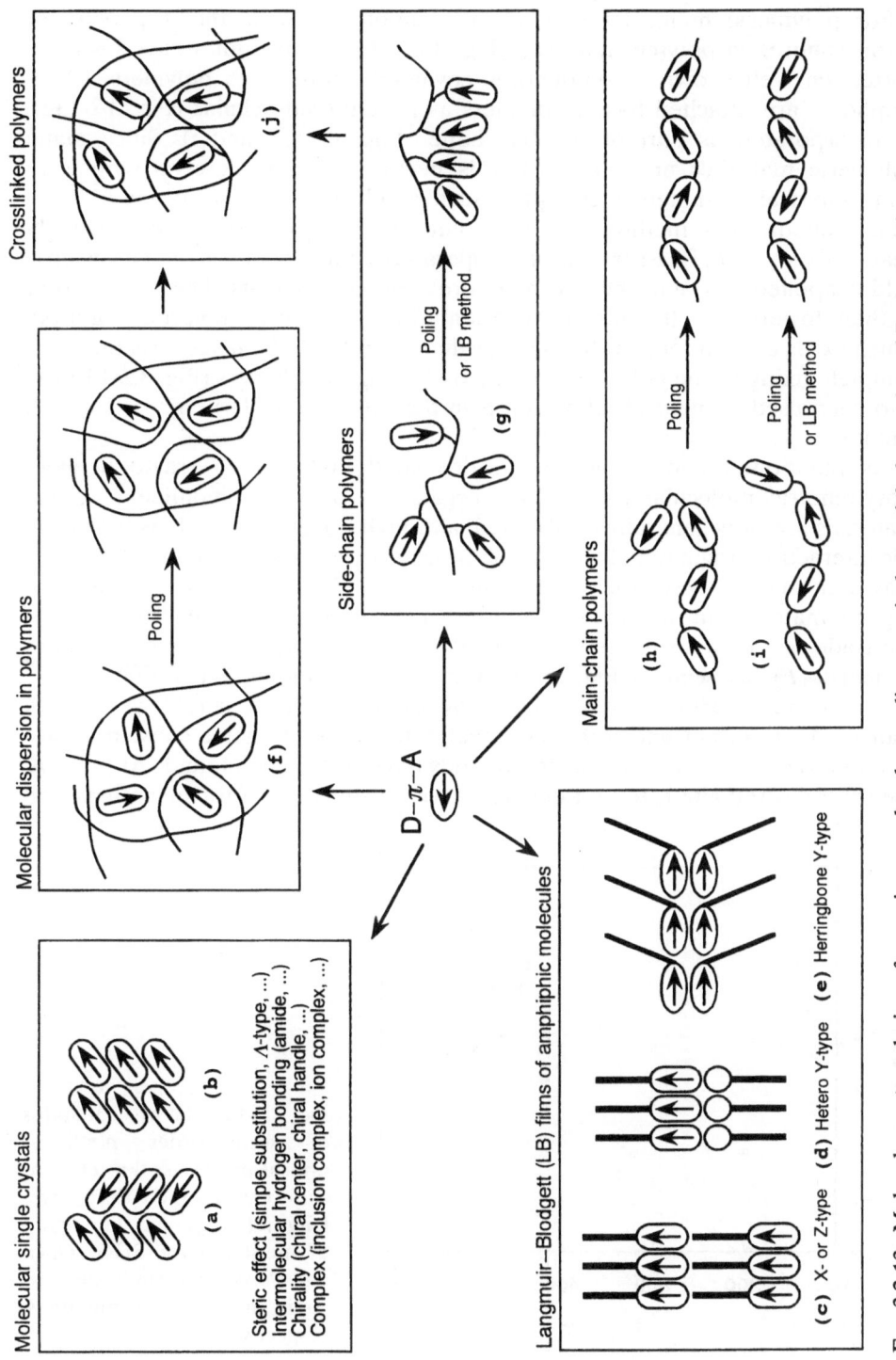

FIG. 3.2-13. Molecular system design of organic second-order nonlinear optical materials for a selected D–π–A molecule

Molecular single crystals

(a)

(b)

Steric effect (simple substitution, Λ-type, ...)
Intermolecular hydrogen bonding (amide, ...)
Chirality (chiral center, chiral handle, ...)
Complex (inclusion complex, ion complex, ...)

Molecular dispersion in polymers

Poling

(f)

Crosslinked polymers

(j)

D–π–A

Side-chain polymers

Poling
or LB method

(g)

Main-chain polymers

Poling

(h)

Poling
or LB method

(i)

Langmuir–Blodgett (LB) films of amphiphilic molecules

(c) X- or Z-type   (d) Hetero Y-type   (e) Herringbone Y-type

For polymeric molecular systems, the simplest case is the dispersion of chromophores in polymer matrices (Fig. 3.2-13f). Many studies have been reported on well-known chromophores covalently bonded to polymers. Chromophores are attached to the polymers as pendant side chains (Fig. 3.2-13g), or incorporated as part of the backbone (Fig. 3.2-13h and i). Since many polymeric materials are isotropic, it is necessary to break centrosymmetry using physical alignment techniques, such as electric-field or corona-induced poling or the LB method. In the poling process, the polymer is usually heated above its T_g (glass transition temperature) and a high voltage DC electric field is applied to orient the chromophore's dipole moments. The temperature is then lowered to fix this polar orientation. However, molecular motion, which exists even at temperatures lower than T_g, relaxes the poled structure. For complete fixing of the polar structure, crosslinking of polymers (Fig. 3.2-13j) has also been used. Temporal relaxation is in the following order: Fig. 3.2-13f, g, h and i, j.

As mentioned above, the preparation methodologies to create noncentrosymmetric molecular systems for organic second-order nonlinear optical materials are now generally established, although novel ideas such as ferroelectric interaction originating from proton transfer in hydrogen bonding arrays [51] may still emerge. Nevertheless, no one is satisfied with the present status of organic materials in the field of useful practical applications. This is because of the trade-off relationship between the absorption cutoff (λ_{cutoff}) and the figure of merit (F), as seen in Fig. 3.2-14 for second-harmonic generation (SHG) [52]. For real SHG devices excited by diode lasers, a cutoff wavelength shorter than 400 nm and a $\chi^{(2)}$ far greater than for inorganic substances are required. However, no suitable compounds have so far been found. At present, the most desired aim is to find molecules with a short cutoff wavelength as well

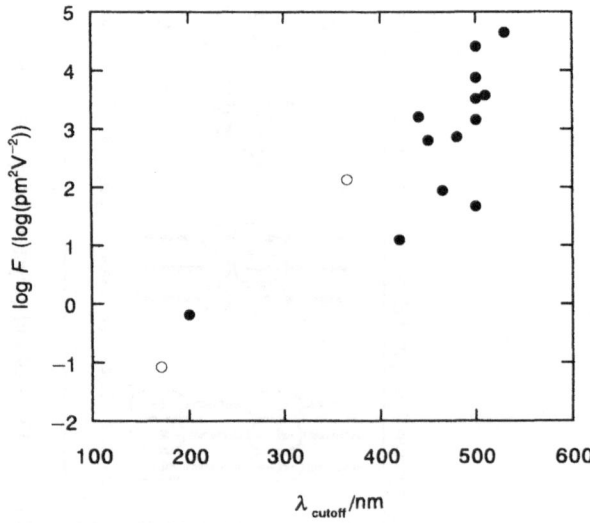

Fig. 3.2-14. The present status of second-order nonlinear optical materials: the relationship between the cutoff wavelength (λ_{cutoff}) and the figure of merit (F) for SHG. *Open* and *closed circles* indicate inorganic and organic compounds, respectively

as a large β. Recently, we found that p-toluenesulfonate anion, which has no absorption at wavelengths longer than 300 nm, has a comparatively large β, i.e., two-thirds that of 4-nitroaniline [53]. Ionic species may be good candidates for the breakthrough.

3.2.3.2 Molecular System Design for Third-Order Nonlinear Optics

The third-order hyperpolarizabilities (γ) of a series of linear π-conjugated molecules without large polarization were evaluated, and the molecules with a longer π-conjugation were found to exhibit larger hyperpolarizability. This was also proved theoretically. An extension of this line of approach led to quasi-one-dimensional π-conjugated polymers such as polydiacetylenes, polyacetylenes, poly(arylenevinylene)s, and so on. The σ-conjugated polysilanes have also been investigated. For these conjugated polymers, it was found that $\chi^{(3)}$ values in the non-resonant region are in inverse proportion to the sixth power of the band-gap energy, and those in the resonant region are proportional to absorbance. The resonant χ^3 values of those polymers, obtained by third-harmonic generation (THG), are in the range from 10^{-12} to 10^{-8} esu, and those evaluated with degenerate four-wave mixing (DFWM) are from 10^{-10} to 10^{-6} esu.

Since the largest $\chi^{(3)}$ component of π-conjugated polymers (Fig. 3.2-15a) is in the main-chain direction, an improvement of the main-chain orientation causes an increase of the raw $\chi^{(3)}$ values: compared with random oriented samples (Fig. 3.2-15b), $\chi^{(3)}$ values with planar and single crystalline orientations (Fig. 3.2-15c) become larger by a factor of 8/3 and 5, respectively. Thus, single crystals are best from the point of view of molecular orientation.

$\chi^{(3)}$ enhancement by the molecular design has also been tried. A reduction of inactive parts for optical nonlinearity, e.g., alkyl groups, increases the susceptibility. Conjugation between the polymer backbone and pendant substituents (Fig. 3.2-15d) extends the π-electron delocalization, resulting in a narrower band gap and larger $\chi^{(3)}$ [54]. New conjugated repeating units, including a ladder structure (Fig. 3.2-15e) and a superlattice structure along the conjugated main chain (Fig. 3.2-15f) are considered to be interesting and worth investigation. From the point of view of a well-defined structure with more homogeneous optical properties, oligomers (Fig. 3.2-15g) or polymers having mono-dispersed molecular weights are also promising. When an oligomer with a π-conjugation length just large enough to show efficient third-order optical nonlinearity per unit volume is packed at high density with perfect orientation, it is expected to give maximum susceptibility for that molecular system.

As the counterpart of the polymers, π-conjugated molecules (Fig. 3.2-15h), e.g., varieties of dyes represented by phthalocyanine, pseudoisocyanine (PIC), and azo dyes, and charge transfer complexes have of course been investigated. Fabrication techniques for these molecules are similar to those for second-order materials, but for third-order nonlinear systems, centrosymmetric structures are also acceptable. Their resonant $\chi^{(3)}$ values for THG are in the range from 10^{-12} to 10^{-10} esu, whereas those for DFWM vary from 10^{-8} to 10^{-7} esu. It is well known

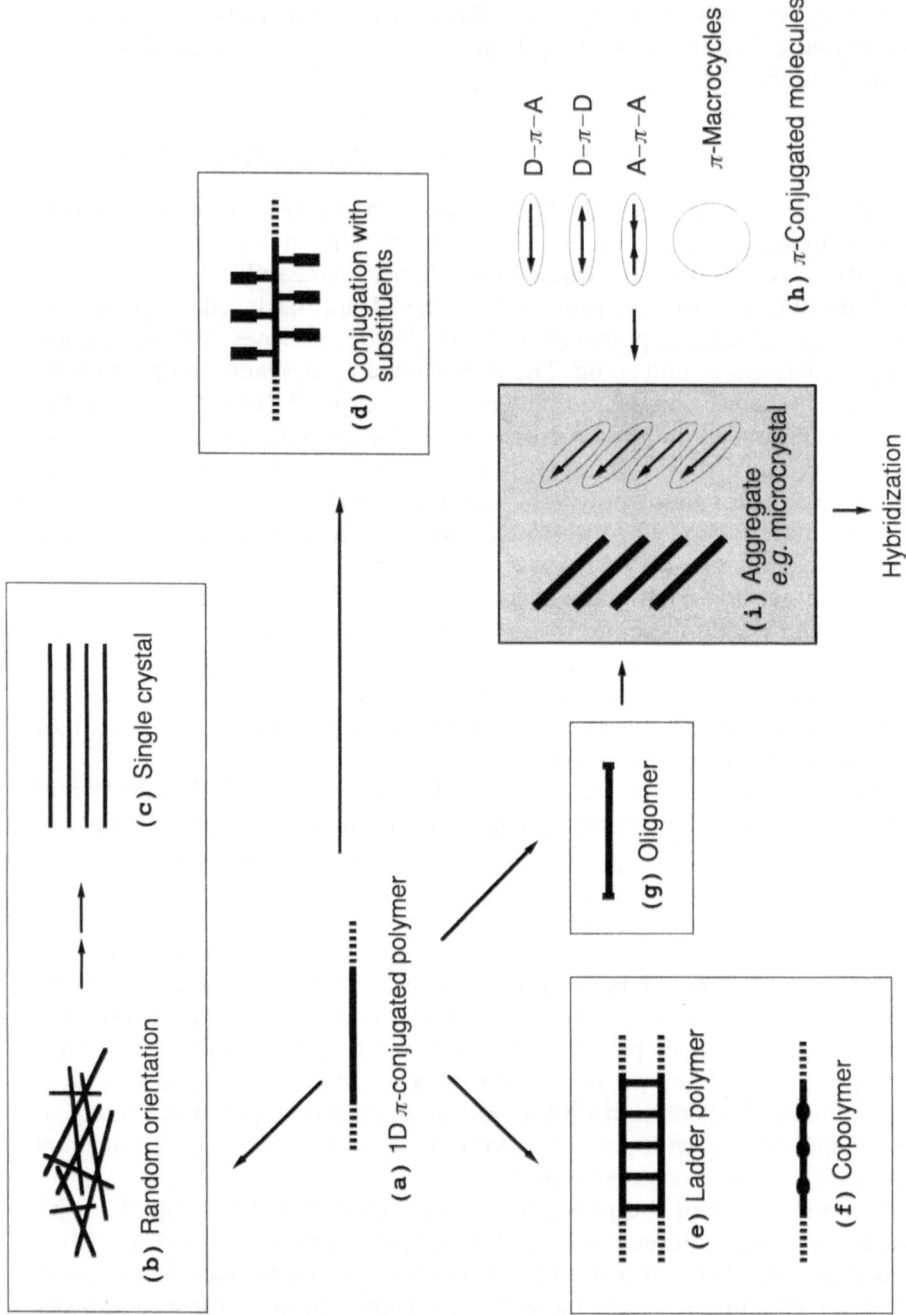

FIG. 3.2-15. Molecular system design of organic third-order nonlinear optical materials

that dyes sometimes show unique absorption spectra due to a specific molecular aggregation (Fig. 3.2-15i). For example, J-aggregated PIC shows extremely sharp absorption and its maximum is shifted to a longer wavelength compared with that of the monomeric spectrum. The $\chi^{(3)}$ is enhanced to 10^{-3} esu, which was evaluated by saturable absorption at 77 K [55]. Although the mesoscopic behavior of semiconductors and metals has been extensively investigated, that of organic compounds has rarely been studied. This is because the fabrication techniques necessary for an organic assembly on the mesoscopic scale have not been established.

Recently, we found a convenient preparation procedure for microcrystals [56], and proposed that microcrystallization is advantageous for optical device fabrication. Microcrystals are a class of molecular system with dimensions far smaller than the wavelengths of their component parts, and when they can be prepared successfully and dispersed uniformly in matrices of high concentration, the optical loss from the materials is very low. In addition, peculiar size effects have been expected for organic compounds, and several interesting phenomena were observed for organic microcrystals. From PIC and an amphiphilic merocyanine, J-aggregated microcrystals are easily obtained at room temperature. In the case of perylene microcrystals, a blue shift in absorption and emission with decreasing size is clearly observed in a size range which is at least one order of magnitude larger than those of metals and semiconductors. Although the reason for this is not yet clear, it is a very challenging subject. Application of the microcrystallization technique to solid-state polymerizable diacetylenes provided us with a molecular system of a single crystalline conjugated polymer with mono-dispersed molecular weight. From such polydiacetylenes, sharp absorption due to the uniform electronic structure of the polymer backbone and real molecular (device) properties are expected. Hybridization of a π-conjugated organic microcrystal as a core with a metal shell has been predicted, and this should also induce enlarged optical nonlinearity [57]. This cored structure type of electronic interaction between π-conjugated microcrystals and their surroundings will not be the only possibility. This kind of research will extend to removing the barriers between organic and inorganic materials, including metals and semiconductors, and creating a new frontier of science, namely pseudo-molecular science, which would deal with a hybridized molecular system covering all kinds of substances.

3.2.4 Optical Computers (Molecular Processors)

Optical computing techniques have been studied extensively by many workers in view of the advantages of optics compared with electronics, such as interconnectivity and the parallel nature of light propagation. Several models have been proposed for optical digital and parallel computers using optics assemblies consisting of photorefractive crystals, optical filters, fibers, and waveguides [58]. As far as switching devices are concerned, it is worth noting that

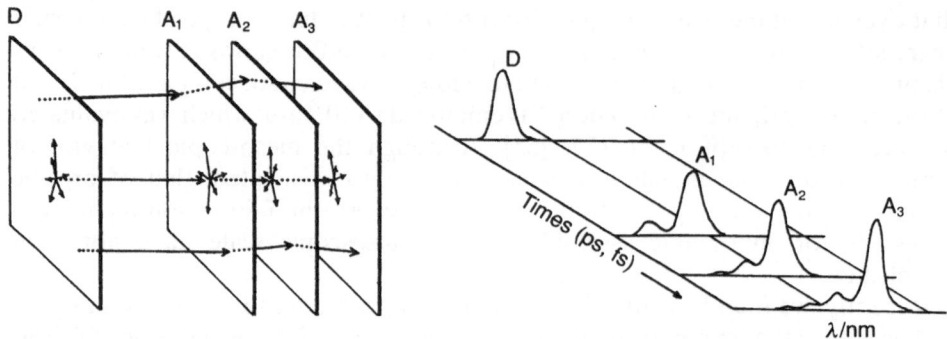

FIG. 3.2-16. Schematic illustration of sequential energy transport in LB multilayers and its observation by means of time-resolved fluorescence spectra. Following 1 ps laser excitation at layer D, the fluorescence spectra associated with excitation energy transport appear sequentially from layer D to layer A_3

photochemically reactive materials can work as optical gates in which molecules, or their electronic states, are converted to other molecules or states in reversible responses to input light.

A Langmuir–Blodgett (LB) film is a mono- or multilayered molecular assembly which is prepared by transferring a compressed monolayer spread on a water surface onto a solid substrate [59]. With a LB multilayer film, one can obtain a molecular stacking architecture in which different kinds of dyes as donor and acceptor molecules are stacked sequentially in such a way that the excitation energy is transported from layer to layer through the Förster dipole–dipole interaction mechanism [60–62]. Figure 3.2-16 illustrates schematically the vectorial transport of excitation energy in LB multilayers. The pulsed laser light is irradiated onto the film at a wavelength corresponding to the absorption band of the first donor layer (D). The excitation energy at D can be transferred electronically to acceptor dyes in the neighboring layer (A_1). When a LB monolayer containing dye molecules capable of a reversible photochromic reaction [63], e.g., spiropyran, is inserted into LB multilayers, as described above, the transport of the excitation energy can be switched, depending on the photochromic reaction, between spiropyran and merocyanine upon irradiation with UV or visible light. We can monitor the switching of energy transportation by measuring the change of fluorescence intensities in particular layers. This type of photochromic LB film may function as a two-dimensional (2D) optical molecular switching device.

3.2.4.1 Excitation Energy Transport Switching in Photochromic LB Films [64]

Spiropyran (SP) and merocyanine (MC) are known to undergo a reversible photochromic reaction [63]: when spiropyran is irradiated with UV light

(~380 nm) SP is converted to MC, while MC is converted to SP under irradiation with VIS light (~560 nm). The quantum yield of this photochemical reaction was estimated to be $\Phi(SP \rightarrow MC) = 0.44$ and $\Phi(SP \leftarrow MC) = 0.04$ in LB monolayer films [65]. In this model of an optical switching device, the LB multilayers consisted of oxacarbocyanine (OCC), SP, and indodicarbocyanine (IDC). The structure and energy-level diagrams are shown in Figs. 3.2-17 and 3.2-18, respectively. The energy levels of the lowest singlet states of OCC, MC, and IDC become lower in this order, and the stacking sequence of OCC–MC–IDC gives substantial spectral overlaps between the donor fluorescence and the acceptor absorption. Therefore, Förster excitation energy transfer can take place in a straightforward way along this pathway. However, the energy level of SP is substantially higher than that of OCC, so energy transport from OCC to SP cannot proceed. Thus the transport of excitation energy is switched depending upon the state of the

FIG. 3.2-17. The structure of the photochromic LB multilayers and the excitation energy transfer pathway oxacarbocyanine (*OCC*) → merocyanine (*MC*)— indodicarbocyanine (*IDC*) under UV irradiation. VIS irradiation switches the photochromic reaction back toward spiropyran (*SP*) and turns off the excitation energy transport

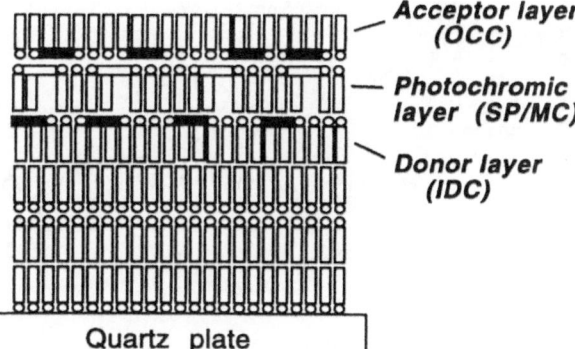

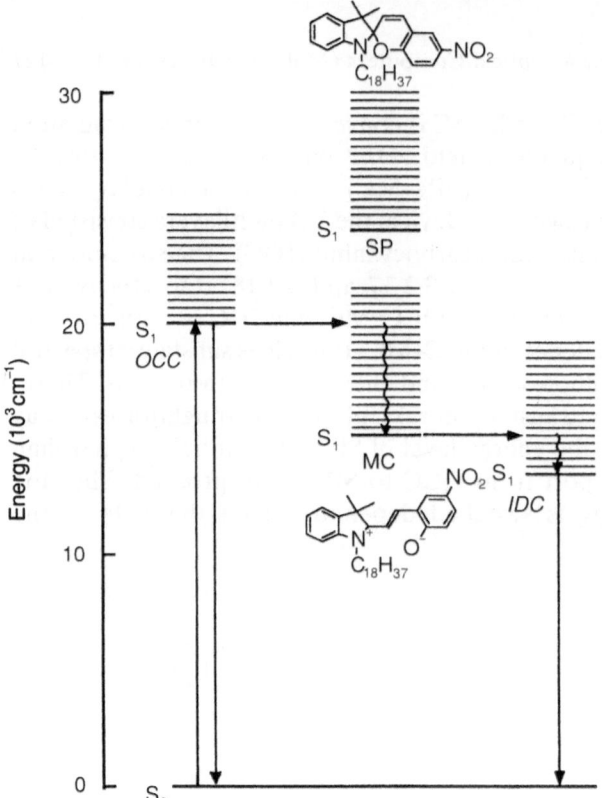

FIG. 3.2-18. Energy-level diagram of the excited singlet states of donor (*OCC*), photochromic dye (*SP/MC*), and acceptor (*IDC*)

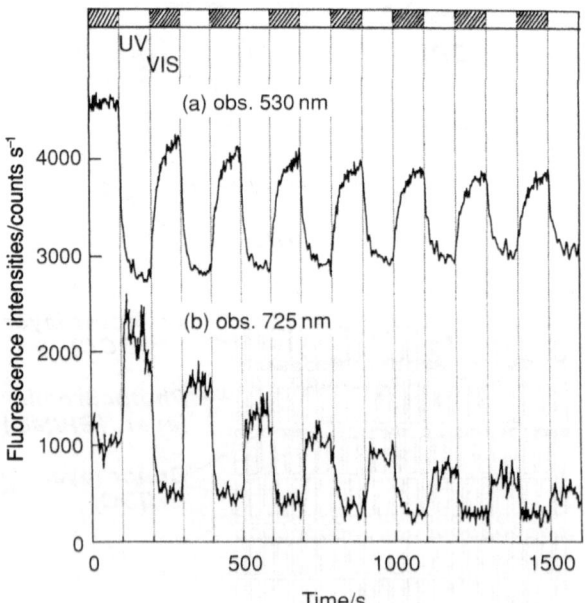

FIG. 3.2-19. Fluorescence intensity changes due to optical switching of the transportation of excitation energy. Fluorescence intensities, monitored at **a** layer D (530 nm, OCC) and **b** layer A (725 nm, IDC), are switched in and out of phase by switching the controlling light between UV (380 nm) and VIS (560 nm)

photochromic molecule, in other words, depending upon whether the irradiation is UV or VIS light.

One can observe this switching behavior in the transport of excitation energy by monitoring the fluorescence emissions from particular layers: OCC and IDC emit fluorescence at 530 and 725 nm, respectively. The experimental results [64] are shown in Fig. 3.2-19. Under UV irradiation (380 nm) where the excitation energy transport circuit is open, the excitation energy is transferred from OCC to IDC and the fluorescence is emitted from IDC. Under VIS light irradiation (560 nm) where the circuit is closed, the fluorescence is emitted from OCC. It can be seen in Fig. 3.2-19 that the fluorescence intensities of OCC and IDC are switched in and out of phase depending on whether the irradiation is UV or VIS light.

3.2.4.2 Applicability to 2D Optical Switching Devices [64]

First, we consider the logicisties of the optical switching device described above, following the diagram shown in Fig. 3.2-20a. Let the inputs $P = 1$ and 0 denote UV and VIS irradiation, respectively, and let the outputs $D = 1$ and 0 mean that the excitation *does* locate and *does not* locate at layer D (OCC), respectively. In other words, this means that the 510-nm fluorescence (D) *does* emit in $D = 1$, and *does not* emit in $D = 0$. Similarly, the outputs $A = 1$ and 0 mean that the excitation *does* and *does not* locate at layer A (IDC), respectively, and that the 720-nm fluorescence (A) *does* emit in $A = 1$ and *does not* emit in $A = 0$. The relationship between the input P and the output D or A is that for $P = 0$ it gives $D = 1$ and $A = 0$, and for $P = 1$ it gives $D = 0$ and $A = 1$. As a result of this relationship, the output D gives NOT P, while the output A gives P, which is the same as input P.

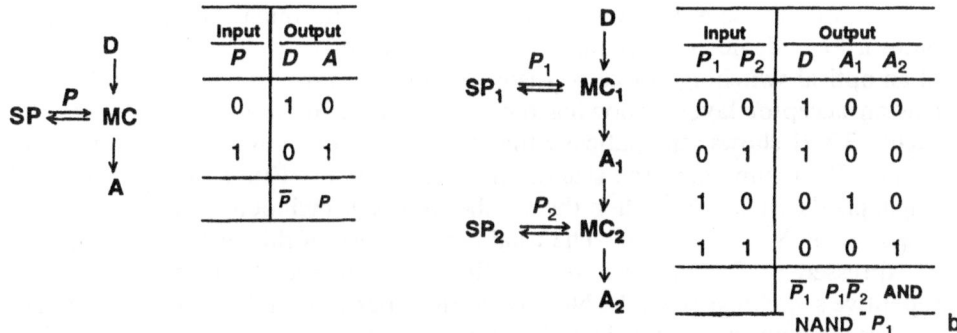

FIG. 3.2-20. Truth tables for photochromic operations in two systems (**a** and **b**). The inputs P, P_1, and $P_2 = 1$ mean that the LB film is irradiated with UV light, and the gate is open for the transport of excitation energy. The outputs D, A_1, and $A_2 = 1$ mean that the excitation energy locates at these layers, and that is where its fluorescence is emitted

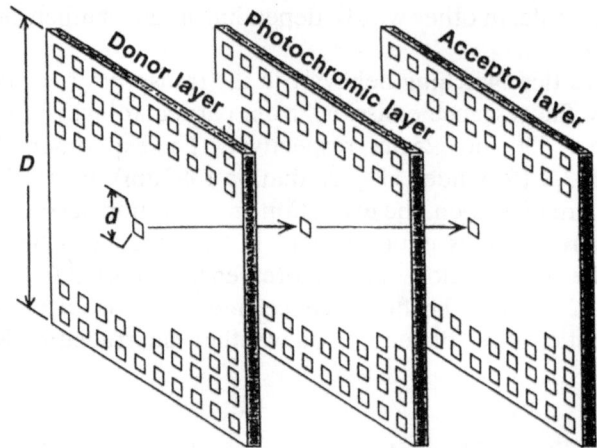

F IG. 3.2-21. Model of an element of an optically switching parallel processor. Photoexcitations at the donor layer are transported or *not* transported to the acceptor layer depending on whether the controlling light irradiation at the photochromic layer is UV or VIS

Now we consider a photochromic LB device consisting of two serially connected elements, each of which has the same structure as shown in Fig. 3.2-20b, but contains different types of photochromic dyes. As responses to the two input signals P_1 and P_2, the output D gives NOT P_1 (which is similar to the previous case), the output A_1 gives $P_1 \times$ NOT P_2, and the output A_2 gives AND where it gives 1 only in the case that both P_1 and P_2 are 1. The molecules for D, SP_1 (MC_1), and A_1 are the same as those given above. The molecules for SP_2 (MC_2) and A_2 are different types of molecules for which we are now testing several candidates.

Since LB film is a 2D film, photochromic LB film might be applicable to a pattern logic device or a parallel processor. Let us consider a simple element of an optical switching device consisting of a donor layer, a photochromic layer, and an acceptor layer, following the method of Tanida and Ichioka [66, 67]. Figure 3.2-21 shows schematically the structure of an element of a parallel processor. We assume here the size of an element (D^2) is 10 mm² and the size of a single pixel (d^2) is 1 μm². Then the number of pixels included in an element (N^2) is written as $N^2 = D^2/d^2 = 10^8$. The total performance of this switching device (P) is expressed as $P = vN^2$, where v is the switching speed, i.e., the number of repetitive switchings (e.g., the NOT operation) per second. Note that the number of photochromic molecules included in a pixel of 1 μm² is 5×10^5 molecules in a LB monolayer film of 10 mol% concentration. The switching threshold was taken to be when one-half of molecules in a pixel had converted to MC. Then the switching speed was calculated to be $v = 10^4 s^{-1}$ by using a *cw* argon-ion laser (5 W) as a controlling light source. The performance of this device was estimated to be

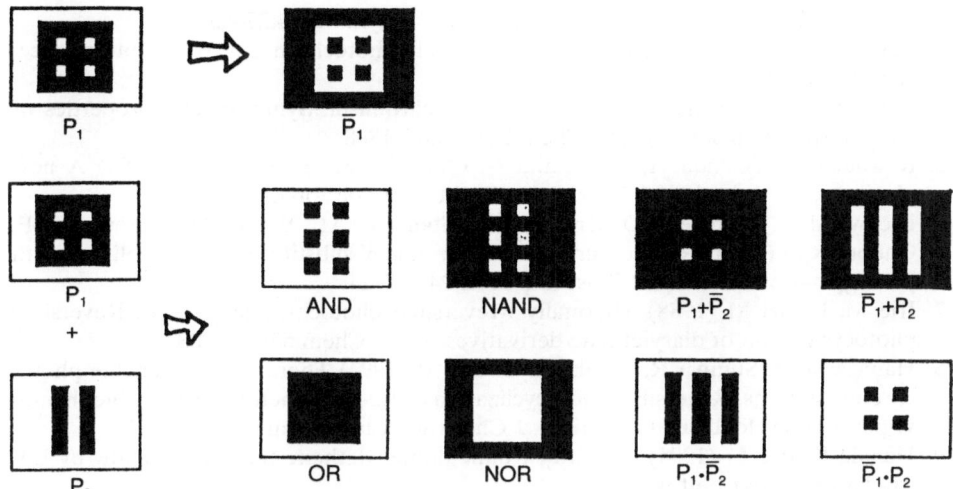

FIG. 3.2-22. 2D pattern logics obtained in two types of photochromic LB multilayers. The input optical patterns P_1 and P_2, given in UV or VIS light, and the output fluorescence patterns at D, A_1, and A_2 are shown

10^{12} NOT operations s^{-1}. This value is comparable to those of other optical switching devices such as a liquid crystal light valve ($P = 7 \times 10^9$ operations s^{-1}), optical bistability devices (7×10^{12}), optoelectronic hybrid devices (6×10^{14}), and an optical computer (OPALS) (1×10^{19}) [66, 67] It is concluded that this type of photochromic LB film can be used in optical switching molecular devices, in particular as an element in an optical computer.

Examples of 2D pattern logics which might be obtained in photochromic LB films are shown in Fig. 3.2-22 for the logic schemes given in Fig. 3.2-20. If we irradiate the element shown in Fig. 3.2-21 with a particular 2D optical pattern (UV or VIS) as input, the output pattern might be obtained by monitoring the fluorescence from D or A on a 2D multichannel photodetector in a 2D pattern of P_1, AND, NAND, OR, NOR, $P_1 + P_2$, etc., as a result of gating the excitation energy transport. All these logics are equivalent to those of electronic devices constructed using standard integrated circuits. It should be noted that in the present photochromic device, the operation in each pixel can be performed independently. Therefore, it is concluded that photochromic LB multilayers are applicable to optically switching parallel processors, in particular a 2D spatial light modulator in optical computers.

References

1. Fischer E, Hirshberg Y (1952) Formation of coloured forms of spirans by low-temperature irradiation. J Chem Soc:4522–4524

2. Irie M (1990) Photoresponsive polymers. Adv Polym Sci 84:27–65
3. Bertelson RC (1971) In: Brown GH (ed) (1971) Photochromism. Wiley Interscience, New York, pp 45–431
4. Arakawa S, Kondo H, Seto J (1985) Photochromism. Synthesis and properties of indolinospirobenzothiopyrans. Chem Lett: 1805–1808
5. Kawauchi S, Yoshida H, Yamashita N, Ohira M, Saeda S, Irie M (1990) A new photochromic spiro [3H-1,4]-oxazines. Bull Chem Soc Jpn 63:267–268
6. Rickwood M, Marsden SD, Ormsby ME, Stannton AL, Wood DW, Hepworth JP, Gabbutt CD (1994) Red colouring photochromic 6'-substituted spiroindolinonaphth [2,1-B][1, 4] oxazines. Mol Cryst Liq Cryst 246:17–24
7. Irie M, Mohri M (1988) Thermally irreversible photochromic systems. Reversible photocyclization of diarylethene derivatives. J Org Chem 53:803–808
8. Hanazawa M, Sumiya R, Horikawa Y, Irie M (1992) Thermally irreversible photochromic systems. Reversible photocyclization of 1,2-bis(2-methylbenzo[b]thiophen-3-yl)perfluorocycloalkene derivatives. J Chem Soc Chem Commun 206–207
9. Uchida K, Irie M (1995) A photochromic dithienylethene that turns yellow by UV irradation. Chem Lett 969–970
10. Irie M, Miyatake O, Sumiya R, Hanazawa M, Horikawa Y, Uchida K (1994) Photochromism of diarylethenes with intralocking arms. Mol Cryst Liq Cryst 246:155–158
11. Gilat SL, Kawai SH, Lehn J-M (1993) Light-triggered electrical and optical switching devices. J Chem Soc Chem Commun 1439–1442
12. Kulisch JR, Franke M, Irmscher R, Buchal Ch (1992) Optooptical switching in ion-implanted poly(methyl methacrylate) waveguide. J Appl Phys 71:3123–3126
13. Tanio M, Irie M (1994) Refractive index of organic photochromic dye-amorphous polymer composites. Jpn J Appl Phys 33:3942–3946
14. Tomlinson WJ, Chandross EA, Fork RL, Pryde CA, Lamola AA (1972) Reversible photodimerization. Appl Opt 11:533–548
15. Nakamura S, Irie M (1988) Thermally irreversible photochromic systems. A theoretical study. J Org Chem 53:6136–6138
16. Yokoyama Y, Kurita Y (1994) Photochromism of fulgides and related compounds. Mol Cryst Liq Cryst 246:87–94
17. Tomoda A, Kaneko A, Tsuboi H, Matsushima R (1992) Photochromism of heterocyclic fulgides. Bull Chem Soc Jpn 65:1262–1267
18. Uchida K, Nakayama K, Irie M (1990) Thermally irreversible photochromic systems. Reversible photocyclization of 1,2-bis(benzo[b]thiophen-3-yl)ethene derivatives. Bull Chem Soc Jpn 63:1311–1315
19. Irie M, Miyatake O, Uchide K (1992) Blocked photochromism of diarylethenes. J Am Chem Soc 114:8715–8716
20. Irie M, Miyatake O, Uchida K, Eriguchi T (1994) Photochromic diarylethenes with intralocking arms. J Am Chem Soc 116:9894–9900
21. Yokoyama Y, Yamane T, Kurita Y (1991) Photochromism of a protonated 5-dimethylaminoindolylfulgide: a model of a non-destructive readout for a photon mode optical memory. J Chem Soc Chem Commun 1722–1724
22. Uchida M, Irie M (1993) Two-photon photochromism of a naphthopyran derivative. J Am Chem Soc 115:6442–6443
23. Walz J, Ulrich K, Port H, Wolf HC, Wonner J, Effenberger F (1993) Fulgides as switches for intramolecular energy transfer. Chem Phys Lett 213:321–324

24. Feringa BL, Jager WF, de Lange B, Meijer EW (1991) Chiroptical molecular switch. J Am Chem Soc 113:5468–5470

25. Moerner WE (1988) Persistent spectral hole burning: science and applications. Springer, Berlin

26. Horie K, Furusawa A (1992) Photochemical hole burning. In: Rabek JF (ed) Progress in photochemistry and Photophysics. CRC Press, Boca Raton, vol 5, Chap 2

27. Moerner WE (1994) Examining nanoenvironments in solids on the scale of a single, isolated impurity molecule. Science 265:46–53

28. Winnacker A, Shelby RM, Macfarlane RM (1985) Photon-gated hole burning: a new mechanism using two-step photoionization. Opt Lett 10:350–352

29. Sakoda K, Kominami K, Iwamoto M (1988) High temperature photochemical hole burning of tetrasodium 5,10,15,20-tetra(4-sulfonatophenyl)porphin in polyvinylalcohol. Jpn J Appl Phys 27:L1304–L1306

30. Horie K, Mori T, Naito T, Mita I (1988) Photochemical hole burning of tetraphenylporphin in phenoxy resin. Appl Phys Lett 53:935–937; J Appl Phys 66:6041–6047

31. Prass B, von Borczyskowski C, Stehlik D (1989) Photochemically burnt holes with high-temperature stability (1989) J Phys Chem 93:8276–8278

32. Jaaniso R, Bill H (1991) Room-temperature persistent spectral hole burning in Sm-doped $SrFCl_{1/2}Br_{1/2}$ mixed crystals. Europhys Lett 16:569

33. Zhang J, Huang S, Yu J (1991) Persistent hole burning at room temperature. Chinese J Lumin 12:181–182

34. Hirao K, Todoroki S, Cho DH, Soga N (1993) Room-temperature persistent hole burning of Sm^{2+} in oxide glasses. Opt Lett 18:1586–1587

35. Kurita A, Kushida T, Izumitani T, Matsukawa M (1994) Room-temperature persistent spectral hole burning in Sm^{2+}-doped fluoride glasses. Opt Lett 19:314–316

36. Bauer R, Osvet A, Sildos I, Bogner U (1993) Room temperature persistent spectral hole burning in neutron-irradiated Iab-type diamond. J Lumin 56:57–60

37. Sauter B, Bräuchle C (1992) Efficient persistent spectral hole-burning of free-base octaethylporphine adsorbed on γ-alumina between 1.6 and 90 K. Chem Phys Lett 192:321–326

38. Ehrl M, Deeg FW, Bräuchle C, Franke O, Sobbi A, Schulz-Ekloff G, Wöhrle D (1994) High-temperature non-photochemical hole-burning of phthalocyanine–zinc derivatives embedded in a hydrated $AlPO_4$-5 molecular sieve. J Phys Chem 98:47–52

39. Machida S, Horie K, Yamashita T (1995) Photochemical hole burning of organic dye doped in inorganic semiconductor. Appl Phys Lett 66:1240–1242

40. Murase N, Horie K, Terao M, Ojima M (1992) Theoretical study of the recording density limit of photochemical hole-burning memory. J Opt Soc Am B 9:998–1005

41. Murase N, Horie K (1993) Mechanism of laser-induced hole filling in photochemical hole burning. Chem Phys Lett 209:42–46

42. Murase N, Horie K (1994) Excitation processes of a dye doped into an amorphous material investigated by photochemical hole-burning. Jpn J Appl Phys 33:1046–1052

43. Murase N, Horie K (1994) Quantitative analysis of wavelength and temperature dependence of laser-induced hole filling in photochemical hole-burning. Chem Phys 183:135–146

44. Vacha M, Machida S, Horie K (1995) Photochemical hole burning in higher excited states (Soret absorption bands) of tetrabenzo- and tetraphenylporphins. J Lumin 64:115–123; Chem Phys Lett 243:297–301; J Phys Chem 99:13163–13167
45. Jankowiak R, Hayes JM, Small GJ (1993) Spectral hole burning spectroscopy in amorphous molecular solids and proteins. Chem Rev 93:1471–1502
46. Machida S, Horie K, Kyono O, Yamashita T (1993) Photon-gated photochemical hole burning by two-color sensitization in an azide polymer matrix. J Lumin 56:85–92
47. Suzuki T, Ikemoto M, Machida S, Vacha M, Horie K, Yamashita T (1994) Relationship between spectral hole burning and Stokes shift of organic dyes in polymer system. Chem Lett 1247–1250
48. Prasad PN, Williams DJ (1991) Introduction to nonlinear optical effects in molecules and polymers. Wiley, New York
49. Chem Soc Jpn (ed) (1992) Organic nonlinear optical materials. Gakkai Shuppan Center, Tokyo (Kikan Kagaku Sosetsu, vol 15)
50. Zyss J (ed) (1994) Molecular nonlinear optics. Academic Press, Orlando
51. Tokura Y (1992) New aspects of organic nonlinear optical materials. In: Chem Soc Jpn (ed) Organic nonlinear optical materials. Gakkai Shuppan Center, Tokyo (Kikan Kagaku Sosetsu, vol 15) pp 191–203
52. Umegaki S (1992) Nonlinear optics and organic materials. In: Chem Soc Jpn (ed) Organic nonlinear optical materials. Gakkai Shuppan Center, Tokyo (Kikan Kagaku Sosetsu, vol 15) pp 1–42
53. Duan XM, Okada S, Oikawa H, Matsuda H, Nakanishi H (1994) Comparatively large second-order hyperpolarizability of aromatic sulfonate anion with short cutoff wavelength. Jpn J Appl Phys 33:L1559–1561
54. Matsuda H, Okada S, Nakanishi H (1992) Molecular and crystal engineering of polydiacetylenes for enlarged nonlinear optical susceptibility. Mol Cryst Liq Cryst 217:43–46
55. Kobayashi S (1992) Structure dependence of the optical nonlinearity for pseudoisocyanine J-aggregates. In: Extended abstracts of the third symposium on photonics materials. Basic technologies for future industries. Japan High Polymer Center, Tokyo, pp 23–32
56. Kasai H, Nalwa HS, Oikawa H, Okada S, Matsuda H, Minami N, Kakuta A, Ono K, Mukoh A, Nakanishi H (1992) A novel preparation method of organic microcrystals. Jpn J Appl Phys 31:L1132–L1134
57. Birnboim MH, Ma WP (1990) Nonlinear optical properties of structured nanoparticle composites. Mater Res Soc Symp Proc 164:277–282
58. Arrathoon R (ed) (1989) Optical computing: digital and symbolic. Dekker, New York
59. Kuhn H, Möbius D, Bücher H (1972) Spectroscopy of monolayer assemblies. In: Weissberger A, Rossiter BW (eds) Techniques of chemistry, vol 1, part 3B. Wiley, New York, pp 577–701
60. Förster TH (1948) Zwischenmolekulare Energiewanderung und Fluoreszenz. Ann Phys 2:55–75
61. Baumann J, Fayer MD (1986) Excitation energy transfer in disordered two-dimensional and anisotropic three-dimensional systems: effects of spatial geometry on time-resolved observables. J Chem Phys 85:4087–4107
62. Yamazaki I, Tamai N, Yamazaki T (1990) Electronic excitation transfer in organized molecular assemblies. J Phys Chem 94:516–525
63. Duerr H, Bouas-Laurent H (eds) (1990) Photochromism. Molecules and systems. Elsevier, Amsterdam

64. Yamazaki I, Okazaki S, Minami T, Ohta N (1994) Optically switching parallel processors by means of Langmuir–Blodgett multilayer films. Appl Opt 33:7561–7568
65. Minami T, Tamai N, Yamazaki T, Yamazaki I (1991) Picosecond time-resolved fluorescence spectroscopy of the photochromic reaction of spiropyran in Langmuir–Blodgett films. J Phys Chem 95:3988–3993
66. Tanida J, Ichioka Y (1986) OPALS optical parallel array logic system. Appl Opt 25:1565–1572
67. Tanida J, Ichioka Y (1988) Programming of optical array logic. 1. Image data processing. Appl Opt 27:2926–2930

3.3 Chemoactive Molecular Systems

MASUO AIZAWA (3.3.1)
MAMORU OHASHI (3.3.2)

3.3.1 Integrated Molecular Systems for Biosensing

Molecular communication is the characteristic information system of biological systems. The endocrine system can be taken to represent the features of molecular communication, where instead of electrons and photons, such molecules as hormones and neurotransmitters serve as information carriers. The molecular information is released from a gland cell into the extracellular fluid, and transported via the blood to the target cells. The receptor, which selectively recognizes the relevant information molecule, is located on the surface of the target cell and, in association with such proteins as GTP-binding protein (G-protein) and phosphokinase, generates a second messenger, in amplified form, in the interior of the cell, as schematically illustrated in Fig. 3.3-1.

The supramolecular mechanisms of these receptor proteins may provide us with the principles to design highly selective and sensitive molecular systems for information transduction in various sensors, specifically biosensors.

A biosensor is defined as a sensor that makes use of biological or living material for its sensing function. Biological materials have been successfully implemented in sensing devices to improve selectivity, which is difficult to attain with other materials such as semiconductor or metals. A biosensor will probably be the first realization of a biomolecular electronic device in which biomolecules play a key role in information transduction.

This section describes the design principles for sensing materials and the molecular systems for information transduction.

3.3.1.1 Molecular Systems for Molecular Recognition and Signal Transduction

Either a biocatalyst or a bioaffinity substance may be used as the major material for molecular recognition to attain extremely high selectivity [1]. The molecular recognition should be coordinated efficiently with the signal transduction when constructing biosensors.

Fig. 3.3-1. Schematic illustration of an integrated molecular system for transducing molecular information

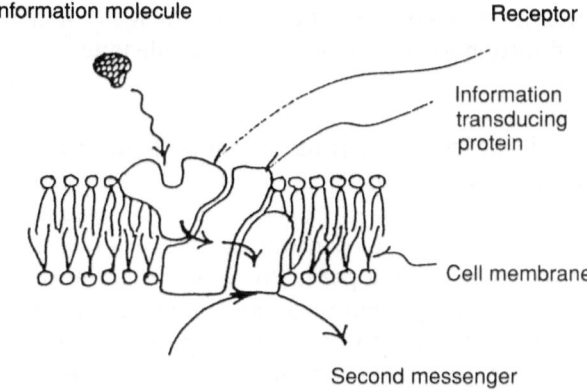

Information molecule

Receptor

Information transducing protein

Cell membrane

Second messenger

A biocatalyst recognizes the corresponding substrate and immediately generates a product by an automatic reaction. The complex of the catalyst and the substrate remains stable only in the transition state. It is so difficult to quantify the transition state of a complex that parameters such as oxygen, hydrogen peroxide, heat, and photons are commonly measured and used to indicate signal transduction.

Redox enzymes are recognized as the major useful biocatalyst for constructing biocatalytic sensors. Biosensors for glucose, lactate, and alcohol utilize glucose oxidase, lactate oxidase/dehydrogenase, and alcohol oxidase/dehydrogenase, respectively. Since these redox enzymes are mostly associated with the generation of electrochemically active substances, many electrochemical enzyme sensors have been developed by linking redox enzymes for molecular recognition with electrochemical devices for signal transduction. These enzymes, however, have been incorporated in an indirect manner with electrochemical devices. The electrochemically active substances generated enzymatically mediate the enzyme reaction to the electrode in the electrochemical device. These sensing systems have often been very complex in structure and with a slow response time.

In contrast, extensive research has been devoted to the direct linking of redox enzymes for molecular recognition with electrochemical devices for signal transduction. Despite efficient electron transfer from redox enzymes to the corresponding substrate molecules, few redox enzymes can transfer electrons directly to metal or semiconductor electrodes even if the energy correlation could be satisfied. Several "molecular interfaces" between a redox enzyme and an electrode to promote electron transfer have been developed, including electron mediators and molecular wires. These have offered a new design principle in the search for an electron transfer type of enzyme sensor.

A bioaffinity protein, which involves an antibody, a binding protein, and a receptor protein, forms a stable complex with the corresponding ligand to result in signal transduction. A change in the bioaffinity protein/ligand complex formation is detected without any label, or with the help of a label such as an enzyme, a fluorophore, or an electroactive substance.

3.3.1.2 Molecular Interfacing of Redox Enzymes on the Electrode Surface

Electrochemical biosensors make use of electrochemical reactions for signal transduction. As a parameter, the potential or conductance between two electrodes or the current through a polarized electrode is measured. A great variety of chemical substances can be measured in this way.

An enzyme sensor for glucose is a well-known example of an electrochemical biosensor. The basic structure of a glucose sensor consists of either an oxygen or a hydrogen peroxide electrode which is covered with a glucose oxidase membrane. Glucose in a sample solution is enzymatically oxidized, with the resulting consumption of oxygen or generation of hydrogen peroxide. Either a decrease in dissolved oxygen or an increase in hydrogen peroxide is sensitively detected with an oxygen or hydrogen peroxide electrode.

These amperometric glucose sensors have long been included in various bench-top analyzers which are widely used for clinical analysis, food analysis, and other tests. In contrast with these continuous-use glucose sensors, a disposable type of glucose sensor has recently become available. They are mainly used for the personal diagnosis of diabetes. Disposable glucose sensors have been developed by innovative technology, including screen-printing fabrication of electrodes and enzyme layers, which has brought economically feasible mass-production.

The pioneering work of Hill has led to the development of fast and efficient electron transfer of enzymes on an electrode surface [2]. They modified a gold electrode with 4,4'-bipyrydyl, which is an electron promoter and not a mediator since it does not take part in electron transfer in the potential region of interst, to accomplish rapid electron transfer of cytochrome c. Their work has triggered intensive investigation of electron transfer of enzymes using modified electrodes [3].

Apart from electron promoters, a large number of electron mediators have been investigated to make redox enzymes electrochemically active on the electrode surface. In this line of research, electron mediators such as ferrocene and its derivatives have successfully been incorporated into an enzyme sensor for glucose. The mediator was easily accessible to both glucose oxidase and an electrode to transfer electrons in an enzyme sensor. Degani and Heller have chemically modified glucose oxidase with an electron mediator [4]. They presumed that an electron tunnelling pathway could be formed within the enzyme molecule. The present authors [5] and Foulds and Lowe [6] used a conducting polymer as a molecular wire to connect a redox enzyme molecule to the electrode surface.

This progress in the electron transfer of enzymes has led us to conclude that a molecular-level assembly should be designed to facilitate electron transfer at the

Fɪɢ. 3.3-2. Three types of molecular interface on an electrode surface

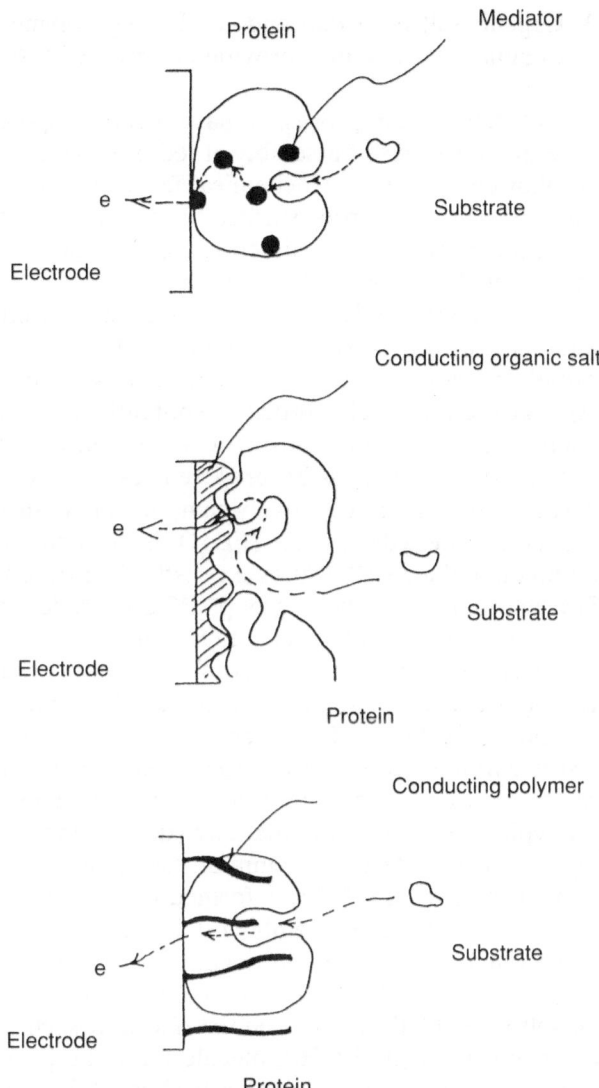

interface between an enzyme molecule and an electrode. Such a molecular-level assembly at the interface may be termed a "molecular interface" [7].

There are several molecular interfaces where redox enzymes can promote electron transfer at the electrode surface [8] (Fig. 3.3-2).

1. Electron mediator: either the electrode or the enzyme is modified with an electron mediator in various ways.
2. Molecular wire: the redox center of an enzyme molecule is connected to an electrode with a molecular wire such as a conducting polymer chain.

3. Organic salt electrode and conducting polymer electrode: the surface of an organic electrode may provide enzymes with smooth electron transfer.

Two fabrication processes have been proposed by our group. One is a potential-assisted self-assembly of redox enzymes on the electrode surface, which is followed by the electrochemical fabrication of a monolayer-scale conducting polymer on the electrode surface for molecular interfacing. The other is the self-assembly of mediator-modified redox enzymes on a porous gold electrode surface through a thiol–gold interaction.

The potential-assisted self-assembly is carried out in an electrolytic cell equipped with a platinum or gold electrode (working electrode) on which a protein monomolecular layer is formed, a platinum counter electrode, and a Ag/AgCl reference electrode. The potential of the working electrode is precisely controlled with a potentiostat with reference to the Ag/AgCl electrode. The protein solution should be prepared taking into account the isoelectric point, because proteins are negatively charged in pH ranges above the isoelectric point.

Fructose dehydrogenase (FDH) is a redox enzyme which has pyrrole–quinoline quinone (PQQ) as a prosthetic group. Upon enzymatic oxidation of D-fructose, the prosthetic group (PQQ) is reduced to $PQQH_2$, and an electron acceptor reoxidizes $PQQH_2$ to PQQ with the liberation of two electrons. FDH is a requisite element of a biosensor for fructose because it can selectively recognize D-fructose as a result of electron transfer from the D-fructose to an electron acceptor in solution. However, it is difficult for FDH to make the electron transfer from fructose directly to an electrode in place of an electron acceptor in solution owing to steric hindrance. Fructose dehydrogenase is therefore one of the typical redox enzymes that have demanding conditions for molecular assembly resulting in electronic communication on the electrode surface.

A monolayer of FDH was formed on platinum or gold electrode surfaces by potential-assisted self-assembly [9]. FDH was dissolved in phosphate buffer (pH 6.0) to make its net charge negative. FDH molecules instantly adsorb on the electrode surface due primarily to electrostatic interaction. FDH molecules may be self-assembled on electrode surfaces in such a manner that the negatively charged side of the FDH molecule faces the positively charged surface of the electrode. An enzyme assay clearly showed that electrode-bound FDH retained its enzymatic activity without appreciable inactivation.

In the next step, a molecular wire of the molecular interface was prepared for the electrode-bound FDH by potential-assisted self-assembly. Polypyrrole was used as the molecular wire of the molecular interface for the electrode-bound FDH and was synthesized by electrochemical oxidative polymerization of pyrrole.

Electrochemical oxidative polymerization of pyrrole was performed on the FDH-adsorbed electrode in a solution containing 0.1 M pyrrole and 0.1 M KCl under anaerobic conditions at a potential of 0.7 V. The thickness of the polypyrrole membrane was controlled by the polymerization electricity. The electrochemical polymerization was stopped when the monolayer of FDH on the electrode surface was presumably covered by the polypyrrole membrane.

The total electricity of electrochemical polymerization was controlled at 4 mC. The molecular interfaced FDH was thus prepared on the electrode surface.

Electronic communication between electrode-bound FDH and an electrode has been confirmed by differential pulse voltammetry. Differential pulse voltammetry of the molecular-interfaced FDH was conducted in a pH 4.5 buffered solution. A pair of anodic and cathodic peaks were observed for the molecular-interfaced FDH. The peaks are attributed to the electrochemical oxidation and reduction of the PQQ enzyme at redox potentials of 0.08 and 0.07 V vs. Ag and AgCl, respectively. In addition, the anodic and cathodic peak shapes and peak currents of the molecular-interfaced FDH were identical, which suggests reversibility of the electron transfer process. On the other hand, FDH exhibited no appreciable peaks in differential pulse voltammetry on the electrode surface without the polypyrrole molecular interface.

In addition to FDH, the potential-assisted self-assembly has been successfully applied to several redox enzymes, including glucose oxidase and alcohol dehydrogenase [10]. The self-assembled redox enzymes have also been molecularly interfaced with the electrode surface via a conducting polymer.

In contrast to the molecular wire of molecular interfaces, electron mediators are covalently bound to a redox enzyme in such a manner that an electron tunneling pathway is formed within the enzyme molecule. Therefore, enzyme-bound mediators work as a molecular interface between an enzyme and an electrode. In 1987, Degani and Heller [4] proposed that the intramolecular electron pathway of ferrocene molecules were covalently bound to glucose oxidase. However, few fabrication methods have been developed to form a monolayer of mediator-modified enzymes on such electrode surfaces. We have succeeded in the development of a new preparation method for an electron transfer system of mediator-modified enzymes by self-assembly in a porous gold–black electrode.

Glucose oxidase (from *Aspergilus niger*) and ferrocene carboxyaldehyde were covalently conjugated by the Schiff base reaction, which was followed by NaBH$_4$ reduction. The conjugates were dialyzed against phosphate buffer with three changes of buffer, and assayed for their protein and ion contents. Porous gold–black was electrodeposited on a microgold electrode by cathodic electrolysis with chloroauric acid and lead acetate. Aminoethane thiol was self-assembled on a smooth gold disk electrode and a gold–black electrode. Ferrocene-modified glucose oxidase was covalently linked to either a modified plain gold or a gold–black electrode by glutaraldehyde.

The ferrocene/glucose oxidase conjugates were characterized by the molar ratios of ferrocene to enzyme in the range from 6 to 11. All the oxidase conjugates retained enzyme activity. Self-assembled ferrocene–glucose oxidase conjugate on the gold disk electrode showed reversible electron transfer. The anodic peak currents in the cyclicvoltammograms were independent on the molar ratio of ferrocene to enzyme in the range from 6 to 11.

It should be noted that the anodic peak current strongly increases with an increase in the molar ratio of ferrocene to glucose oxidase, whilst the amount of enzyme self-assembled on the electrode surface is fixed. This indicates that each modified ferrocene may contribute to electron transfer between the enzyme and the electrode in the case of a gold–black electrode, and the ferrocene-modified enzyme could form multi-electron transfer paths on the porous gold–black electrode.

The substrate concentration dependence of the response current of the gold–black electrode was compared with that of the gold disk electrode. The ferrocene-modified glucose oxidase which was used in this measurement had 11 ferrocenes per glucose oxidase. The electrode potential was controlled at 0.4 V vs. Ag/AgCl. The response current was recorded when the output reached a steady state. The response current was enhanced when ferrocene-modified glucose oxidase was self-assembled on a porous gold–black electrode.

The porous matrix of the gold–black electrode has allowed ferrocene-modified glucose oxidase to perform a smooth electron transfer by means of easy access between self-assembled molecules and the electrode surface.

3.3.1.3 Potentiometric Immunosensing on a Solid Surface

Three types of potential immunosensing have been proposed [11]. The first is based on the determination of the transmembrane potential across an antibody (or antigen) membrane that specifically binds the corresponding antigen (or antibody) in solution. Concentrations of either the target antigen or the antibody can be determined by measuring the change in the transmembrane potential which occurs when the immunocomplex forms on the membrane surface.

The second type is based on a determination of the electrode potential. The surface of an electrode is modified by an antibody or antigen to bind specifically the corresponding antigen or antibody. Immunocomplex formation causes the electrode potential to vary, primarily as a result of a change in surface charge resulting from the concentration of the analyte in solution.

The third type is based on the determination of the surface potential of the gate of a field effective transistor (FET) covered by a thin antibody-binding membrane. The surface potential of the FET gate may change with the concentration of the corresponding antigen in solution.

Results for potentiometric immunosensors for syphilis and blood typing have been reported [12]. They are based on the determination of the transmembrane potential across an immunoresponsive membrane. For blood typing, for instance, the immunoresponsive membrane incorporates blood group substances and a pair of reference electrodes for measuring the transmembrane potential [13]. A, B, and O blood samples have been typed using this potentiometric immunosensor.

Yamamoto et al. [14] proposed a potentiometric immunosensor with an antibody against human chorionic gonadotropin (HCG) hormone, covalently bound to the surface of an electrode. The electrode is reported to respond to HCG in solution.

3.3.1.4 Amperometric Immunosensing on a Solid Surface

A variety of labeled immunosensors have been proposed [11]. Enzyme labels can enhance the sensitivity of immunosensors owing to chemical amplification. Since an enzyme rapidly converts a substrate into a product, the product that accumulates can be detected with various electronic and optoelectronic devices [15–18].

Some enzyme labels are electrochemically detected with high sensitivity. These include catalase, glucose oxidase, and peroxidase. Both catalase and glucose oxidase may be associated with an oxygen electrode.

When catalase, which catalyzes the decomposition of hydrogen peroxide into oxygen, is the labeling enzyme for an antigen, an enzyme immunosensor can be constructed by assembling an antibody-bound membrane and an oxygen-sensing electrode. In heterogeneous enzyme immunoassays, the labeling enzyme activity is measured by amperometry with an oxygen-sensing device. Because only a short time is required for measuring the labeling enzyme, a rapid and highly sensitive enzyme immunoassay can be achieved. An enzyme immunosensor was prepared by attaching the antibody-bound membrane to a Clark-type oxygen electrode involving an oxygen-permeable synthetic (e.g., Teflon) membrane on the cathode surface.

Enzyme labels are efficiently detected with optoelectronic devices in a manner similar to that used for electrochemical devices. Peroxidase and luciferase, for instance, catalyze luminescent reactions of luminol and luciferin, respectively, to generate photons. These enzymes may be incorporated as labels to form optical enzyme immunosensors.

3.3.1.5 Optical Immunosensing

If the solid surface is sensitive enough to allow changes in its optical properties with immunocomplex formation, optical immunosensors without label may be constructed [19]. Surface plasmon resonance (SPR) is so sensitive that an immunocomplex may be detected on the surface of a silver-coated solid. It should be noted that an immunosensing system based on SPR is already on the market, primarily for basic research use [20, 21].

Acoustic immunosensors have attracted many researchers because of their simplicity of operation. The surface of a piezoelectric material is coated with antibodies or antigens. The intrinsic frequency of the piezoelectric material is expected to shift when responding to the corresponding antigen or antibody in solution.

Since homogeneous immunoassays require no bound/free species separation process, they may have advantages over heterogeneous immunoassays. Although intensive efforts have been concentrated on the development of homogeneous immunosensors, a very limited number of papers has been published on the subject.

An optical fiber structure has been proposed for designing a homogeneous immunosensor. An antibody is immobilized on the surface of an optical fiber core, and a fluorescence compound is used as the label. Both the labeled antigen and the free antigen to be determined react competitively with the bound antibody to form an immunocomplex on the core surface. The surface-bound label can be excited by an evanescent wave that passes through the optical fiber core. However, fluorescence labels in the bulk solution cannot be excited even if the excitation beam comes through the optical fiber core. Labels attached to the surface-bound immunocomplex are thus distinguished from labels in solution. Although the details have not been reported, the principle is elegant and homogeneous immunoassay may be possible.

Some electrochemically active substances able to generate photons on an electrode surface can be used as labels for homogeneous reactivity and to generate luminescence, but when the labeled antigen is immunochemically complexed it loses its electrochemiluminescence property. An optical immunosensor for homogeneous immunoassay was assembled by spattering platinum on the end surface of an optical fiber [22, 23]. Spattered platinum maintains optical transparency and works as an electrode. An optical fiber electrode efficiently collects photons generated on the surface of the transparent electrode. Luminol as a label offers excellent characteristics for designing a homogeneous immunosensor and generates luminescence by two different types of electrochemical excitation. One is based on a two-step electrochemical excitation, i.e., cathodic excitation of hydrogen peroxide that causes the anodically generated luminol radical to emit photons. The other is a single-step electrochemical excitation which provides a very high sensitivity with a limit of detection equal to $10^{-13}\,\text{mol}\,\text{L}^{-1}$ of luminol. Homogeneous immunoassay with immunoglobulin G (IgG) as a model antigen, labeled with luminol, was thus performed by two-step electrochemical excitation using an optical fiber electrode. As with free luminol, labeled IgG generates electrochemical luminescence in the presence of hydrogen peroxide by anodic oxidation. Electrochemical luminescence sharply decreases by immunocomplexation with anti-IgG antibody. The addition of $10^{-13}\,\text{g}\,\text{L}^{-1}$ of antibody results in an appreciable suppression of luminescence. The lower limit of detection may then be close to $10^{-13}\,\text{g}\,\text{L}^{-1}$ of antibody.

3.3.1.6 Self-Assembled Antibody Protein Array on a Protein A Monolayer

Immunosensors are characterized by a single step of determination and high selectivity as well as high sensitivity. The responses of these immunosensors, however, result from averaging the physicochemical properties of the antibody-bound solid surface. We have succeeded in fabricating an ordered array of antibody molecules on a solid surface, and in quantitating individual antigen molecules which are complexed with the antibody array.

Protein A is a cell wall protein from *Staphylococcus aureus* and has a molecular weight of 42 kDa. Since protein A binds specifically to the Fc region of IgG from various animals, it has been widely used in immunoassays and affinity chromatography. We found that protein A could be spread over a water surface to form a

FIG. 3.3-3. Antibody assembling on a protein A monolayer prepared by the LB method

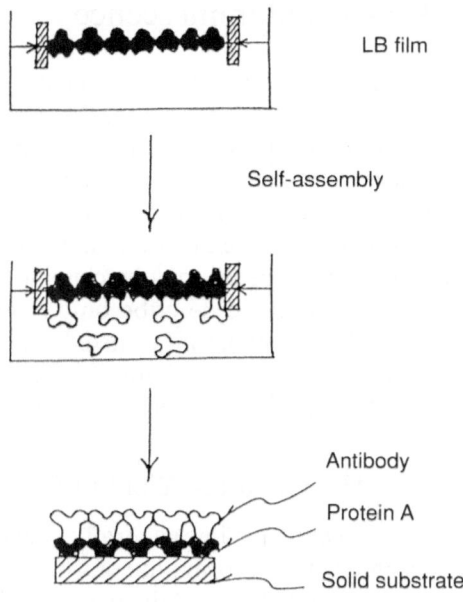

LB film

Self-assembly

Antibody

Protein A

Solid substrate

monolayer membrane using Langmuir–Blodgett (LB) methods [24]. On the basis of this finding, an antibody array on a solid surface can be obtained by the two following steps. The first step is fabrication of an ordered protein A array on the solid surface by the LB method. The second step is self-assembling of antibody molecules on the protein A array by biospecific affinity between protein A and the Fc region of IgG.

For preparation of an antibody protein array, a monolayer of protein A which was compressed at a surface pressure of $11\,mN\,m^{-1}$ was transferred to a compartment containing antiferritin antibody in 10 mM phosphate buffer (pH 7.0). The antibody molecules were self-assembled onto the protein A layer (Fig. 3.3-3). The protein A/antibody molecular membrane was transferred to a compartment containing ultrapure water for rinsing, and was transferred onto the surface of a highly ordered pyrolitic graphite (HOPG) plate by the horizontal method. Atomic force microscopy (AFM) imaging of the protein A array deposited on the HOPG plate showed an ordered alignment of protein molecules when the measurement was made in phosphate buffer at a controlled force of $4 \times 10^{-11}\,N$ and a scanning rate of 0.6 Hz. However, an ordered structure was not observed unless protein A molecules were not cross-linked by glutaraldehyde. The antibody array which was self-assembled on the protein A array was also visualized in molecular alignment by AFM.

The antibody array was soaked in different concentrations of ferritin solutions for 1 h, and was assayed for AFM imaging in solution. Individual ferritin molecules on the antibody array can be selectively quantitated by AFM.

3.3.2 Chemiluminescence

There are many natural and artificially produced lights all around us. The sun, moon, stars, lightning, aurora, fire light, fireflies, noctilucae, luminous animalcules, luminous bacteria and fungi, candlelight, electric light, fluorescent lamps, and laser light are just a few examples with which we are all familiar. Incandescence is the emission of light by a body heated to a high temperature, while luminescence is the emission of light at a temperature below that of incandescent bodies. Luminescence may be classified as shown in Table 3.3-1 [25].

In this section we describe only luminescence related to chemical reactions, i.e., chemi- and bioluminescence. Since bioluminescence is a type of chemiluminescence carried out in biological systems, such as luciferin–luciferase reactions, we can treat bioluminescence as a part of chemiluminescence.

3.3.2.1 The Quantum Yield of Chemiluminescence

A luminescent reaction generally consists of three processes: (a) the processes associated with the formation of a high-energy intermediate, efficiency Φ_r; (b) the processes associated with the decomposition of this intermediate to give a singlet excited state light emitter, Φ_s; (c) the processes of fluorescence from the excited emitter, where the quantum yield of fluorescence of the excited state of the emitter is Φ_f. The overall efficiency of chemiluminescence, Φ_{cl}, is provided by the product of the three terms:

$$\Phi_{cl} = \Phi_r \times \Phi_s \times \Phi_f \qquad (3.3\text{-}1)$$

TABLE 3.3-1. Various types of luminescence

---

1. Heating
   (a) Thermoluminescence: luminescence of solids on mild heating
   (b) Pyroluminescence: luminescence of metal atoms in flames

2. Irradiation
   (a) Photoluminescence: fluorescence and phosphorescence
   (b) Radioluminescence: irradiation by γ- or X-rays, aurora
   (c) Laser: light amplification by stimulated emission of radiation

3. Electricity
   (a) Electroluminescence: luminescence related to electric discharge, lightning, fluorescent lamps
   (b) Electrochemiluminescence: luminescence on electrodes in solution or solid

4. Mechanical transformations
   (a) Triboluminescence: luminescence on crushing crystals
   (b) Sonoluminescence: luminescence from ultrasonic waves in solution

5. Chemical reactions
   (a) Chemiluminescence: luminescence from chemical reactions
   (b) Bioluminescence: luminous organisms

The quantum yield of chemiluminescence of luminol, the standard for luminescence measurements, is 0.0124 [26], while those of the bioluminescence of the firefly and the sea firefly are 0.88 [25] and 0.3 [25], respectively.

3.3.2.2 Ground State Reactions Leading to an Excited State

Chemical reactions leading to an excited state are not often found. A reaction whose transition state energy is greater than the singlet excited state energy of one of the products is a candidate for chemiluminescence. For concerted reactions, the electronic configuration of the excited state is controlled according to the rules of conservation of orbital symmetry.

There are three possible pathways leading to an excited state product from a ground state reaction: (a) thermal opening of a four-membered ring, such as Dewar benzene or 1,2-dioxetanes, in which the reaction process is controlled by the rules of conservation of orbital symmetry; (b) an electron-transfer reaction such as electrochemiluminescence or chemically initiated electron exchange luminescence (CIEEL); (c) a highly exothermic reaction through a charge-transfer transition state. We will discuss each of these cases in detail.

Thermal Opening of a Four-Membered Ring of 1,2-Dioxetane. 1,2-Dioxetanes are typical compounds related to luminescence as studied to date. Furthermore, many examples of bio- and chemiluminescence are expected to involve 1,2-dioxetane intermediates in their luminescent processes. 1,2-Dioxetane contains a strained four-membered ring with a weak O–O bond. For example, since the sum of the activation energy for decomposition and the heat of formation of tetramethyl-dioxetane (TMD, $90 \, \text{kcal} \, \text{mol}^{-1}$) is larger than the singlet ($85 \, \text{kcal} \, \text{mol}^{-1}$) and triplet energy ($78 \, \text{kcal} \, \text{mol}^{-1}$) of the product, acetone (ACE), the formation of an excited state of acetone is possible. However, numerous experiments have shown that the major product is the triplet state of acetone. The ratio of the triplet state to the singlet state is about 140/1 [27, 28]. The predominant formation of the triplet state may suggest that the reaction does not take place in the concerted process, but proceeds via the biradical

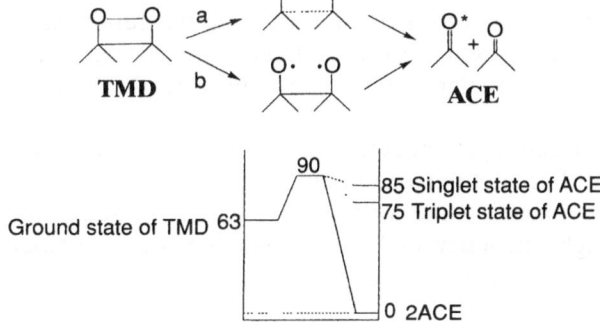

Fig. 3.3-4. The energies related to the thermal decomposition of tetramethyl dioxetane (TMD): ACE acetone

Electrogenerated chemiluminescence (ECL)

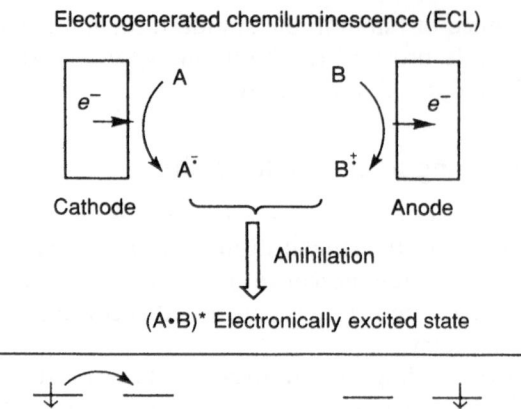

FIG. 3.3-5. The way that electron-transfer produces electrochemiluminescence

intermediate produced from the initial cleavage of the weak O–O bond, as shown in Fig. 3.3-4 [29].

Formation of an Excited State Through Electron-Transfer (CIEEL Mechanism). As shown in Fig. 3.3-5, what is known as electrochemiluminescence (ECL) occurs when electron-transfer from the radical anion to the radical cation produced on the electrodes yields an excited state of either the electron donor or the acceptor and emits light. If the radical cation and the radical anion are produced in a thermal reaction, an excited state of neutral species would be formed through electron-transfer. On the basis of this idea, Koo and Schuster [30] proposed a mechanism known as chemically initiated electron exchange luminescence (CIEEL), and suggested that the high efficiency of bioluminescence might be associated with this mechanism. A typical example is the thermolysis of diphenoyl peroxide (DPP) in the presence of an activator such as perylene, whose fluorescence is observed during the reaction. Figure 3.3-6 summarizes the series of reactions which produce this luminescence. The high efficiency of the oxidative luminescence of oxalic acid derivatives in the presence of an activator clearly demonstrates the participation of the intermolecular CIEEL mechanism in the formation of excited state processes [31].

Recently, Catalani and Wilson [32] claimed that the CIEEL in the reaction of diphenyl peroxide is inefficient (2×10^{-5}), and does not necessarily produce an excited state with a high yield, as suggested by Schuster. Although the high efficiency of bioluminescence was explained by a mechanism associated with CIEEL by Schuster and Thorn [31] and McCapra [33], these is still

FIG. 3.3-6. Reactions involved in the
CIEEL of diphenoyl peroxide (DPP):
BEC benzocumarin

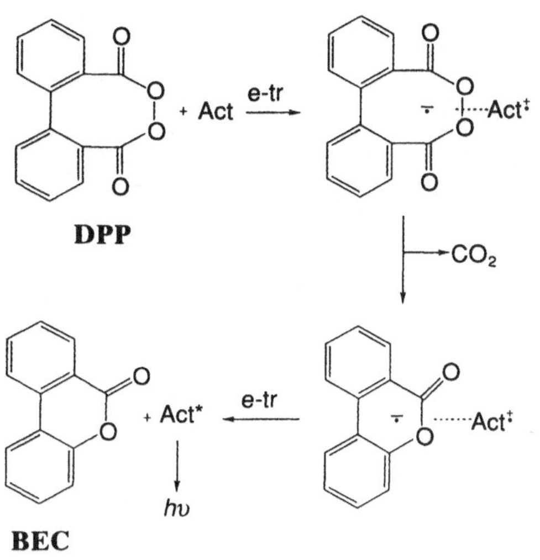

debate about what factors would make an intermolecular CIEEL more efficient.
The most critical argument is concerned with how it is possible to produce
more than 25% of the singlet state under conditions where intermolecular
electron-transfer between a radical cation and a radical anion occurs, since
the statistical probability of producing the singlet state should not exceed
25%. To answer this point, the following charge-transfer mechanism has been
proposed.

Formation of an Excited State Through a Charge-Transfer Intermediate. Instead
of the CIEEL mechanism, McCapra [34], Wilson, and White [35] have independently proposed the idea of the charge-transfer transition state mechanism, in
which the energy surface of the excited state of the product fluorescer may cross
over that of the charge-transfer transition state formed from a ground state inter-
or intramolecular charge-transfer species. However, Yang and Yang [36] suggested that in pericyclic chemiluminescence the introduction of polar groups may
lead the charge-transfer intermediate, whose decomposition pass may cross the
energy surface of the excited state of the product, as shown in Fig. 3.3-7. Recently, Kimura et al. [37] demonstrated that there is a relationship between the
efficiency of singlet excited state formation and the σ value of substituents in
the chemiluminescence of lophine peroxides (LOP), and suggested the perturba-
tion of a charge-transfer character in the process of singlet excited state
formation. Although those arguments have attracted considerable interest, the
following question still remains to be answered: how can we design a charge-
transfer transition state with simultaneous generation of a singlet excited
fluorescer?

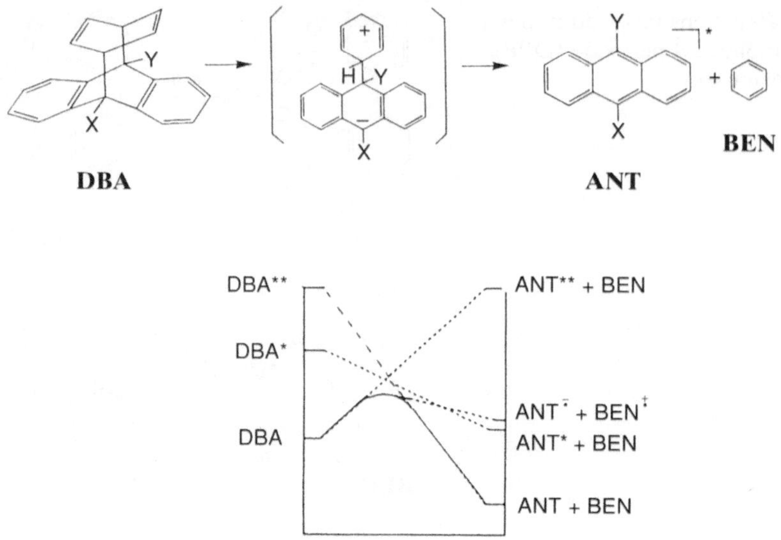

FIG. 3.3-7. Pericyclic chemiluminescence and the energy surface of anthracene (*ANT*) pericyclic chemiluminescence: *DBA*, 9,10-dihydro-9,10-0-benzeno-anthracene

3.3.2.3 Examples of Chemiluminescent Reactions

Some well-known chemiluminescent compounds and their reactions are summarized in Fig. 3.3-8 [25, 38, 39]. In these examples the emitter is an oxidation product associated with dioxetan intermediates. Schaap et al. [40] synthesized dioxetane derivatives (ADO) containing an adamantane skeleton, and demonstrated that the efficiency of the formation of an excited state of product (MCP) is about 0.57, which is close to that of bioluminescence. The highest efficiency so far reported for this step in chemiluminescence is 0.79 [37].

In the examples shown in Fig. 3.3-9 the emitter is the activator, and is associated with intermolecular electron-transfer.

In some cases of chemiluminescence the emitter is the reactant and is associated with dimeric intermediates. An example is shown in Fig. 3.3-10.

3.3.2.4 Bioluminescence

The famous luciferin–luciferase reaction was discovered by Dubois in 1887. He showed that an extract from the luminous beetle *Pyrophorus* with cold water initially emitted light, but this faded away gradually. An extract with hot water did not give light, but when this extract was added to the dimmed cold-water extract, light was emitted again. From this result he concluded that bioluminescence, i.e., the emission of light by living organisms, is the result of a chemical reaction

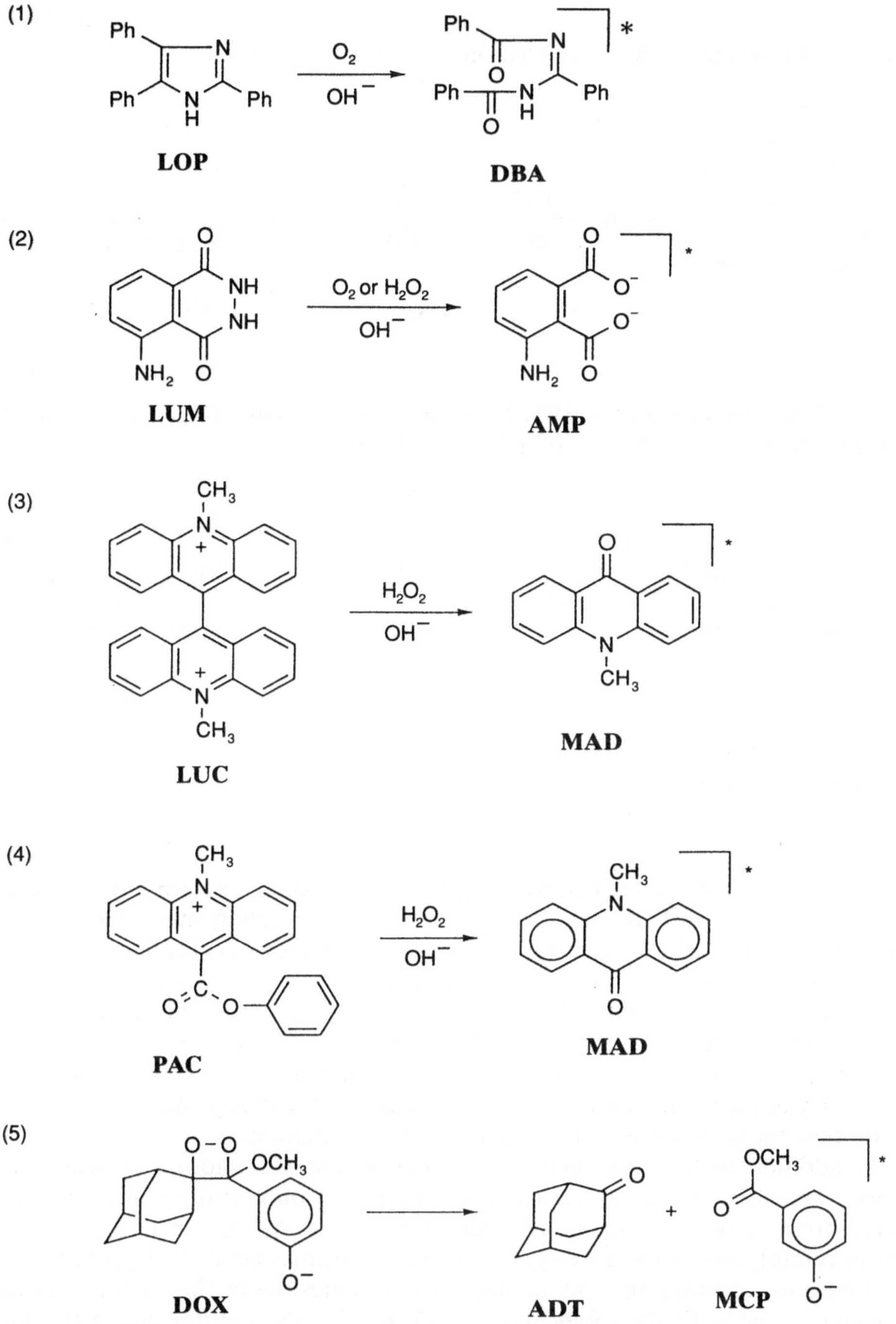

FIG. 3.3-8. Examples of chemiluminescence in which the emitter is an oxidation product: *LOP*, lophine; *DBA*, N^1,N^2-dibenzoyl benzamidine; *LUM*, luminol; *AMP*, 3-aminophthalate; *LUC*, lucigenin; *MAD*, N-methylacridone; *PAC*, phenyl 10-methylacridinium-9-carboxylate; *DOX*, 3-(2'-spiroadamantane)-4-methoxy-4-(3-hydroxy)phenyl-1,2-dioxetane; *ADT*, adamantanone; *MCP*, m-methoxycarbonylphenolate

151

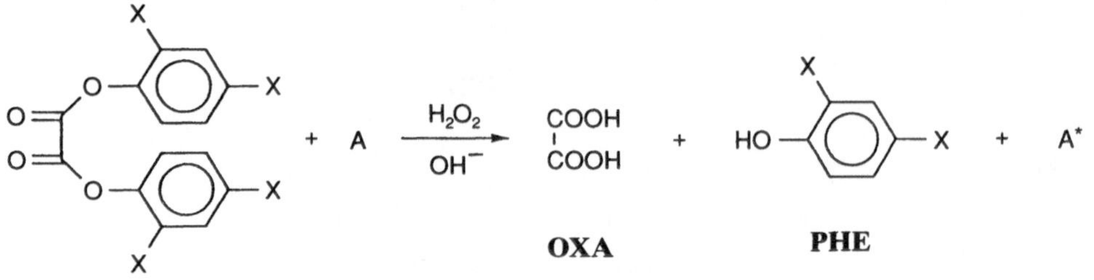

FIG. 3.3-9. Chemiluminescence in which the emitter is an activator: *DPO*, diphenyloxalate derivatives; *OXA*, oxalic acid; *PHE*, phenol derivatives

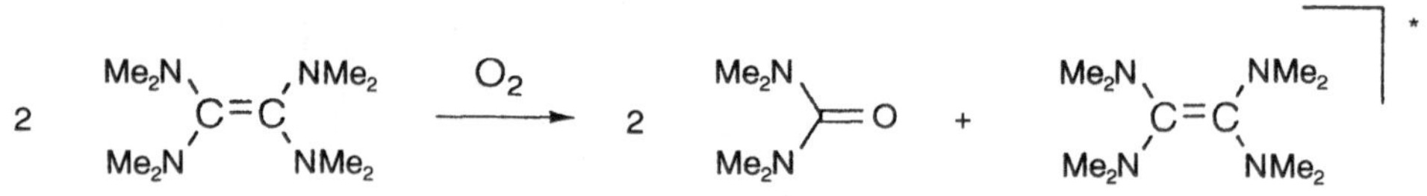

FIG. 3.3-10. Chemiluminescence in which the emitter is the reactant: *TDE*, tetrakis(dimethylamino)ethylene; *TMU*, tetramethylurea

between the heat-stable luciferine and the heat-labile luciferase [25]. Later, Harvey [41] established this general concept by revealing many luciferin–luciferase reactions in a wide variety of luminous organisms (Table 3.3-2).

Although the quantum yield of bioluminescence is usually quite high (for example, firefly 0.88, sea firefly 0.3 [25]), the key step in bioluminescence is the formation of an excited state of oxidation product via a dioxetane or peroxide intermediate. The CIEEL mechanism was first assumed to account for the high efficiency of the formation of the excited state, but the charge-transfer transition state concept now seems to be a more likely explanation.

In addition to luciferin–luciferase reactions, there is another system called photoproteins in bioluminescence. For example, a photoprotein aequorin, isolated from jellyfish, consists of coelenterazine (chromophore), apoaequorin (apoprotein), and molecular oxygen. The luminescent reaction is triggered by the presence of calcium ions, and apoaequorin becomes the luciferase that oxidizes coelenterazine with the molecular oxygen to give the emitter called the blue fluorescent protein in which the excited coelenteramide is bound to the apoprotein in a noncovalent manner [42].

Recent developments in genetic techniques have opened the way to cloning the genes of luciferase, and the primary structures of several luciferases have been established (apoaequorin, vargula luciferase, bacterium *Vibrio harveyi*,

TABLE 3.3-2. Examples of luciferins and approximate molecular weights of luciferases

Structure	Species	M of luciferase
1. Imidazopyrazine	Sea firefly (*Vargula*)	68 000
	Jellyfish (*Aequorea*)	24 000 (Phot. pr.)
2. Benzothiazole	Firefly (*Photinus*)	62 000
3. Tetrapyrole	Dinoflagellate (*Gonyaulax*)	135 000
	Millipode (*Luminodesmus*)	104 000 (Phot. pr.)
4. Flavin	Fungi (*Lampteromyces*)	
	Bacteria (*Photobacterium*)	80 000
5. Aldehyde	Gastropod (*Latia*)	170 000
	Earthworm (*Phylum*)	

firefly *Photinus pyralis*, etc.). Through this cloning technique the preparation of luciferase by expression of the complementary DNA in *Escherichia coli* became possible, and a new area of bioluminescence applications in various fields is developing. Recently it was reported that point mutation of the luciferase of firefly [43] and click beetle [44] changed the colors of their luminescence, and that a single amino acid change in the 548 amino acids of firefly luciferase is enough for the color to change from yellow–green to red, orange, green, or yellow [43].

Figure 3.3-11 shows several bioluminescent reactions. Recently, Nakamura et al. [45] succeeded in establishing the structure of dinoflagellate luciferin.

3.3.2.5 Ultraweak Chemiluminescence

In addition to the bioluminescence due to luciferin–luciferase reactions or photoproteins, there is another type of luminescent reaction in which the quantum yields are 10^{-10} less than the bioluminescence, and the intensity is less than 10^{-14} W. This luminescence is not visible to the naked eye and is called ultraweak chemiluminescence or low-level chemiluminescence. It originates mainly from excited carbonyls or singlet oxygen produced in radical reactions of peroxides in organs.

Clinical and pharmaceutical applications for this ultraweak chemiluminescence are developing rapidly. A discussion of these applications is outside the scope of this book, but see the excellent review by Cambell [25].

3.3.2.6 Applications of Chemi- and Bioluminescence

Numerous applications of chemiluminescence have been proposed in analytical and biological fields. Applications may be classified into three categories, as described below [25, 38, 39].

1. Inorganic systems. Detection of metal ions using a luminol–hydrogen peroxide–metal catalyst. A well-known, classical method of determining blood group is based on the luminol–hydrogen peroxide–Fe (III) system. Ag (lucigenin), Bi (lucigenin), Ce (luminol), Co (luminol, lucigenin), Cr(III) (luminol), Cu (luminol), Fe (luminol), Hg (luminol), Mn (lucigenin), Ni(II)

(1) Imidazopyrazines (jellyfish, sea firefly)

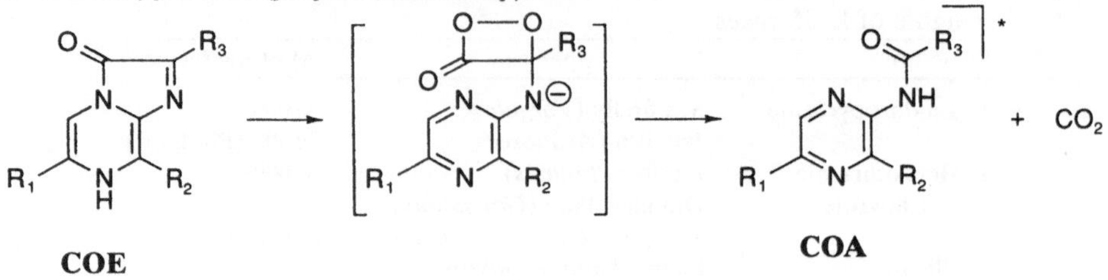

COE   COA $+$ CO_2

(2) Benzothiazole (firefly, beetle)

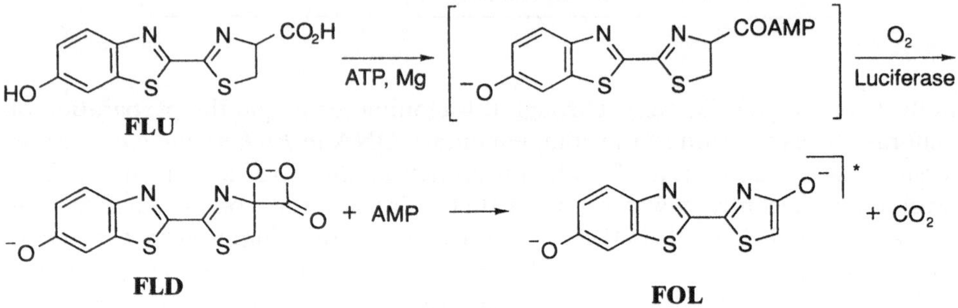

FLU

FLD   FOL $+$ CO_2

(3) Flavin (bacteria)

FLA   DHF

HFL $+$ $R'CO_2H$

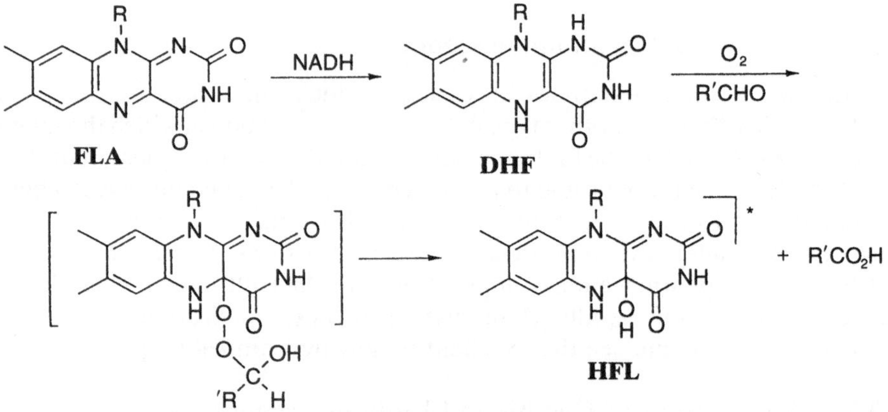

FIG. 3.3-11. Some bioluminescent reactions: *COE*, coelenterazine analogs; *COA*, coelenteramide analogs; *FLU*, firefly luciferin; *FLD*, firefly luciferin–dioxetanone; *FOL*, firefly oxyluciferin; *FLA*, flavin; *DHF*, dihydroflavin; *HFL*, 4a-hydroxyflavin; *DFL*, dinoflagellate luciferin; *DFO*, dinoflagellate oxyluciferin; *LAL*, latia luciferin; *DIO*, dihydro-β-ionone (purple protein, pp)

(4) Tetrapyrrole (dinoflagellate)

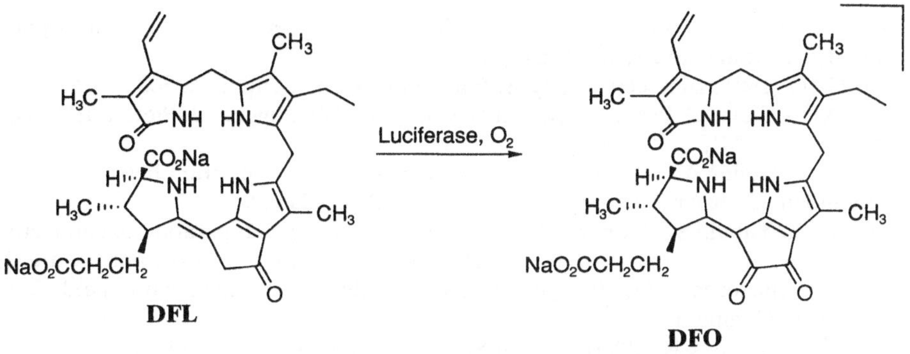

Luciferase, O₂ →

DFL

DFO

(5) Aldehyde (Latia)

LAL   + (PP) ⟶   DIO   + (PP) *

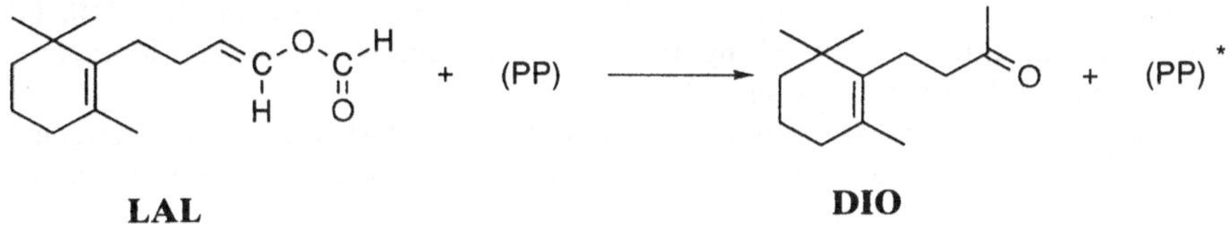

Fig. 3.3-11. *Continued*

(luminol), Os (lucigenin), Pb (lucigenin), T1 (lucigenin), V (lucigenin), and Zr (luminol) have all been determined by the catalytic oxidation of luminol or lucigenin. The photoprotein aequorin emits light when calcium ions are added. Since this reaction is very sensitive, it has been used to measure the concentration of intracellular Ca^{2+} ion in gene technologies.

2. Organic systems. Peroxyoxalate chemiluminescence has been used for emergency lighting, fishing, and toys. Chemiluminescence detection with HPLC is now becoming popular.

3. Biological systems. The major biomedical applications of chemiluminescence are analyses of ATP (firefly luciferin–luciferase) and NADH (bacteria luciferase), of reactive oxygen (ultraweak chemiluminescence, luminol, lucigenin), and of Ca^{2+} (aequorin). Although radioimmunoassay is widely used in clinical laboratories, there are serious problems connected with the use of radioactive isotopes. To replace radioactive isotopes, chemiluminescent compounds may be used as non-isotopic labels in immunoassay. Bioluminescence in combination with recombinant DNA technology is also developing, and many examples are being reported. Bacterial and firefly luciferase genes have been used as a probe for genetic expression, and firefly luciferin derivatives and admantane 1.2-dioxetane derivatives have been developed as a sensor for measuring the activities of enzymes such as alkaline phosphatase and oxygenase [46].

References

1. Aizawa M (1991) Principles and applications of electrochemical and optical biosensors. Anal Chim Acta 250:249–256
2. Hill HAO (1981) Bioelectrocatalysis. Phil Trans R Soc London A, 302:69–75
3. Cass AEG, Devis G, Francis GD, Hill HAO, Aston WJ, Higgins IJ, Plothin EV, Scott LDL, Turner APF (1984) Anal Chem 1880
4. Degani Y, Heller AJ (1987) Direct electrochemical communication between chemically modified enzymes and metal electrodes. J Phys Chem 91:6–12
5. Aizawa M, Yabuki S, Shionhara H (1987) Electrochemical preparation of conductive enzyme membrane. In: Torii S (ed) Proc 1st Int Symp Electroorganic Synthesis. Elsevier, Amsterdam, 353–360; (1989) Electroconductive enzyme membrane. J Chem Soc, Chem Commun 945–946
6. Foulds NC, Lowe CR (1986) J Chem Soc, Faraday Trans 1, 82:1259
7. Aizawa M, Yabuki S, Shinohara H, Ikariyama Y (1989) Molecular wire and interface for bioelectronic molecular devices. In: Aviram A (ed) Molecular electronics: science and technology. Engineering Foundation, New York, pp 139–150; (1988) Biomolecular interface. In: Hong FT (ed) Molecular electronics. Biosensors and biocomputers. Plenum Press, New York, pp 269–276
8. Aizawa M, Khan GF, Shinohara H, Ikariyama Y (1994) Molecular interfacing for electron transfer of redox enzymes. In: Aizawa M (ed) Chemical sensor technology, vol 5. Kodansha, Tokyo, pp 157–175
9. Khan GF, Kobatake E, Shinohara H, Ikariyama Y, Aizawa M (1992) Molecular interface for an activity controlled enzyme electrode and its application for the determination of fructose. Anal Chem 64:1254–1258
10. Ishizuka T, Kobatake E, Ikariyama Y, Aizawa M (1991) Amperometric biosensor for alcohol using enzyme–NAD–mediator electron transferring system. Technical Digest 10th Sensor Symp pp 73–76
11. Aizawa M (1994) Immunosensors for clinical analysis. Adv Clin Chem 31:247–27512.
12. Aizawa M, Suzuki S, Nakagawa Y, Shinohara R, Ishiguro I (1977) An immuno sensor for specific protein. Chem Lett 779–782
13. Aizawa M, Kato S, Suzuki S (1977) Immunoresponsive membrane. I. Membrane potential change associated with an immunochemical reaction between membrane-bound antigen and free antibody. J Membrane Sci 2:125–132
14. Yamamoto N, Nagasawa Y, Sawai M, Suda T, Tsubomura H (1978) Potentiometric investigations of antigen–antibody and enzyme–enzyme inhibitor reactions using chemically modified electrodes. J Immunol Methods 22:309–317
15. Aizawa M, Morioka A, Matsuoka H, Suzuki S, Nagamura Y, Shinohara R, Ishiguro (1976) An enzyme immunosensor for IgG. J Solid-Phase Biochem 1:319–328
16. Aizawa M, Morioka A, Suzuki S, Nagamura Y (1979) Enzyme immunosensor. III. Amperometric determination of human chorionic gonadotropin by membrane-bound antibody. Anal Biochem 94:22–28
17. Aizawa M, Morioka A, Suzuki S (1980) An enzyme immunosensor for the electrochemical determination of the tumor antigen a-fetoprotein. Anal Chim Acta 115:61–67
18. Aizawa M, Suzuki S, Kato T, Fujiwara T, Fujita Y (1980) Solid-phase luminescent enzyme immunoassay of IgG and anti-IgG using a transparent and nonporous antibody-bound plate. J Appl Biochem 2:190–195

19. Schaffar BPH, Wolfbeis QS (1991) In: Blum LJ, Coulet PR (eds) Biosensors—principles and applications. Marcel Dekker, New York, pp 163–194
20. Liedberg B, Nylander C, Lundstrom I (1983) Sensors Actuators 4:299
21. Lofas S, Johnson B (1990) A novel hydrogel matrix on gold surface plasmon resonance sensors for flat and efficient covalent immobilization of ligands. J Chem Soc, Chem Commun 1526–1528
22. Ikariyama Y, Kunoh H, Aizawa M (1987) Electrochemical luminescence-based homogeneous immunoassay. Biochem Biophys Res Commun 128:987–992
23. Aizawa M, Tanaka M, Ikariyama Y, Shinohara H (1989) Luminescence biosensors. J Biolumin Chemilumin 4:535–542
24. Owaku K, Goto M, Ikariyama Y, Aizawa M (1995) Protein A LB film for antibody immobilization and its use in optical immunosensing. Anal Chem 67:1613–1616
25. Campbell AK (1988) Chemiluminescence. Ellis Horwood, Chichester
26. Lee J, Seliger HH (1972) Quantum yields of the luminol chemiluminescence reaction in aqueous and aprotic solvents. Photochem Photobiol 15:227–237
27. Adam W, Beinhauser A, Hauser H (1989) Activation parameters and excitation yields in 1,2-dioxetane chemiluminescence. In: Scaiano JC (ed) Handbook of organic photochemistry, vol II. CRC Press, Ottawa, pp 271–328
28. Turro NJ (1978) Modern molecular photochemistry. Benjamin, Menro Park, pp 597–602
29. Reguero M, Bernardi F, Bottoni A, Olivucci M, Robb MA (1991) Chemiluminescent decomposition of 1,2-dioxetanes: an MC-SCF/MP2 study with VB analysis. J Am Chem Soc 113:1566–1572
30. Koo J-Y, Schuster GB (1977) Chemically initiated electron exchange luminescence. A new chemiluminescent reaction path for organic perroxides. J Am Chem Soc 99:6107–6109; (1978) Chemiluminescence of diphenoyl peroxide. Chemically initiated electron exchange luminescence. A new general mechanism for chemical production of electronically excited state. J Am Chem Soc 100:4496–4503
31. Schuster GB (1979) Chemiluminescence of organic peroxides. Conversion of ground-state reaction to excited-state products by the chemically initiated electrom-exchange luminescence mechanism. Acc Chem Res 12:366–373
32. Catalani LH, Wilson T (1989) Electron transfer and chemiluminescence. Two inefficient systems: 1,4-dimethoxy-9,10-diphenylanthracene peroxide and diphenoyl peroxide. J Am Chem Soc 111:2633–2639
33. McCapra F (1977) Alternative mechanism for dioxetan decomposition. J Chem Soc Chem Commun 946–947
34. McCapra F (1990) Chemiluminescence and bioluminescence. J Photochem Photobiol A 51:21–28
35. White EH, Roswell DF, Dupont AC, Wilson AA (1987) Chemiluminescence involving acidic and ambient ion light emitters. The chemiluminescence of the 9-acridinepercarboxylate ion. J Am Chem Soc 109:5189–5196
36. Yang NC, Yang X-Q (1987) A new type of pericyclic chemiluminescence. J Am Chem Soc 109:3804–3805
37. Kimura M, Nishikawa H, Kura H, Lim H, White EH (1993) Maximization of the chemiluminescence efficiency of 1,4,5-triarylhydroperoxy-4H-isoimidazoles. Chem Lett 505–508
38. Van Dyke K (ed) (1985) Bioluminescence and chemiluminescence: instruments and applications. CRC Press, Baca Raton

39. Gundermann KD, McCapra F (1987) Chemiluminescence in organic chemistry. Springer, Berlin, Chap. 13
40. Schaap AP, Chen T-S, Handley RS, DeSilva R, Giri BP (1987) Chemical and enzymatic triggering of 1,2-dioxetanes. 2. Fluoride-induced chemiluminescence from tert-butyldimethylsilyloxy-substituted dioxetanes. Tetrahedron Lett 28:1155–1158
41. Harvey EN (1952) Bioluminescence. Academic Press, New York
42. Shimomura O, Musicki B, Kishi Y (1988) Semi-synthetic aequorin. Biochem J 251:405–410
43. Kajiyama N, Nakano E (1991) Isolation and characterization of mutants of firefly luciferase which produce different colors of light. Protein Eng 4:691–695
44. Wood KV, Lam YA, Seliger HH, McElroy WD (1989) Complementary DNA coding click beetle luciferases can elicit bioluminescence of different colors. Science 244:700–702
45. Nakamura H, Kishi Y, Shimomura O, Morse D, Hastings JW (1989) Structure of dinoflagellate luciferin and its enzymatic and nonenzymatic air-oxidation products. J Am Chem Soc 111:7607–7611
46. Beck S, Koester H (1990) Application of dioxetane chemiluminescent probes to molecular biology. Anal Chem 62:2258–2270

4. Molecular Systems and Their Applications to Energy Conversion

IWAO YAMAZAKI

A number of molecular energy conversion systems can be seen in biological organisms, e.g., photosynthesis, vision, muscular movements, and photophobic/phototactic responses. A particular form of energy, e.g., photonic energy, electric energy, mechanical energy, or chemical energy, is received as a stimulus from the external environment by sensor molecules and then converted to another form of energy. The initial steps in these energy conversions are driven by the transfer of an electron, excitation energy, or a soliton through a molecular channel in which functional molecules are arranged in a specific spatial configuration within polypeptide networks. These processes are characterized by high efficiencies and ultrafast reaction rates. To interpret the mechanisms of such sequential and cooperative reactions, it may be necessary to develop a new theoretical description of the intermolecular interactions which spread the functional molecules along the reaction channel. This description is being sought by detailed analyses of the biological systems, and also by research into artificial molecular systems based on synthesized supramolecules or planned solid complex superstructures. This chapter first considers sequential reactions in biological molecular systems, and then those in artificial systems.

4.1 Photoactive Molecular Systems

Iwao Yamazaki

Photoactive responses in molecular systems are initiated by light absorption by molecules generating an electronically excited state. Then the excited molecule A* may undergo various relaxation processes with rate constant k.

$$A* \xrightarrow{k_F} A + h_\nu \qquad \text{Fluorescence} \qquad (4.1\text{-}1)$$

$$A* + B \xrightarrow{k_{ET}} A + B* \qquad \text{Energy transfer} \qquad (4.1\text{-}2)$$

$$A* + B \xrightarrow{k_{CS}} A^+ + B^- \qquad \text{Charge separation} \qquad (4.1\text{-}3)$$

$$AH* + B \xrightarrow{k_{PT}} A + BH \qquad \text{Proton transfer} \qquad (4.1\text{-}4)$$

$$A* \xrightarrow{k_{ISO}} A' \qquad \text{Photoisomerization} \\ \left(\text{reversible or irreversible}\right) \qquad (4.1\text{-}5)$$

Photoinduced charge separation (Eq. 4.1.3) is a primary step for energy conversion from light to electrical energy, while photoisomerization (Eq. 4.1.5) is involved in the initial step of the conversion of light to chemical energy. Thus the energy conversion efficiency is determined predominantly by the quantum yield of an initial step, which is given by the ratio of the rate constant of a particular process to the sum of all the rate constants of competing relaxation processes. The extraordinarily efficient energy conversion found in biological systems is believed to result from a sequential and cooperative chain reaction in a supramolecular system. In this system different kinds of functional molecules are located close together along the reaction pathway in such a way that a particular process is completely dominant.

4.1.1 Molecular Systems in Biological Photoreceptors

Various kinds of photoactive responses are known in animals and plants in which one particular photophysical process from among those listed in Eqs. 4.1.1–4.1.5 is involved as a primary step. Table 4.1-1 summarizes the biological photorecep-

TABLE 4.1-1. Functional molecules and proteins in biological photoreceptors

Photoreceptor systems	Functional protein	Functional molecules
Photosynthetic systems	Antenna pigment systems	Antenna chlorophyll
	Phycobilisomes	Phycobilin
	Chlorophyll complexes	
	Reaction centers	Chlorophyll dimer
		Pheophytin
		Quinone
		Manganese complex
Visual recognition systems	Rhodopsin	Retinal
Phototactic photophobic movement	Stentorin	Hypericin
Biogenetic control	Phytochrome	Tetrapyrrole (open form)

tor proteins and functional chromophores related to the photoactive response. This section reviews recent research into the mechanisms and structures of organized molecular systems in the biological photoreceptors involved in photosynthesis, biogenetic control, and phototactic/photophobic movement.

4.1.1.1 Photosynthesis: Solar Energy Conversion to Electrical and Chemical Energy

Photosynthesis in plants is a process of energy conversion from solar energy to chemical (electrical) energy. The initial step is the highly efficient absorption of light and the transfer of excitation energy to the reaction center. The transfer of excitation energy occurs in the light-harvesting chromoprotein, in which several different pigment chromophores are arranged in specific sequences or channels. The reaction center is also an organized molecular system in which an electronically excited dimer of chlorophyll molecules, called a "special pair," generates electrons, and the directional transfer of electrons occurs across a biological membrane. This electron transfer takes place through a channel of electron donor–acceptor molecules, and creates a potential difference that drives the subsequent biochemical reactions. The initial steps in the energy and electron transfer are ultrafast (1–100 ps time-scale) and highly efficient (quantum yield of almost unity). The dynamic aspects of the photochemical processes of the reaction center have been reviewed by Fleming and Van Grondelle [1].

In 1985, Deisenhofer et al. [2] revealed the structure of the protein scaffolding in the reaction center and the spatial arrangement of functional chromophores by X-ray crytallographic studies, together with a determination of the sequences of the constituent subunits of a purple bacterium *Rhodopseudomonas viridis*. Figure 4.1-1 shows a schematic illustration of the arrangement of functional chromophores in *R. visidis*. There appear to be two similar paths (labelled A and B in the figure) for electrons to travel, which are arranged with a local two-fold rotation axis. The B branch is regarded as an inactive branch. Along the active A branch, chromophores are located in the order of a heme group of cytochrome

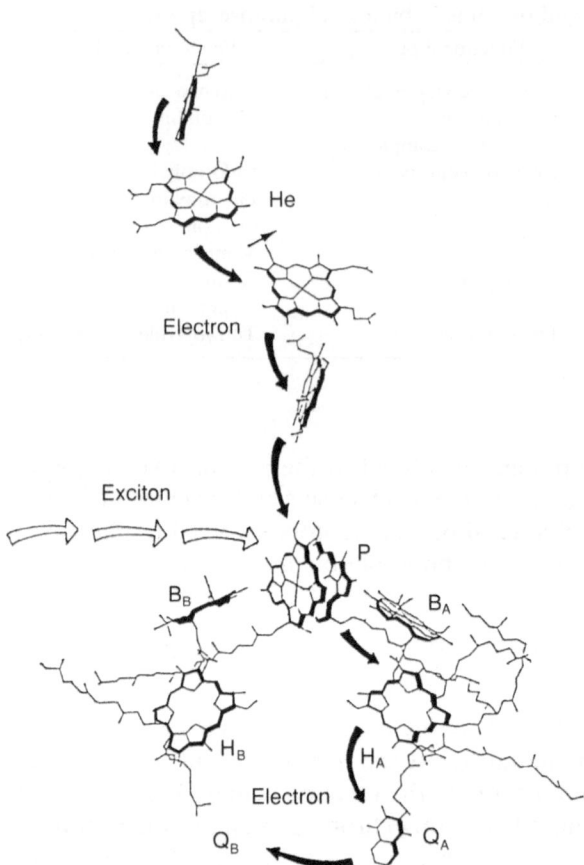

Fɪɢ. 4.1-1. Spatial arrangement of functional chro-
mophores and the sequential
transport of excitation energy
and electrons in the photosynthetic reaction center of a
purple photosynthetic
bacterium, *Rhodopseudomonas viridis*

(He), a pair of bacteriochlorophylls (P), a single bacteriochlorophyll (B_A),
bacteriopheophytin (H_A), the first quinone (Q_A), and the second quinone (Q_B).
The center-to-center distances are 11 Å between P and B_B, 11 Å between B_A and
H_A, and 17 Å between P and H_A. This geometrical arrangement of chromophores
can be rationalized by considering that the π-electron orbitals explicitly, while the
medium is viewed as a dielectric acting as a potential barrier.

The sequence of events in the reaction center can be expressed as

$$P^*B_AH_A \rightarrow P^+B_A^-H_A \rightarrow P^+B_AH_A^- \tag{4.1-6}$$

$$\rightarrow P^+Q_A^-Q_B \rightarrow P^+Q_AQ_B^- \tag{4.1-7}$$

First, the charge separation occurs at the special pair P within about 3 ps, to give
the oxidized special pair of P^+ and a pheophytin anion H_A^- (Eq. 4.1.6). The
observed electron transfer rate between the two molecules is at least 1000 times
faster than the expected rate over that distance in a vacuum. The electron then

Fig. 4.1-2. Energy level structure of the electron transfer pathway in the photosynthetic reaction center of a purple bacterium

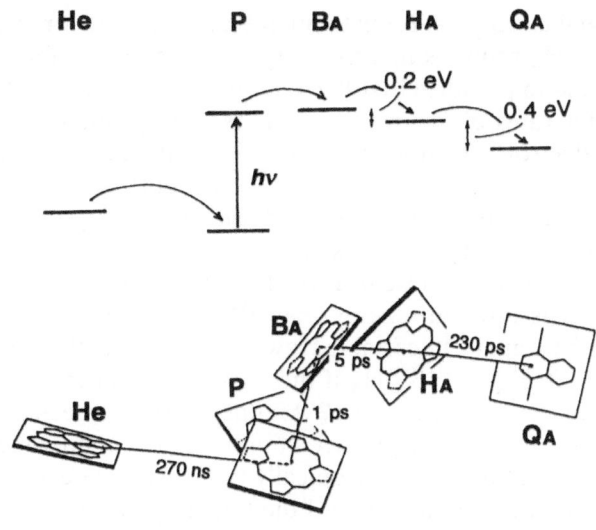

hops to a quinone molecule Q_A in about 200 ps, and on to a second quinone Q_B in about 100 μs (Eq. 4.1.7). This entire sequence is repeated, and the Q_B molecule leaves the protein as $Q_B H_2$ to take part in the chemical reactions that lead to the generation of an electrochemical potential gradient across the biological membrane. Eventually the two electrons of the reduced quinone molecule are returned to the cytochrome, leading to the transport of four protons across the membrane. The resultant electrochemical energy is sufficient for green plants and bacteria to synthesize adenosine triphosphate and other molecules used as energy sources by living organisms.

The energy level structure of the chromophores and the electron transfer rate at each step are shown diagramatically in Fig. 4.1-2 following a reaction model proposed by Kuhn [3]. In the first step of electron transfer, an energy barrier separating He from P is required to prevent the electron in P* from moving to an acceptor other than B_A. However, this barrier should be transparent to an electron moving from donor He to P^+ in a slower process. The way out of this dilemma is a barrier as high as the barrier of a hydrocarbon portion, and sufficiently thin to allow quantum mechanical electron tunneling from He to P^+ within a reasonable time, but not so thin that it will allow tunneling of the photoexcited electron through the barrier instead of moving to B_A. In the sequence of electron transfer, the energy levels of B_A^-, H_A^-, and Q_A^- must be high to keep the photoexcited electron at the highest possible potential, but sufficiently low to avoid back-transfer by thermal activation, i.e., Q_A^- should be about 0.7 eV below the level in the photoexcited P in order to keep the electron in Q_A^- for a few milliseconds at room temperature. The need for transient trapping of the electron in H_A requires that its energy level in H_A^- is about 0.2 eV below its level in B_A^- and in photoexcited P. Consequently, the forward and backward electron transfer rates differ by many orders of magnitude. For example, if Q_B is removed, electron recombination from Q_A to P takes 100 ms, which is six orders of magnitude slower than the forward process P $\rightarrow$ Q_A (via H_A).

The above argument for the initial step of charge sepration $P^*B_AH_A \rightarrow P^+B_AH_A^-$ (Eq. 4.1.6) is based on the sequential transfer process participating an intermediate state of $P^+B_A^-H_A$. This mechanism must operate if the state $P^+B_A^- H_A$ is lower in energy than $P^*B_AH_A$. The energy levels of the relevant states still remain uncertain. How do we rationalize the extraordinarily fast process $P^*B_AH_A \rightarrow P^+B_A^-H_A$ if the energy level of the intermediate state is higher than that of the initial state? Another possible mechanism is a virtual process called the superexchange interaction mechanism, in which the three states P, B_A, and H_A are coupled electronically to form a time-evolutionary state. Coherence must then be maintained until the final population state is reached. Detailed analysis suggests that the spectral features of the intermediate state might be observable in the fully virtual process, which means that the sequential and superexchange mechanisms may be difficult to distinguish experimentally. In fact, Holzapfel et al. [4] concludes that a two-step mechanism is consistent with his data, while Kirmaier and Holten [5] find no evidence for an intermediate state and so favor the superexchange process. The truth of the proposed mechanisms depends on both the energetics of the system and its interaction with its protein environment.

Electron transfer along the "inactive" branch to H_B is interpreted in terms of the energy level structure [6], i.e., $P^+B_B^-H_B$ lies significantly above $P^*B_BH_B$, and intermediates such as $P^+B_A^-$ and $P^+H_A^-$ are lower in energy than $P^+B_B^-$ and $P^+H_B^-$. Experiments by Middendorf et al. [6] exploited the sensitivity of the electronic absorption band to external electric fields, and probed the effective dielectric constant in the vicinity of the chromophores, suggesting that the dielectric constant is higher along the functional or active pathway, and arguing that this determines the directionality of electron transfer. The effect appears to be a collective one involving many amino acids, so that changing a small number of amino acids would be unlikely to alter the direction of electron flow.

Chan et al. [7] have investigated the role of two particular amino acis, L181 (phenylalanine) on the inactive branch, and M208 (tyrosine) on the active branch (Fig. 4.1-1). In their study they presumed that tyrosine controlled led the directionality of the electron transfer, and they therefore reversed the locations of the two amino acids. They found that the electron transfer rate was essentially unchanged by this modification, and still proceeded along the active branch. Even more surprisingly, the electron transfer rate increased slightly when "symmetry" was restored to the reaction center by making both amino acids tyrosine residues, in spite of the fact that this change modified only the "inactive" side. Studies of a set of ten different mutations on these two sites revealed that the amino acids in these positions affect the redox potential of the special pair, that is, the difference in free energy between P and P^+.

4.1.1.2 Photosynthesis: Excitation Energy Transfer in Light-Harvesting Antenna Systems

Photosynthetic reaction centers are surrounded by light-harvesting chromoproteins, the functions of which are to absorb light over a broad range of wavelengths and to transfer the excitation energy to the special pair at the reaction center. The light-harvesting system, often called the antenna, allows the cell to greatly improve the absorption cross section of each reaction center and make optimal use of its energy-converting capacity. Within chromoprotein, the excitation energy is transferred sequentially among different kinds of pigments, such as phycobilin, chlorophyll, and carotenoid molecules, which are bound covalently to proteins and form stacking sequences. The stacking structure of chromoproteins is classified into the four types illustrated schematically in Fig. 4.1-3. It depends on both evolution and environmental conditions. In particular, the intensity of sun-light on a living organism changes the constitution of the structure with regard to its stacking and the number of antenna chromophores.

The excitation energy transfer (Eq. 4.1.2) in the antenna pigment system can be described basically as a Förster dipole–dipole resonance mechanism. The rate of energy transfer varies with the inverse of the sixth power of the distance between pigments, as well as a factor that accounts for the orientation of the dipoles in space, and another factor that measures the overlap between the emission spectrum of the excitation donor and the absorption spectrum of the excitation acceptor.

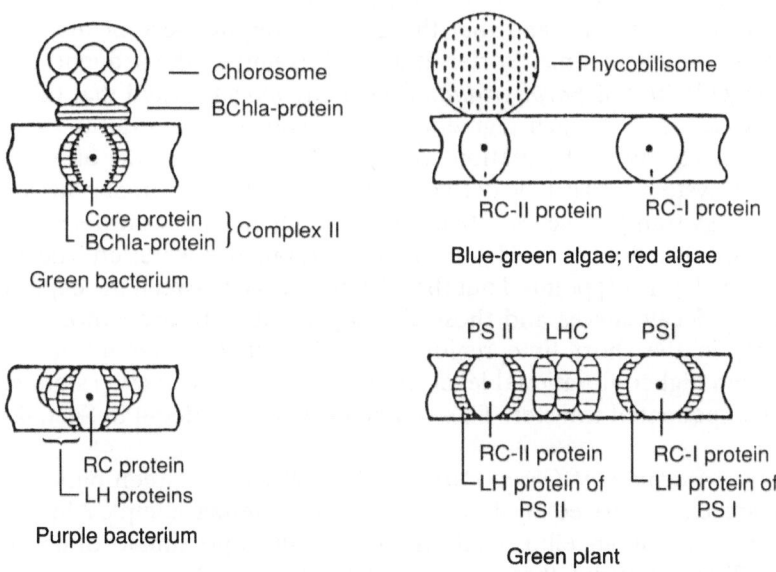

FIG. 4.1-3. Stacking structures of chromoproteins in the light-harvesting antenna of photosynthetic bacteria and green plants

Phycobilisome–Chlorophyll Antennas of Algae. A light-harvesting antenna system of red and blue-green algae consists of chromoproteins, called phycobilisome, attached on the core antenna inside the thylakoid membrane. The phycobilisome is a supramolecular unit involving several kinds of phycobiliproteins, phycoerythrin (PE), phycocyanin (PC), and allophycocyanin (APC), which contain in each protein open-chain tetrapyrroles as chromophores at distances of 20–40 Å [8, 9]. The structure of the phycobilisome is presented schematically in Fig. 4.1-4. In a particular phycobiliprotein, PC for example, three monomers (molecular weight ca. 10^5) are assembled with a C_3 symmetry to form an acyclic trimer with a central hole. Absorption and fluorescence spectra for several proteins are shown in Fig. 4.1-5. The band position depends on the conformation of tetrapyrroles in the protein scaffolding, and the shift to red in the order PE, PC, and APC from the outer surface to the inner core. The Förster critical transfer distances are 61 Å (PE–PC) and 63 Å (PC–APC), so that the energy transfer to the reaction center is straightforward. The dynamic aspect of sequential energy transfer has been studied by using picosecond time-resolved fluorescence spectroscopy [10–12].

Figure 4.1-6 shows the time-resolved fluorescence spectra of intact cells of *Synechocystis* sp. It can be seen that the fluorescence bands of various pigments appear sequentially in the following order: (1) the PE spectrum appears in the time-region 0–100 ps; (2) the PC spectrum first appears at 0–100 ps, with its peak being shifted gradually to the red; (3) the APC spectrum becomes dominant at 100–400 ps; (3) the Chl *a* spectrum appears clearly at 500 ps and remains the same after 700 ps. The fluorescence growth and decay curves show that fluorescence in particular pigments appears with a delay time which gets longer from the outer surface to the inner core, and that the growth time in every pigment is much shorter than the decay time. Porter et al. [10] examined the sequential fluorescence decay kinetics of *Porphyridium cruentum*, and suggested that the fluorescence time behavior of each pigment can be expressed by a decay function of $\exp(-2k\,t^{1/2})$. This type of equation corresponds to those for the original three-dimensional system, and can be derived from Förster kinetics under the condition that the energy transfer occurs with extremely high efficiency between the donor and acceptor chromophores. However, some anomalies were left unexplained. Yamazaki et al. [11, 13] pointed out that the difference between the experimental fluorescence decay curves and those developed theoretically is not negligible, particularly in the short time region, suggesting that nonequilibrium energy transfer from higher vibrational levels in the electronically excited state contributes to this sequential transfer. This will be described in detail in Sect. 4.1.2.1.

Chlorosome Antennas of Green Photosynthetic Bacteria. Green photosynthetic bacteria are characterized by having a unique antenna complex known as a chlorosome. This is an ellipsoidal structure with approximate dimensions of 100 nm × 30 nm × 10 nm that is attached to the cytoplasmic side of the cell membrane [14]. The overall architecture of their pigment composition is shown schematically in Fig. 4.1-7a. Chlorosomes contain approximately 10 000 mol-

Phycobilisome

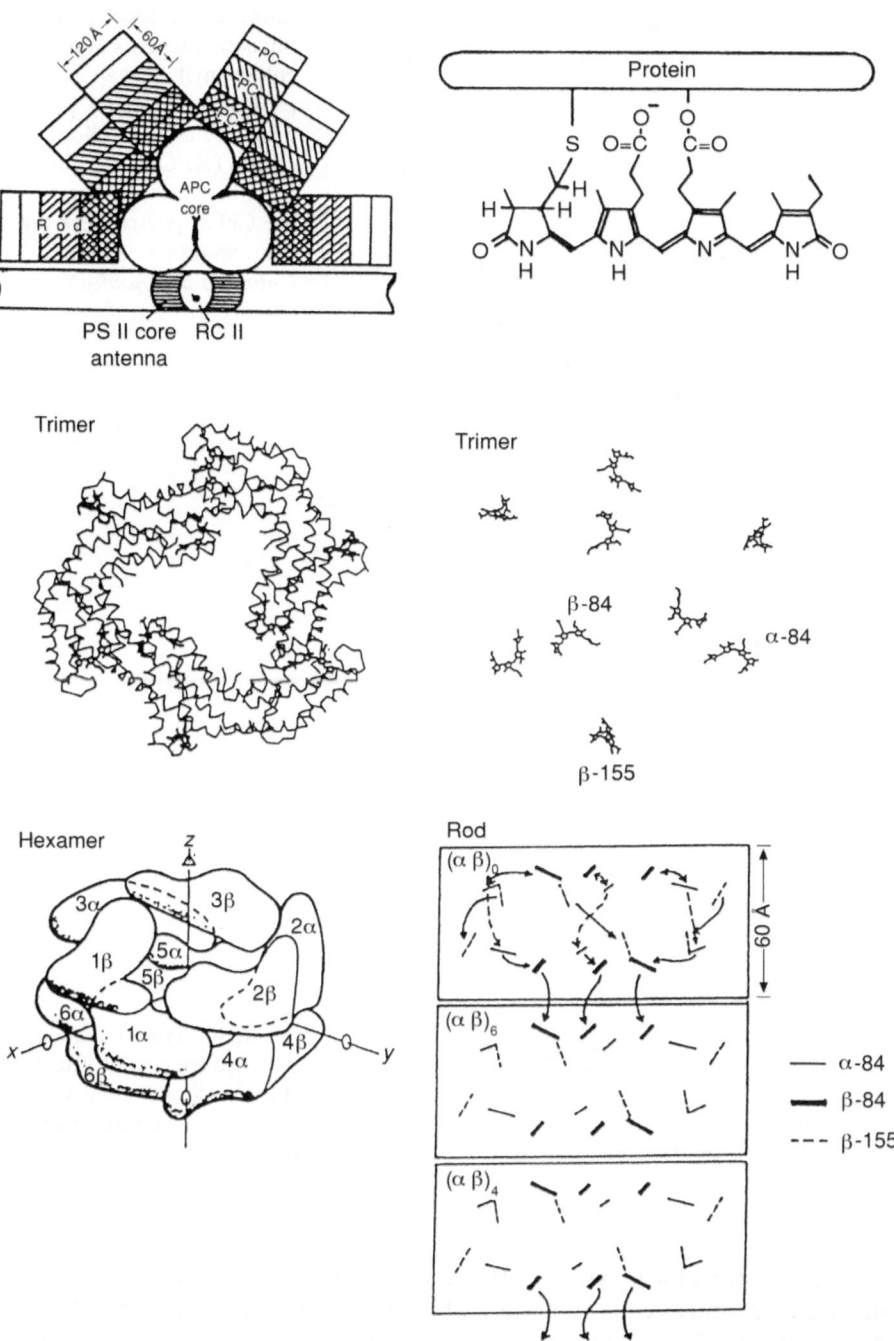

FIG. 4.1-4. Stacking structure of light-harvesting antenna chromoproteins, phycobiliprotein, in blue-green algae

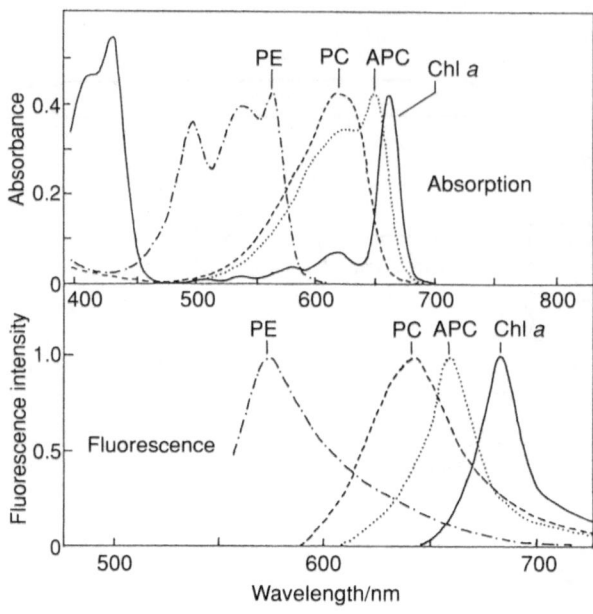

FIG. 4.1-5. Absorption and fluorescence spectra of phycobiliproteins and chlorophyll. The spectra of phycoerythrin (*PE*), phyco-cyanin (*PC*), and allophyco-cyanin (*APC*) were obtained from the hexamer forms, and the Chl *a* spectra were taken from an intact cell of *Chlorella pyrenoidosa*

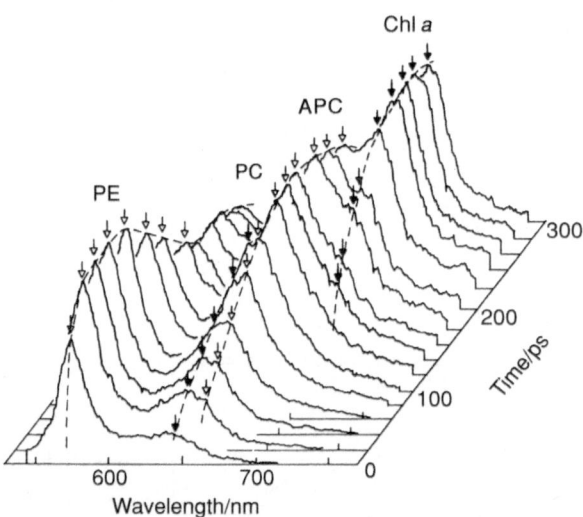

FIG. 4.1-6. Time-resolved fluo-rescence spectra of intact cells of algae, *Synechocystis* sp., ob-tained by 2-ps laser excitation at 540 nm

ecules of bacteriochlorophyll (BChl) *c*, *d*, or *e* (depending on the species) as their major pigment, and also carotenoids and small amounts of BChl *a* [15]. The BChl *c* molecules form large oligomers in vivo by means of direct pigment–pigment interactions, with the geometrical arrangement and large absorption red shifts typical of *J* aggregates [16]. Two of several possible structures for these BChl *c*

FIG. 4.1-7. **a** Schematic model of an antenna pigment system in the photosynthetic green bacterium *Chloroflexus aurantiacus*. **b** Proposed structural models for the oligomeric organization of antenna BChl *c* in chlorosomes. *i*, asymmetric 6-coordinate model; *ii*, antiparallel-chain model with 5-coordinate Mg but involving both the 2-hydroxyethyl and 9-keto groups

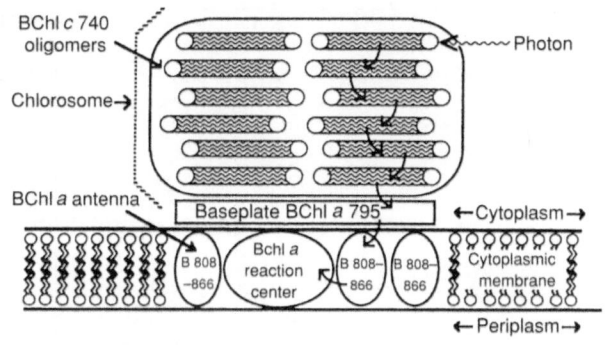

oligomers are shown in Fig. 4.1-7b. This organizational principle is in sharp contrast to the situation in other photosynthetic antennas, where pigment–protein interactions are of dominant importance in determining the structural arrangement of the chromophores, and direct pigment–pigment contacts are rare.

Chloroflexus Aurantiacus. is a thermophilic filamentous green photosynthetic bacterium found in hot springs throughout the world [17]. Its absorption and fluorescence spectra are shown in Fig. 4.1-8. The major antenna pigment is BChl *c*, with an absorption maximum located at 740 nm (see Fig. 4.1-7a). The BChl *a* in the chlorosome absorbs maximally at 795 nm (B795) and is located in the baseplate, i.e., the region of the chlorosome in contact with the cytoplasmic membrane [14]. It is a minor component (BChl *c* 740:B795 = 25:1) and is not apparent in the absorption spectrum. The two other absorption maxima at 808 and 866 nm are due to an integral membrane BChl *a* pigment–protein complex called B808–866. Reaction-center absorption is not discernible in the spectra of membranes, but it is suggested that the reaction-center photoactive pigment P865 exhibits an absorption at 865 nm [18].

Excitation energy flow in *C. aurantiacus* has been studied by steady-state and time-resolved fluorescence spectroscopy [17]. In the steady fluorescence spectrum (Fig. 4.1-8), when BChl *c* is excited, most of the emission is observed at 883 nm and arises from B866. Weak fluorescence maxima are observed at 750 and

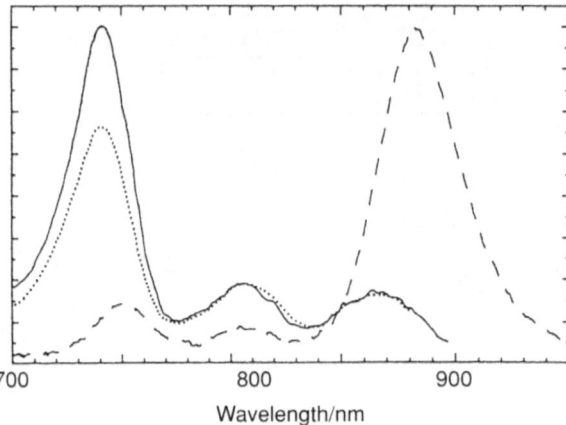

FIG. 4.1-8. Steady-state spectra of *C. aurantiacus* intact cells at 50°C: *solid line*, absorption; *dashed line*, fluorescence emission with excitation at 460 nm; *dotted line*, fluorescence excitation of 900-nm emission

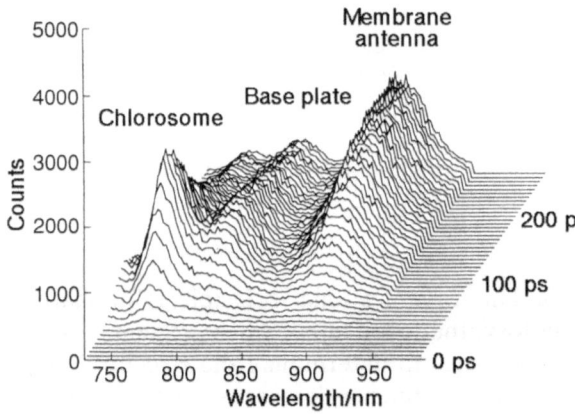

FIG. 4.1-9. Time-resolved fluorescence spectra of *C. aurantiacus* intact cells at 50°C, obtained with 2-ps laser excitation at 715 nm

805 nm, arising from BChl *c* and B795, respectively. No emission from B808 is observed in measurements on whole cells due to an extremly rapid energy transfer to B866. The time-resolved fluorescence spectra are shown in Fig. 4.1-9. Following excitation of BChl *c* at 715 nm, BChl *c* fluorescence appears at 750 nm just after the excitation pulse, the second fluorescence band of BChl *a* in the baseplate appears at 800 nm after 25 ps but shifts to the red (810 nm) within 50 ps, and the third fluorescence band of B808–866 appears in the membrane at 883 nm. Reaction-center fluorescence is extremely weak and was not detected.

The decay time of the BChl *c* emission is 16 ps. The apparent rise time of the baseplate emission is <3 ps. For the transfer from the baseplate BChl *a* to the membrane-bound BChl *a* complex, the baseplate BChl *a* emission decays in 41 ps, which corresponds to the rise time found for the membrane-bound BChl *a* emission. From kinetic analysis, a model is proposed for the structure and func-

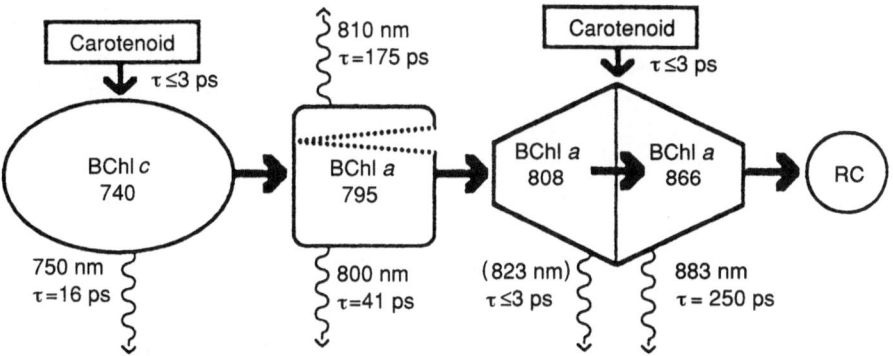

FIG. 4.1-10. Model for the energy flow in the antenna complexes of *C. aurantiacus*. The *broken lines* separate groups of pigments of the same type of antenna complex but with different kinetics. *RC*, reaction center

tion of the chlorosome antenna system [17] as shown in Fig. 4.1-10. The data suggest that the excitation transfer process may utilize a novel mechanism that takes advantage of the photophysical properties of aggregated pigments. BChl *c* molecules form naturally occuring aggregates with oligomeric structures similar to *J* aggregates, but which are very different from the organization of antenna pigment-proteins from other photosynthetic organisms. These oligomers absorb light and transfer excitation energy to a small amount of BChl *a* antenna protein in the baseplate. The baseplate acts as an interface in the energy transfer between the chlorosome and the antenna protein complexes located within the membrane. The integral membrane antenna complexes in turn deliver the excitations to the reaction center, where photosynthesis is initiated by electron transfer reactions.

Chlorophyll Antennas Surrounding the Reaction Center. The excitation energy transfer in the photosynthetic antenna occurs very rapidly, particularly in the antenna pigments located nearby the reaction center inside the thylakoid membrane. Recent studies by X-ray and/or electron crystallography revealed the molecular structure of the bacterial light-harvesting complex LH1 and LH2 [19, 20]. The structure of LH2 of the purple bacterium *Rhodopseudomonas acidophila* has been demonstrated by McDermott et al. [20] and is shown in Fig. 4.1-11. It has nine identical units, each consisting of two fairly short α-helical polypeptides and associated pigment molecules, combine into a ring. The helices are perpendicular to the plane of the membrane, or slightly tilted, as was found in three other α-helical membrane proteins. Within one unit, each pair of polypeptides coordinates three bacteriochlorophylls and one carotenoid. Two bacteriochlorophylls are held near the periplasmic, carboxy terminal end of the helices, and one is suspended between the helices near the middle. Within the complex, there are thus two rings of bacteriochlorophylls, one set of 18 close to

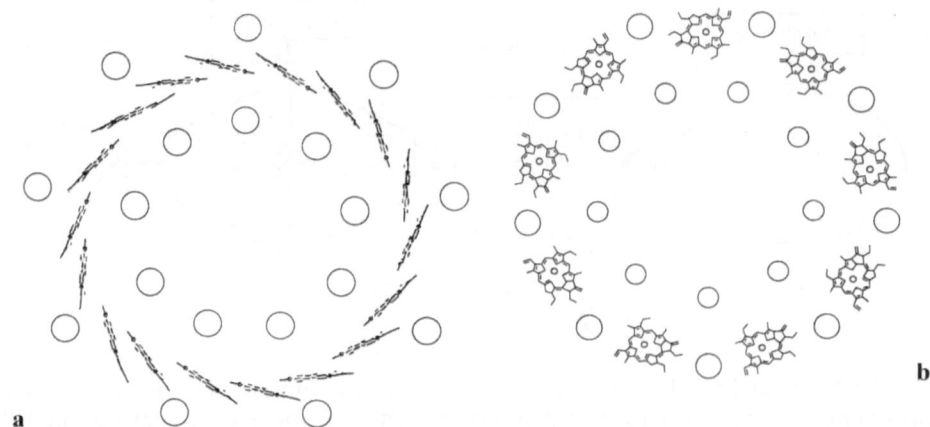

Fig. 4.1-11. Molecular arrangement in the antenna protein LH2 of the purple bacterium *R. acidophila*. **a** The ring of B850 Bchl *a* molecules. **b** The positions of the B800 Bchl *a* molecules between the helices of the β-apoproteins forming the outer cylindrical wall of the complex. The α-apoproteins are colored yellow and the β-apoproteins are green

the membrane surface, arranged as in the wheel of a turbine, and another set of nine in the middle of the bilayer.

However, a projection map of LH1 by Karrasch et al. [19] at 8.5 Å resolution shows that this complex also has a ring-like structure, and is built up from the same type of molecular units as LH2. The LH1 ring contains 16 rather than nine units, and is therefore considerably wider to accommodate the reaction center. Kühlbrandt [21] proposed a model of the excitation energy transfer in the membrane, as shown in Fig. 4.1-12. When excitation energy reaches one of the chlorophylls, it spreads extremely rapidly (in less than one picosecond) to the others in the ring as a result of their favorable spacing and orientation. Where the rings touch in the close-packed membrane, the energy can easily jump the short distance to an adjacent complex where it again spreads into the ring. LH1 absorbs at a longer wavelength, and hence lower energy, than LH2. It serves as an energy funnel for the reaction center which, because it has the most red-shifted absorption maximum, acts as an energy sink. In this way, the energy contained in a single photon is transmitted in a very short time, and with minimal loss, from the point where it is absorbed to the point where it is needed.

In time-resolved experiments, van Mousik et al. [22] demonstrated that equilibration of the initial distribution of excitation energy occurs on a timescale of a few hundred femtoseconds in the light-harvesting antenna of purple bacteria. Similarly, Du et al. [23] observed ultrafast energy transfer processes for LHC I and II of higher plants by measuring time-dependent fluorescence depolarization. In both cases, depolarization occurs on a timescale of 150–300 fs. Depolarization of the fluorescence results from excitation energy jumping from the

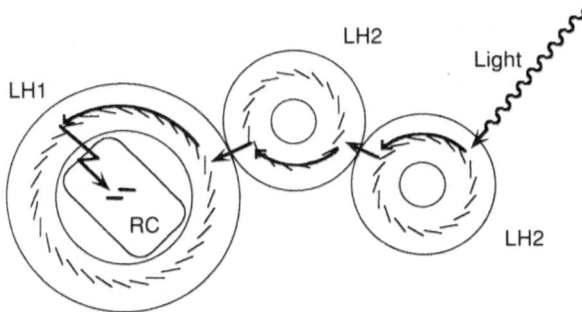

FIG. 4.1-12. Model of the excitation energy flow in the bacterial light-harvesting complexes LH2 and LH1 to the special pair of bacteriochlorophylls in the reaction center. The diagram shows a view from above the membrane

initially excited chlorophyll molecule to other molecules with different spatial orientations. In the LHC II system, further depolarization is observed on a 5 ps timescale, possibly corresponding to energy transfer between different units of the whole assembly. Effective localization of excitation energy on the reddest fraction of the piments causes the dynamic red shift of the spectrum and the increase in excited state absorption.

For these short distances and fast rates, Fleming and Van Grondelle [1] pointed out that a Förster-type mechanism described in terms of the point–dipole approximation has only limited applicability, as in the case of electron transfer, and that the excitation energy transfer may occur from a vibrationally unrelaxed state. In addition, in densely packed chlorophyll protein systems, excitonic interactions between pigments leads to a distinct splitting of the excited state energies. In this case it is more appropriate to describe the short time-period of the energy transfer as a relaxation between the different exciton levels of the system. Thus, energy transfer in the complete photosynthetic apparatus spans a time-range of three orders of magnitude, from perhaps a few tens of femtoseconds up to hundreds of picoseconds. The process may initially involve coherent migration in which the excitation energy is delocalized over several molecules, while the longer timescales correspond to incoherent hopping from molecule to molecule.

4.1.1.3 Phytochrome: Reversible Phototransformation Regulating Photo-morphogenesis in Plants

Phytochromes in plants mediate light signal transduction processes such as the expression of light-responsive genes, seed germination, stem growth, and many other developmental and morphogenic photoresponses [24, 25]. Phytochromes function according to a reversible photoisomerization (photochromic) reaction (Eq. 4.1.5) between the two forms of phytochromes, Pr and Pfr.

$$\text{Pr} \xrightarrow[h_\nu(\text{far-red})]{h_\nu(\text{red})} \text{Pfr} \longrightarrow \begin{array}{c} \text{Responses} \\ \left(\text{e.g. gene expression}\right) \end{array} \qquad (4.1\text{-}8)$$

The absorption spectra of the two forms are shown in Fig. 4.1-13. Accompanying the phototransformation from the Pr to the Pfr form, there is a red shift of the absorption band maximum from 666 to 730 nm. The functional chromophore in phytochrome is a tetrapyrrole with semi-extended conformation, The phototransformation mechanism is illustrated schematically in Fig. 4.1-14. The overall conformation of the tetrapyrrolic chromophore is retained during the transformation, with conservation of the exocyclic dihedral angle at ring D by a chromophore–apoprotein interaction.

The representative phytochrome is phytochrome A (Phy A), which is the most abundant form involved in hypocotyl growth/inhibition in etiolated plant tissues

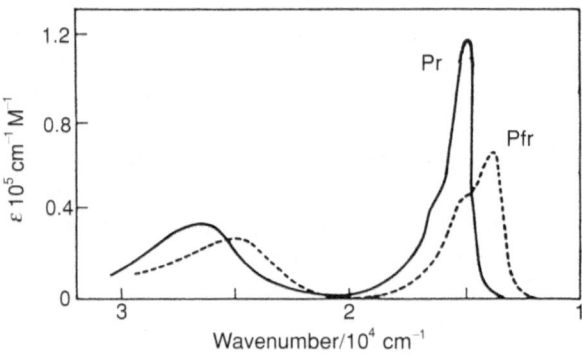

FIG. 4.1-13. Absorption spectra of 124-kDa oat phytochrome at room temperature. *Solid line,* Pr; *broken line,* Pfr (87% at photostationary equilibrium reached upon red light irradiation)

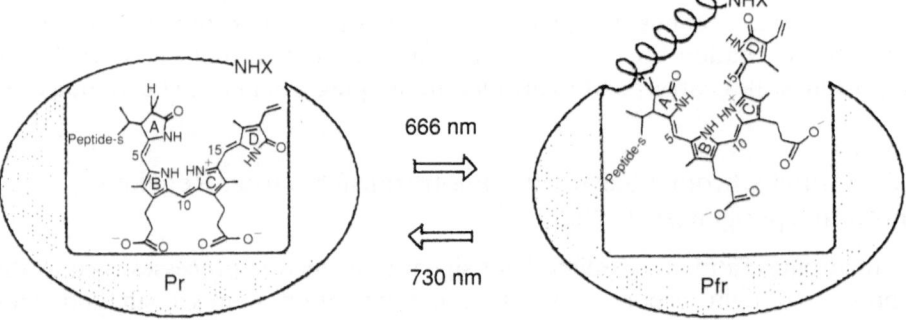

FIG. 4.1-14. Phototransformation between the Pr and Pfr forms of tetrapyrrolic chromophore in phytochrome

[26]. The molecular weights of phytochromes from different plant species are in the range 121 000–129 000 daltons (Da): e.g., 124 kDa *Avena* protein is a typical monocot phytochrome species (hereafter referred to as phytochrome unless specified otherwise), and 121 kDa *Pisum* protein is a representative dicot phytochrome species. Using current knowledge of the primary structures of several phytochrome species, the secondary structure of phytochrome A was analyzed by Song and co-workers [27, 28]. Striking features are extensive α-helical folding and apparent lack of β-sheet conformation in native holoproteins (both Pr and Pfr forms).

Light Activation of Phytochrome. Pr → Pfr phtotransformation involves Z(15) → E(15) isomerization [29], which occurs within a few picoseconds [30]. Quantum efficiencies for the Pr → Pfr phototransformation of phytochrome are 15–17%, and those for the Pfr → Pr photoreversion are 6–7%. Although the forward reaction is fairly efficient, the quantum efficiency of the reverse reaction is significantly lower than that for the primary reaction, leading to the formation of some intermediates. This suggests that the Pr species is regenerated from these intermediates via phtochemical and thermal routes, thus partially accounting for the relatively low quantum efficiencies for the Pr → Pfr phototransformation.

Numerous photolysis studies have been carried out with various phytochrome preparations. Fig. 4.1-15 shows the generally accepted scheme of phytochrome phototransformation derived from such studies. However, a recent 100 ns to 800 ms time-resolved spectroscopic study of Pr → Pfr phototransformation kinetics, using 7-ns laser pulse excitation and transient absorbance difference measurements over the UV–visible range, revealed more complex kinetics at 283 K for the forward reaction pathway than those represented in Fig. 4.1-15. [31] Thus, a global analysis fitting of the time-resolved absorption spectra of phytochrome entails at least five kinetic intermediates.

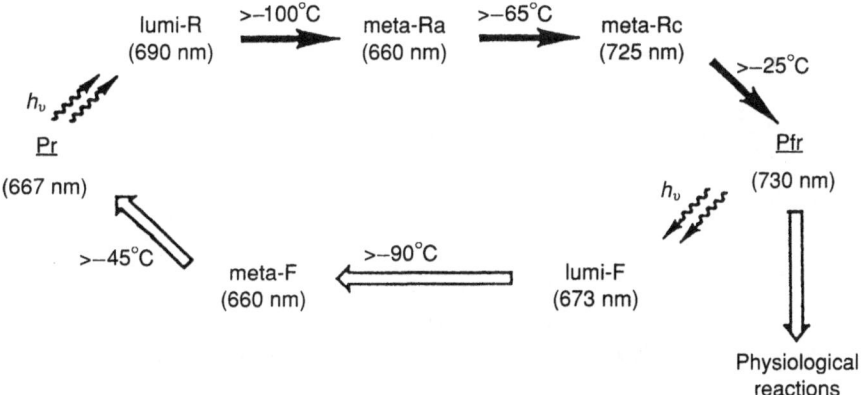

Fig. 4.1-15. Reaction scheme of reversible phototransformation Pr ↔ Pfr through intermediates determined from time-resolved spectroscopy

4.1.1.4 Photomechanical Responses in Unicellular Ciliates

The unicellular ciliary protozoans *Stentor coeruleus* and *Blepharisma japonicum* exhibit photophobic and phototactic (light-avoiding) responses to visible light stimuli. *S. coeruleus* is a typical example of a unicellular organism, and trajectories of its movements are shown in Fig. 4.1-16. The primary photoreceptor for the step-up photophobic and negative phototactic responses is a chromoprotein called stentorin [32, 33]. Stentorin forms a large molecular assembly (stentorin-2, up to 800kDa) comprised of the 50kDa protein and additional non-chromophore-bearing protein subunits (stentorin-2A), These chromoproteins are localized in the pigment granules (0.3–0.7 µm diameter) of the cell, and apparently bound to the membrane. The chromophore of stentorin is identified as a derivative of hypericin, octahydroxy diisopropylnaphthodianthrone, with two possible structures [34].

Thus the ciliate photosensor pigments represent a novel class of photoreceptor molecules with a chromophore structure significantly different from the other well-known photoreceptors listed in Table 4.1-1.

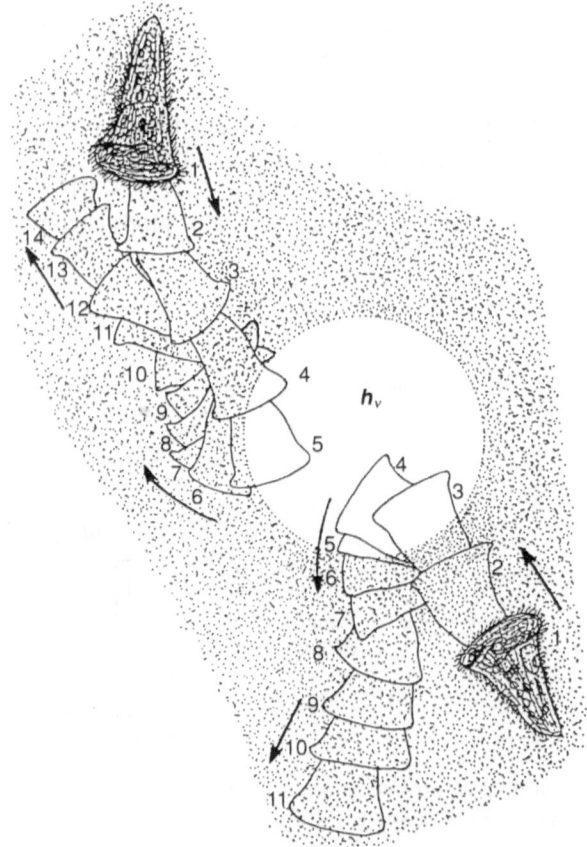

FIG. 4.1-16. Trajectories of the movement of the unicellular ciliary protozoan *S. coeruleus*. The cells exhibit a stop–turn response upon encountering an illuminated area (*center*)

FIG. 4.1-17. Tentative scheme for a step-up photophobic response in *S. coeruleus*. The light signal cascade is represented by four steps

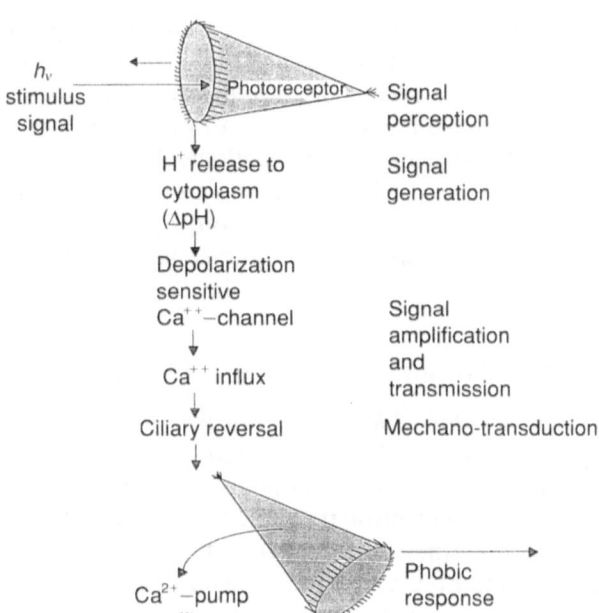

The reaction scheme is shown in Fig. 4.1-17. The stimulus light signal in *Stentor* and *Blepharisma* is transduced in the form of an intracellular pH gradient, which is eventually amplified by a transient influx of calcium ions into the cell. The transient ΔpH generated may activate the Ca^{2+} influx through voltage-sensitive Ca^{2+} channels, eventually eliciting the ciliary stroke reversal and stop–reverse swimming. Preliminary studies also suggest that the photosignal transduction in both organisms utilizes G-protein(s) as an initial transducer and a cGMP-phosphodiesterase as the effector system, which is analogous to the visual system of higher animals [35].

Time-resolved fluorescence decay studies of the stentorin and blepharismin chromoproteins indicated that a primary event occurs from the excited singlet state of the photoreceptor molecules within a short time period: a few picoseconds for stentorin, and 200–500 PS for blepharismin [36]. The ultrafast process in stentorin was confirmed by a recent pump–probe spectroscopy study, which also suggested that intermolecular proton transfer to an appropriately situated amino acid residue(s) of the apoprotein represents the radiationless decay mode in the primary photoprocess of the pigment molecule [37].

4.1.2 Artificial Molecular Systems of Photoinduced Energy Conversion

As is seen in the previous section, the energy conversions in biological molecular systems are characterized by significantly high efficiency and a fast reaction rate.

These features may come from sequential and cooperative reactions among functional molecules linked in polypeptide chains. There have been numerous attempts to fabricate organized molecular arrangements similar to biological molecular systems with a view to obtaining a deeper understanding of the sequential and cooperative reactions. This work has been done by synthesizing supramolecules or by assembling molecular modules on a two-dimensional base to form three-dimensional structures [3, 38]. This section reviews recent studies on electron transfer and excitation energy transfer in synthesized supramolecules and Langmuir–Blodgett (LB) films.

4.1.2.1 Photoinduced Electron Transfer

Intramolecular Electron Transfer in Donor-Acceptor Linked Molecular Systems. Covalently linked electron donor–acceptor molecules have been extensively studied in order to obtain information about the factors governing interchromophore electron transfer. An important requirement for these molecules to become an efficient photocatalyst for charge separation is that they allow very fast photoinduced electron transfer, and minimize the thermally wasteful backward transfer. Such a molecular organization is actually realized in the bacterial photosynthetic reaction center.

Osuka et al. [39] investigated intramolecular electron transfer by using specially designed molecular systems consisting of zinc porphyrin (ZnP) as an electron donor and ferric porphyrin (Fe(III)P) as an acceptor. They synthesized a series of molecules, as shown in Fig. 4.1-18, in which donor and acceptor molecules were linked with spacer groups of different lengths, and examined the distance-dependence of the electron transfer rate. The fluorescence spectra of ZnP–Fe(III)P are the same as those of the donor, unperturbed ZnP, but their fluorescence quantum yields decrease significantly due to photoinduced electron transfer from the singlet excited state of ZnP to Fe(III)P. Time-resolved fluorescence and absorption measurements revealed the ultrafast reaction dynamics (in the picosecond time range). In the transient absorption spectra of P1 in DMF, for example, a $S_n \leftarrow S_1$ absorption of ZnP at 460 nm decays with a lifetime $\tau = 50$ ps, followed by a rise of the broad absorption in the 480–510 nm and 600–750 nm regions which can be ascribed to the formation of ZnP^+–Fe(II)P. The time constant of the absorption decay agreed with the fluorescence lifetime of the ZnP ($\tau = 50$ ps). At longer time region (3–5 ns), the absorption of ZnP–Fe(II)P decayed slowly by charge recombination. By analyzing these data, the rate constants for charge separation (k_{CS}) and recombination (k_{CR}) were determined in a series of diporphyrin molecules. Figure 4.1-19 shows plots of the rate constants k_{CS} and k_{CR} as a function of the center-to-center distance, r, between two porphyrins.

It can be seen in Fig. 4.1-19 that the k_{CS} value decreases with increasing r. The results fit an electron tunneling equation,

$$k_{CS} = A_0 \exp(-\beta r) \qquad (4.1\text{-}9)$$

FIG. 4.1-18. Structure of ZnP–Fe(III)P hybrid diporphyrins. Abbreviations are indicated to the *left* of the aromatic spacers

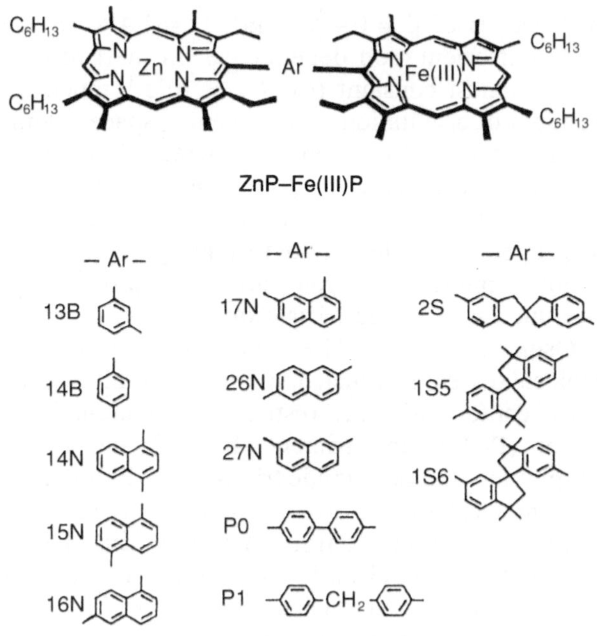

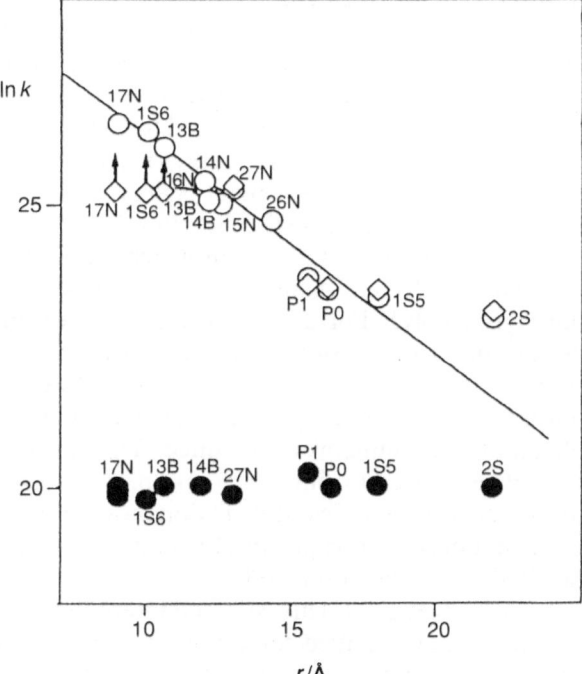

ZnP–Fe(III)P

FIG. 4.1-19. Plots of k_{cs} and k_{cr} vs. the center-to-center distance, r, of two porphyrins. *Open circles*, k_{cs} determined by the fluorescence lifetime of ZnP in CH$_2$Cl$_2$; *open squares*, k_{cs} determined by transient absorption measurements in DMF; *solid circles*, k_{cr} determined by transient absorption measurements in DMF

with values of $\beta = 0.4\,\text{Å}^{-1}$ and $A_0 = 1.4 \times 10^{13}\,\text{s}^{-1}$. Surprisingly, k_{CS} shows no apparent orientation dependence. In marked contrast, and more surprisingly, k_{CR} is almost constant (ca. 4–6 $\times\ 10^8\,\text{s}^{-1}$, much smaller than k_{CS}), and is independent of orientation, of intervening spacers, and even of distance in the range $23\,\text{Å} > r > 8\,\text{Å}$. As a result, the increase in the ratio k_{CS}/k_{CR} is brought about by decreasing the center-to-center distance. For example, the value of the ratio k_{CS}/k_{CR} increases ca. 36 times from 18 in 2S to 650 in 1S6. The constant nature of k_{CR} suggests that the rate-determining step in the charge-recombination process is not an intersite electron transfer but a reaction within the Fe(II)PCl complex site, such as a change of ligation.

Osuka et al. [40, 41] extended this study by synthesizing highly organized molecular systems which had a greater similarity to the reaction center in a plant: (1) a conformationally restricted quinone-linked porphyrin monomer (P–Q), dimer (P$_2$–Q), and trimers (P$_2$–P–Q and P$_2$–P–P–Q), and (2) carotenoid–porphyrin–pyromellitimide triads (C–H$_2$P–Im). Several examples of their synthesized molecules are shown in Fig. 4.1-20. In the first group, the monomeric model exhibits very rapid charge separation and charge recombination, the dimeric model shows significantly suppressed charge recombination, and the trimeric model undergoes an efficient quenching of the excited singlet state of the coplanar diporphyrin by the attached quinone and long-lived charge-separated states. These behaviors are regarded as electron transfer from the diporphyrin part to the monomeric porphyrin part competing with the rapid charge recombination reaction. In the second group, interchromophore energy transfer and electron transfer were investigated. Excitation of C at 532 nm led to formation of C$^+$–H$_2$P–Im$^-$ within several tens of picoseconds. This ultrafast formation was interpreted in terms of long-distance electron transfer from C* to Im mediated by a superexchange interaction involving the π-electronic orbital of the intervening H$_2$P. Selective excitation of the triad to C–H$_2$P*–Im at 585 nm led to much slower formation of C$^+$–H$_2$P–Im$^-$, probably via a C*–H$_2$P–Im state which may be formed by intramolecular singlet–singlet energy transfer.

Segawa et al. [42] investigated photoinduced electron transfer with a view to synthesizing a molecular wire in which an electron is transferred long-distance through a molecular channel. They synthesized oligothiophenes coupled with porphyrin, Poly1, Poly2, and Poly3, as shown in Fig. 4.1-21. Porphyrins were used as the photosensitizer, because they have strong oxidizing powers and are stable to oxidation. After polymerization of Poly1, porphyrin polymers in which porphyrins take a face-to-face configuration and are separated by ordered oligothiophene units were obtained. The conductivity of Poly2 and Poly3 was strongly enhanced by photoirradiation. It is suggested that photoinduced carrier formation occurs efficiently in donor–acceptor polymers. In this study, they found not only appreciable photoconductivity, but also nonlinear optical properties in the polymers obtained.

Yonemura et al. [43] investigated interchromophore electron transfer by using a donor–acceptor linked compound in which the conformation was restricted by complex formation with α- and β-cyclodextrins (CD). The compound consists of

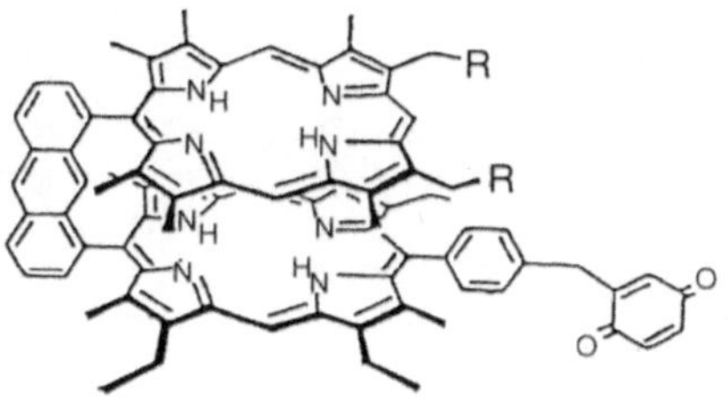

P_2–Q

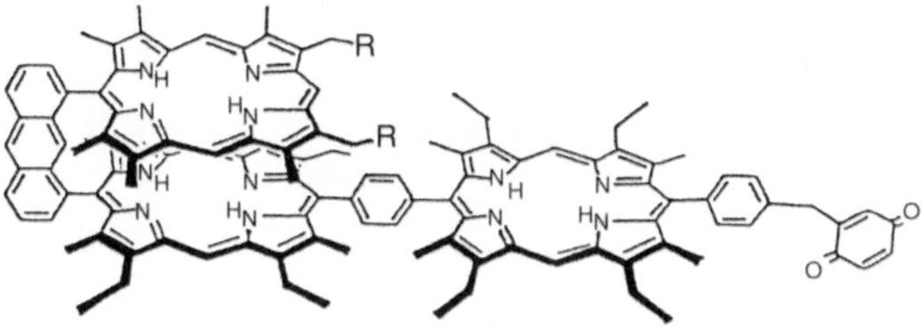

P_2–P–Q

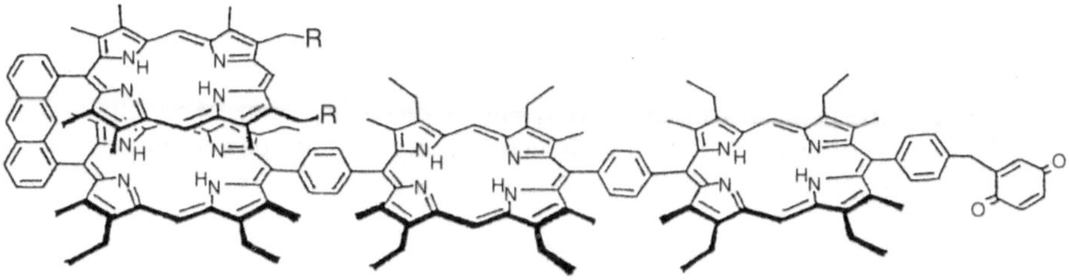

P_2–P–P–Q

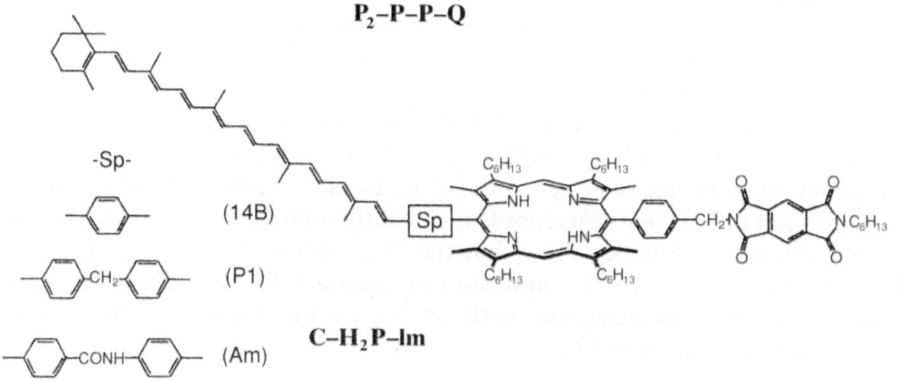

C–H_2P–Im

FIG. 4.1-20. Molecular systems of excitation energy and electron transport synthesized by Osuka et al. [39–41]

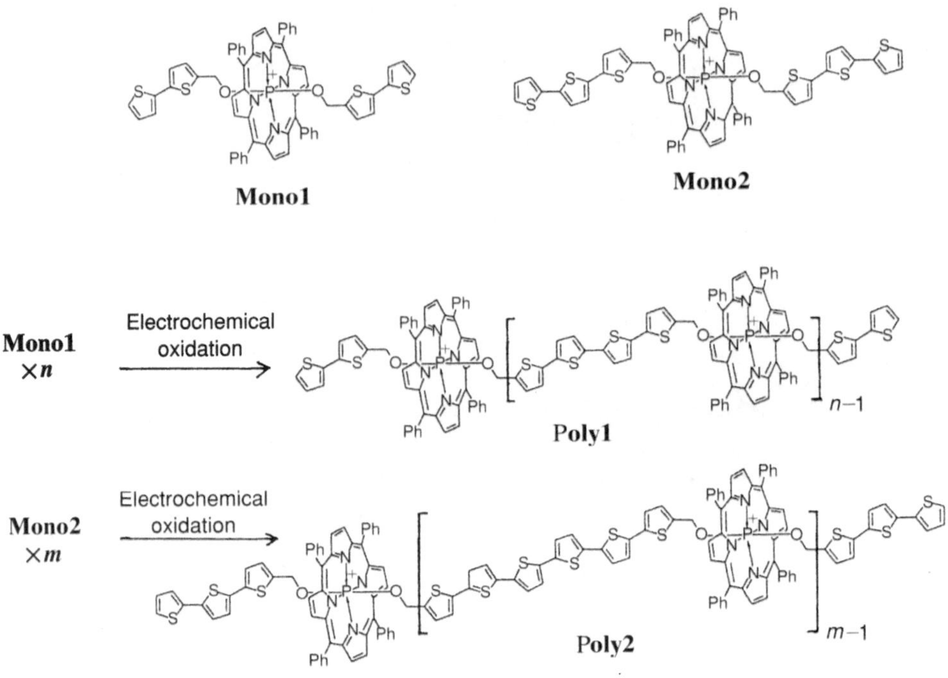

Mono1   **Mono2**

Mono1 **Electrochemical oxidation** → **Poly1**
×*n*

Mono2 **Electrochemical oxidation** → **Poly2**
×*m*

FIG. 4.1-21. Electrochemical synthesis of polymeric molecular systems of one-dimensional electron transfer, consisting of oligothiophenes and phosphorous porphyrins

carbazole as donor and viologen as acceptor linked with the spacers, alkyl chain, and/or biphenyl group. In an acetonitrile solution containing α- or β-CD, compounds with a long alkyl chain form a stable complex with cyclodextrin in which the spacer chain is encased in the cavity of CD. Model structures are shown in Fig. 4.1-22. A spatial restriction resulting from the complex formation with CD arranges the donor and acceptor in a linear configuration. The carbazole–viologen linked compounds generate a radical pair, effectively following laser excitation in the presence of either α- or β-CD. In the case of α-CD, the yield of photogenerated radical pairs increased significantly relative to the case of β-CD. Transient absorption kinetics show that the intramolecular radical pair forms with a rate constant of $\sim 10^{-8}\,\mathrm{s}^{-1}$, and the rate constant for the reverse process (charge recombination) decreases with an increasing number of methylene groups as spacers (from 6 to 12).

Interlayer Electron Transfer in LB Multilayer Films. A LB film is a mono- or multi-layered molecular assembly which is prepared by transferring a compressed monolayer spread on a water surface onto a substrate [44]. One can prepare a LB film in a molecular stacking architecture containing functional chromophores, with their number densities and interlayer distances being variable over a wide range. Thus a LB film provides us with a unique molecular

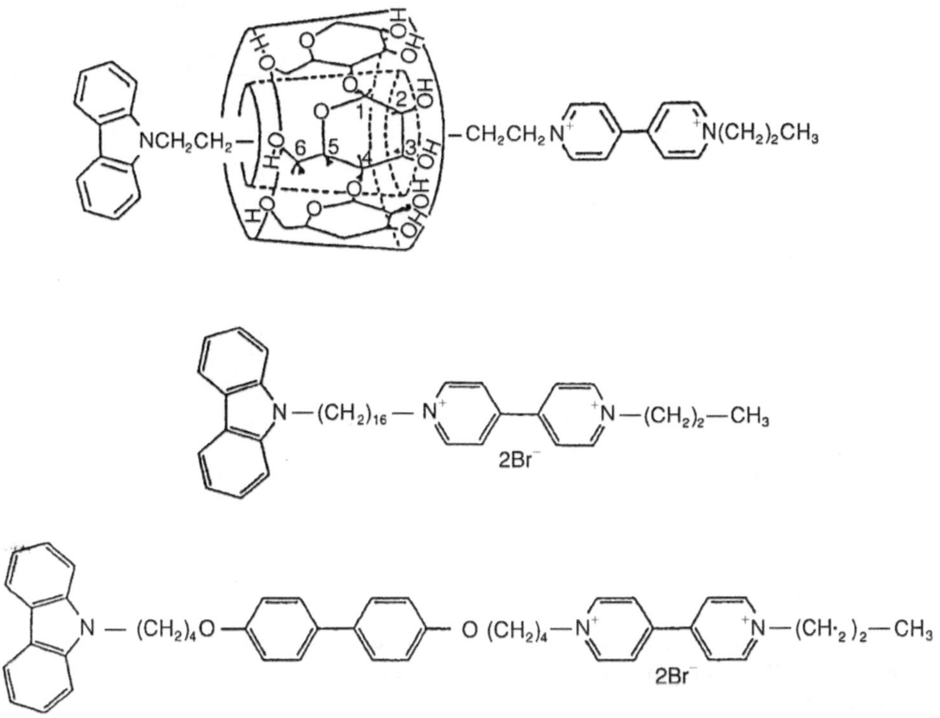

Fig. 4.1-22. Geometrically restricted molecular system of electron transfer enclosed by cyclodextrin

system for investigating sequential photophysical processes. In the 1970s, Prof. H. Kuhn's group [3, 44] extensively studied photophysical processes in LB films, particularly for photoinduced electron transfer and excitation energy transfer. This section reviews their work and othes recent studies of LB films.

The study of photoinduced electron transfer in LB films was started with a three-layer system containing in each layer carbocyanine as an electron donor (D), fatty acid as a spacer, and viologen as an acceptor (A) [45]. The stacking structure of a multilayer film is illustrated in Fig. 4.1-23. Hereafter, this system is referred to as System I. In the donor layer, the concentration of carbocyanine is sufficiently low that the dye is predominantly in the monomer form and is strongly fluorescent. In the spacer layer, the length of the methylene chain in the fatty acid was varied to change the D–A distance. The sample is excited with UV light absorbed by the dye. The fluorescence is quenched from original intensity I_0 to intensity I. The ratio $(I_0–I)/I$ gives the ratio k_{ET}/k, where k_{ET} is the rate of electron transfer from D to A, and k is the decay rate of the dye. Then the electron transfer rate can be derived from the fluorescence intensity, which should decrease exponentially with the interlayer distance d in the case of tunnel-

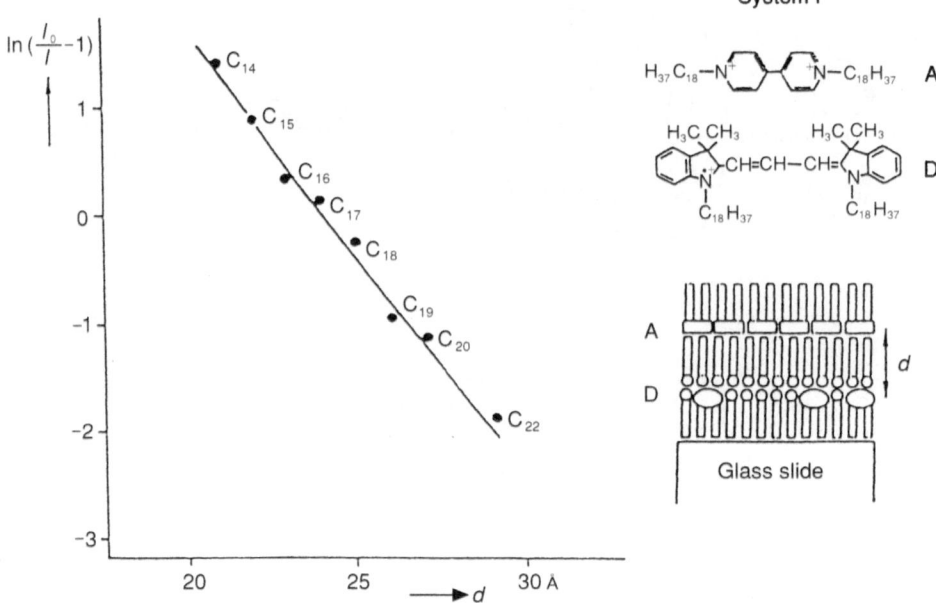

FIG. 4.1-23. Oriented molecular system of electron donor and acceptor incorporated in LB multilayer films, and plots of fluorescence intensity change vs. distance between donor and acceptor

ing. Figure 4.1-23 shows that the plot of $\ln[(I_0-I)/I]$ is indeed on a straight line against d, and the following equation of electron tunneling holds:

$$k_{ET} = A_0 \, e^{-\beta d} \qquad (4.1\text{-}10)$$

with $A_0 = 10^{13}\,\mathrm{s^{-1}}$ and $\beta = 0.41\,\text{Å}^{-1}$. The radical cation A^- is actually building up when it is illuminating the sample, but with a small quantum yield (1%). This follows from the appearance of the absorption band of the viologen radical. The electron in A^- is then usually tunneling backwards to D^+, forming D. Otherwise, A^- is stabilized with a small probability by environmental reorganization.

Subsequent studies revealed that the relation is Eq. 4.1.10 can be applied generally to electron transfer in organized molecular systems. Figure 4.1-24a shows two sets of experimental results using thiacarbocyanine (D)–viologen (A), and thiacarbocyanine (D)–quinolinium (A) [45]. Hereafter, these are referred to as Systems II and III, respectively. Figure 4.1-24b shows the result obtained by Mooney and Whitten [46] for donor stilbene (System IV). In this case, molecules were synthesized with the stilbene chromophore built into the hydrocarbon chain of a fatty acid at various positions, and the molecules were incorporated in a fatty acid monolayer. Thus, in this case the transition moment of the donor is perpendicular to the layer plane, while in the previous three cases of Systems I, II, and III it is parallel.

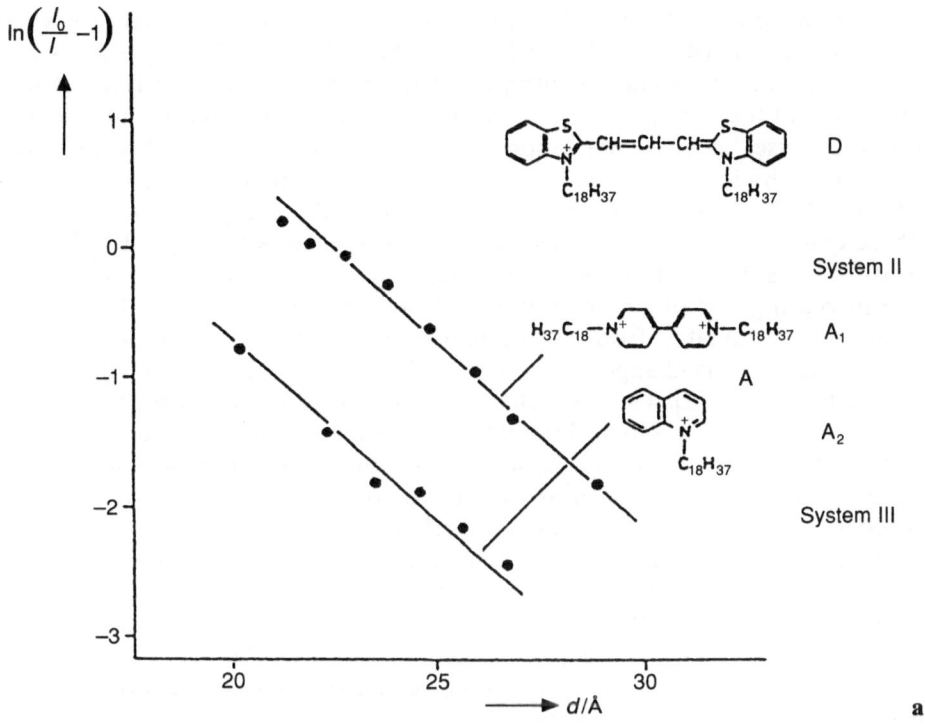

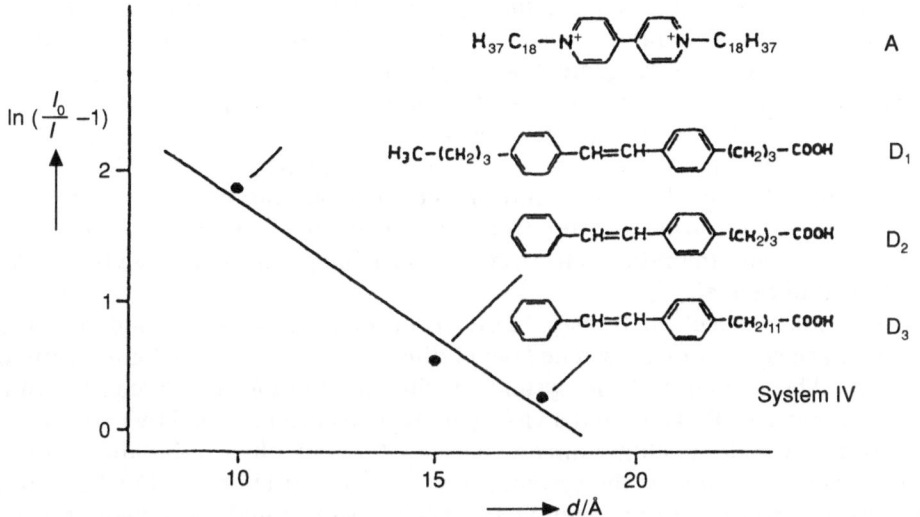

FIG. 4.1-24. Plots of fluorescence intensity changes associated with one-directional electron transfer as a function of D–A distance in LB films. The stacking structure is the same as in Fig. 4.1-23

As seen in Figs. 4.1-23 and 4.1-24, the value of β in Eq. 4.1.10 decreases on going from System I ($\beta = 0.41\,\text{Å}^{-1}$) to Systems II and III ($\beta = 0.27\,\text{Å}^{-1}$) and System IV ($\beta = 0.22\,\text{Å}^{-1}$). This can be interpreted from the fact that the level of the excited state of the donor, as estimated from the oxidation potential and excitation energy, rises in the same sequence, and the barrier height therefore decreases. The driving force (decrease in free energy in the electron transfer reaction ($-\Delta G^0$)) increases in this sequence, since the acceptor is identical in all three cases. The value of A_0 decreases from $A_0 = 10^{13}\,\text{s}^{-1}$ to $A_0 = 10^{12}\,\text{s}^{-1}$ and $A_0 = 10^{10}\,\text{s}^{-1}$ in going from System I to System IV. This is explained as being due to the increasing loss of free energy ($-\Delta G^0 = 0.7\,\text{eV}$ in System I and $1.7\,\text{eV}$ in System IV, as estimated from the reduction potential and excitation energy). When the acceptor is changed, the height of the barrier does not change, and therefore factor β in Eq. 4.1.10 should not change. This is indeed observed as we substitute the acceptor bipyridinium for chinolinium (Fig. 4.1-24a). The preexponential factor A_0 decreases by a factor of three, and when bipyridinium is replaced by pyridinium it diseases by a factor of 50. Both changes can be accounted for as being due to changes in the driving force and vibronic coupling [3].

These results show that electron tunneling can occur over distances of the order of 20 Å, except that efficient removal of the electron from the excited dye molecule is not achieved. The acceptor should be at a greater distance to avoid back-reaction, but then the removal of the electron from the dye is too slow to compete with fluorescence and radiationless deactivation. The energy barrier between donor and acceptor should be lowered by bringing in a π-electron system in the space between donor and acceptor. The π-electron system acts as a molecular wire [3]. An attempt to realize this idea is shown in Fig. 4.1-25 [3]. Polythiophenes (W_1, W_2) are incorporated between the carbocyanine (D) and viologen (A). The fluorescence of D is strongly quenched ($I/I_0 = 0.4$), while in the corresponding arrangement where D and A are separated by two pure arachidic acid layers, no quenching of fluorescence takes place. An acceptor radical A^- is slowly formed (quantum yield 1%). In the absence of A (arrangement D, W_1, W_2), the fluorescence of D is partly quenched ($I/I_0 = 0.6$). These findings indicate that the rate of intramolecular deactivation of D, the rate of deactivation by electron transfer from photoexcited D to W_1, W_2 and back-transfer to D^+, and the rate of electron transfer from photoexcited D via W_1, W_2 to A are about equal.

In relation to electron transfer, Polymeropoulos et al. [47] found that the photovoltage generation is switched depending on whether the UV light is turned on or off. The sequence of this system, as illustrated in Fig. 4.1-26, consists of a carbocyanine dye (D) layer and a viologen (W_2) layer sandwiched by two evaporated thin electrodes of aluminum and barium. A photovoltage is building up and disappearing reversibly as the system is photoexcited at D or not. Furthermore, the fatty acid monolayer between D and W_2 was substituted for a mixed monolayer of fatty acid and a π-electron system W_1 (Fig. 4.1-26b). The long axis of this π-electron system is parallel to the fatty acid chains. It acts as a molecular wire supporting electron transfer, as indicated by the higher photovoltage. In both

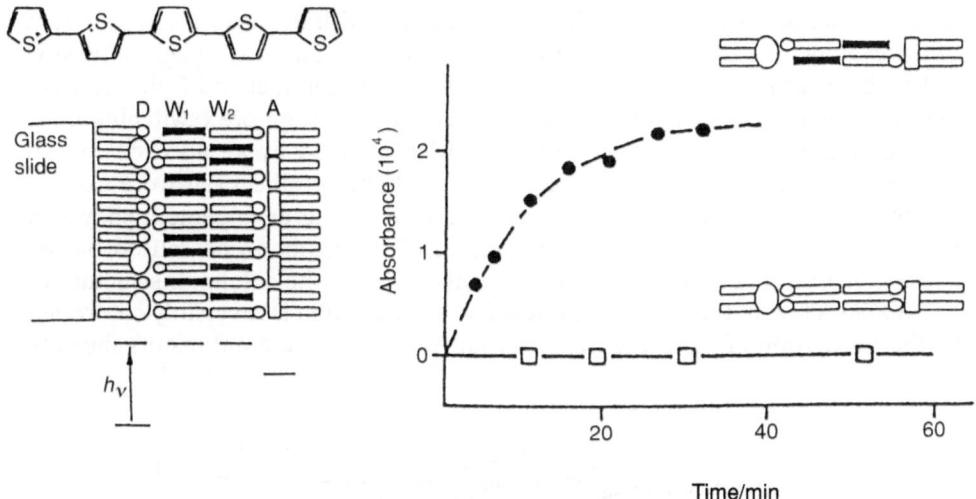

FIG. 4.1-25. Electron transfer molecular system containing a molecular wire. The donor (D) and acceptor (A) are separated by two mixed monolayers of fatty acid and polythiophene (W_1 and W_2) to act as the molecular wire. The plots are of the absorbance of radical A^- (at 400 nm) against time of illumination: *solid circles*, stacking multilayer containing polythiophene; *open squares*, multilayer without polythiophene

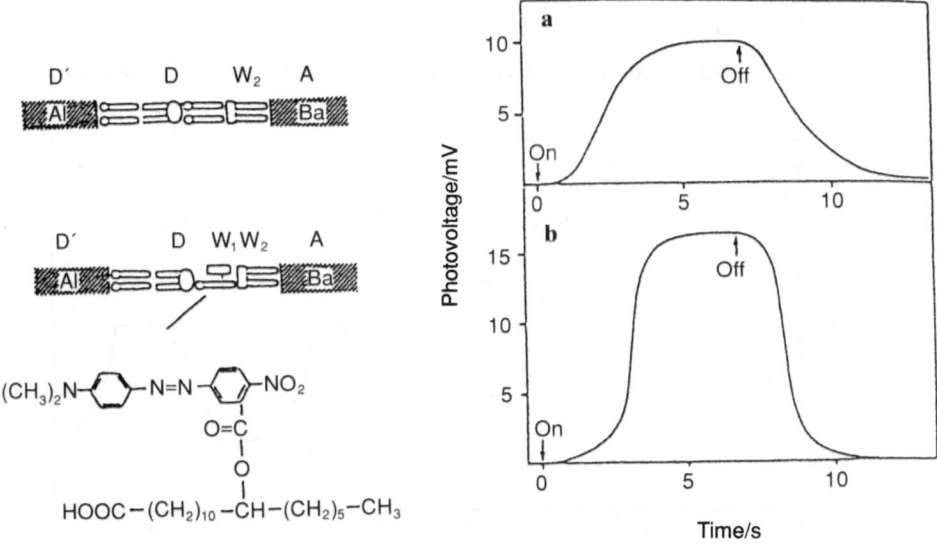

FIG. 4.1-26. Photovoltage in a LB multilayer between aluminium and barium electrodes. **a** D, carbocyanine, and W_2, bipyridinium dication; **b** multilayer assembly in **a** with an additional component of W_1 which conducts electrons to W_2. Photovoltage is enhanced

cases, the action spectra of the photovoltage agree with the absorption spectrum of the cyanine dye D, and the photovoltage is proportional to the light intensity.

Tachibana et al. [48] investigated a photoswitching conductive LB film consisting of linear molecules containing photochromic chromophore (switching unit) and conductive chromophore (working unit) connected by an alkyl chain. An azobenzene derivative and a pyridinium–TCNQ anion radical salt were used as the switching and working units, respectively. With the arrangement shown in Fig. 4.1-27, the *cis–trans* photoisomerization of the azobenzene caused by the irradiation of light induces a conformational change in the columnar structure of TCNQ. As a result, the lateral conductivity of the LB film is reversibly controlled by the irradiation of light. Figure 4.1-27 illustrates experimental results showing

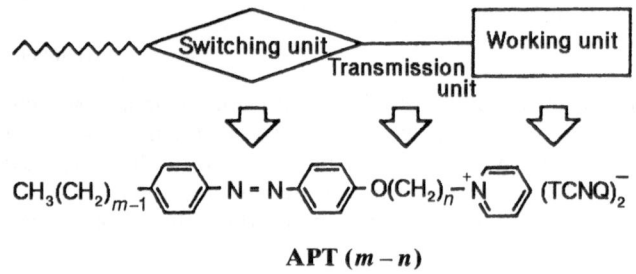

$$\text{APT } (m-n)$$

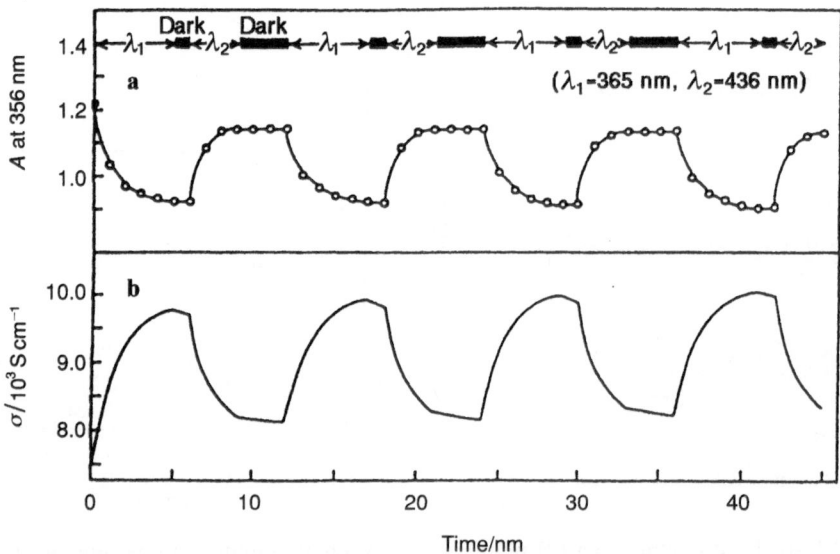

FIG. 4.1-27. Photoswitching conductive LB films. Conductivity is changed light irradiation which causes a conformation change of the columnar structure of TCNQ. **a** Absorbance (A) change at 356 nm due to the *trans*-azobenzene, and **b** conductivity (σ) change on alternating irradiation with UV (365 nm) and visible (436 nm) light

a change in the lateral conductivity of the LB film upon alternate irradiation with UV (365 nm) and visible (436 nm) light. The reversible changes can be repeated more than 100 times.

4.1.2.2 Excitation Energy Transfer

In the light-harvesting antenna of biological photoreceptors, photonic energy is transferred from molecule to molecule along a particular channel toward the reaction center. An artificial analogue of such a molecular channel can be obtained by means of LB multilayer films prepared by successive deposition of monomolecular layers contained dye molecules. As an example of these studies [11, 47], Yamazaki and co-workers [11, 49] demonstrated the dynamic behavior of the interlayer one-directional energy transfer by probing the fluorescence emitted from each layer. A model of the sequential energy transfer and the time course of the fluorescence spectrum are illustrated in Fig. 4.1-28.

Figure 4.1-29 shows the stacking structure of LB multilayer films. Several monolayers containing different dyes in each layer are stacked sequentially, as denoted by $N_1, N_2 \ldots$, in order of decreasing S_1 energy level of the dye molecule. The concentration of pigment molecules is 5 mol% in each layer, and the distance between adjacent layers is 25 Å. The Förster critical transfer distance for the

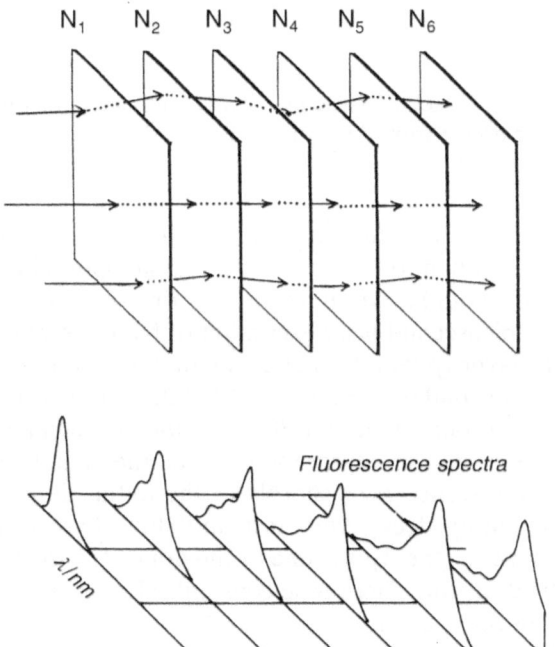

FIG. 4.1-28. Schematic model of sequential excitation energy transfer in LB multilayers, and time-resolved fluorescence spectra as a probing method

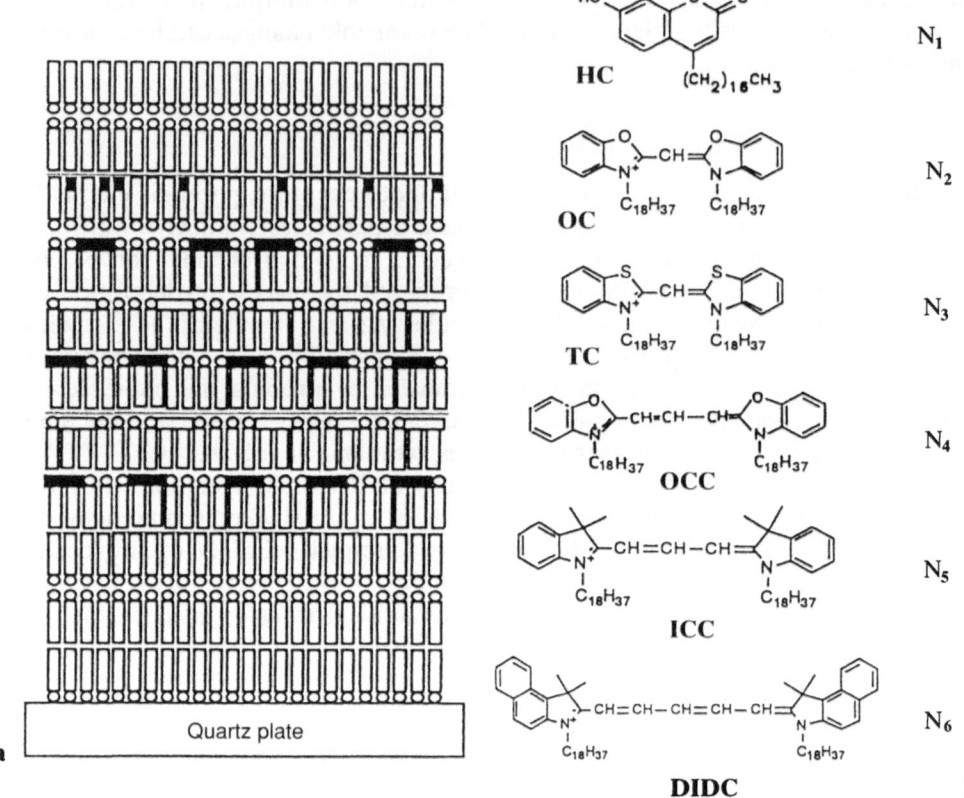

FIG. 4.1-29. Stacking structure of LB multilayers in which sequential excitation energy transfer occurs from the outer surface to the inner layer

donor and acceptor employed here falls between 50 and 70 Å. Taking an interchromophore distance of 25 Å as an optimal D–A pair, the excitation energy transfer should occur in ~25 ps. The time-resolved fluorescence spectra, obtained with a picosecond laser excitation of N_1 at 295 nm, are shown in Fig. 4.1-29b for a four-layer system. It can be seen that the fluorescence bands of N_1–N_4 appear in the picosecond time-range, and that the energy transfer occurs sequentially in the stacking order. Figure 4.1-30 shows the fluorescence growth and decay curves for each layer of a six-layer system. In all the emissions of N_1–N_6 it can be seen that the fluorescence rises sharply in the initial time region, and then slowly later, while during decay both fast and slow decays appear. For the short kinetic components, the decay time of the donor (15–20 ps) corresponds to the rise time of the acceptor. These time constants also correspond to the fluorescence anisotropy decay constant.

These observations indicate that an ultrafast energy transfer pathway is involved in the sequentially stacking molecular system, as was expected from a

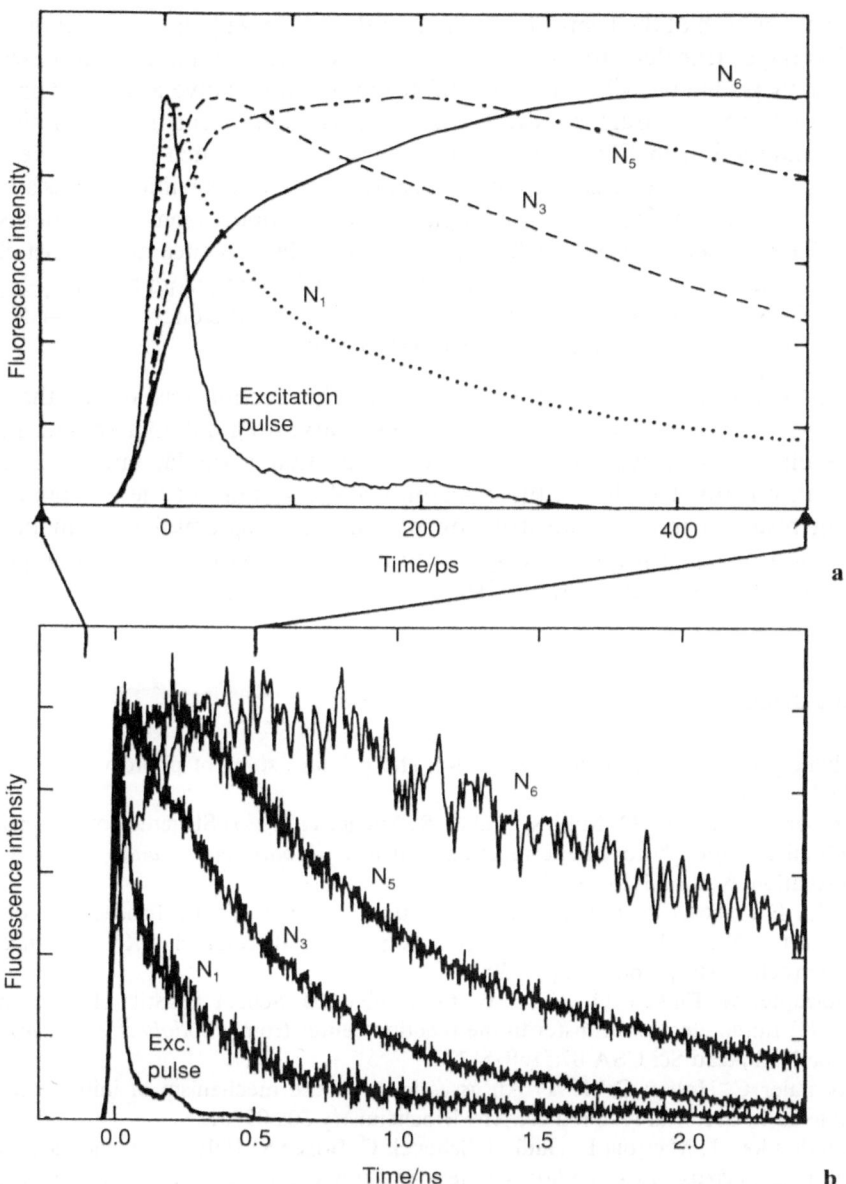

FIG. 4.1-30. Fluorescence growth and decay curves of dye molecules in the six-layer LB films shown in Fig. 5.1-29: **a** 0–400 ps; **b** 0–2 ns time-ranges after 2-ps laser excitation at 415 nm

rough estimation of the transfer rate. Yamazaki and Ohta [49] suggested that this fast interlayer transfer process might compete with vibrational relaxation within the intially populated vibrational manifolds in S_1, and involve a transfer process from vibrationally unrelaxed levels. This process might occur through a dipole–dipole interaction in Förster's medium coupling case, which has considerably more energy than the weak coupling case, and involve a reversible process among D and A molecules [50]. This model gains strong support from experiments on decay kinetics and energy transfer efficiency. The fluorescence decay (fast component) and anisotropy decay of N_1 become faster in going from two- (N_1–N_2) to four- (N_1–N_2–N_3–N_4) layer systems, and the transfer efficiency of $N_1 \rightarrow N_2$ increases from the two- to the four-layer system [49].

Goal of Constructing Multilayer Assemblies. This section has shown that LB multilayer films can provide an artificial light-harvesting antenna system. From the results of these studies, it can be expected that a similar energy transfer mechanism is involved in the biological molecular systems of the antenna. One goal in research into the construction of multilayer assemblies might be the fabrication of an artificial photosynthetic molecular system by combining the antenna system with the system of sequential electron transfer.

References

1. Fleming GR, Van Grondelle R (1994) The primary steps of photosynthesis. Phys Today 47:48–55
2. Deisenhofer J, Epp O, Miki K, Huber R, Michel H (1985) Structure of the protein subunits in the photosythetic reaction centre of *Rhodopseudomonas viridis* at 3 Å resolution. Nature 318:618–624
3. Kuhn H (1986) Electron transfer in organized membranes. In: Proceedings of the Robert A. Welch Foundation Conference on Chemical Research XXX. Advances in Electrochemistry. Houston, pp 339–368
4. Holzapfel W, Finkele U, Kaiser W, Oesterheldt D, Scheer H, Stilz HU, Zinth W (1990) Initial electron-transfer in the reaction center from *Rhodobacter sphaeroides*. Proc Natl Acad Sci USA 87:5168–5172
5. Kirmaier C, Holten D (1991) An assessment of the mechanism of initial electron transfer in bacterial reaction centers. Biochemistry 30:609–613
6. Middendorf T, Mazzola L, Gaul D, Schenck C, Boxer S (1991) Photochemical hole-burning spectroscopy of a photosynthetic reaction center mutant with altered charge separation kinetics: properties and decay of the initially excited state. J Phys Chem 95:10142–10151
7. Chan CK, DiMagno TJ, Chen LXQ, Norris JR, Fleming GR (1991) Mechanism of the initial charge separation in bacterial photosynthetic reaction centers. Proc Natl Acad Sci USA 88:11202–11206
8. Glazer AN (1984) Phycobilisome, a macromolecular complex optimized for light energy transfer. Biochim Biophys Acta 768:29–51
9. Huber R (1989) A structural basis of light energy and electron transfer in biology. EMBO J 8:2125–2147

10. Porter G, Tredwell CJ, Searle GFW, Barber J (1978) Picosecond time-resolved energy transfer in *Porphyridium cruentum*. Biochim Biophys Acta 501:232–245

11. Yamazaki I, Tamai N, Yamazaki T, Murakami A, Mimuro M, Fujita Y (1988) Sequential excitation energy transport in stacking multilayers: comparative study between photosynthetic antenna and Langmuir–Blodgett multilayers. J Phys Chem 92:5035–5044

12. Mimuro M, Yamazaki I, Tamai N, Katoh T (1989) Excitation energy transfer in phycobilisomes at −196°C isolated from the cyanibacterium *Anabaena variabilis* (M-3): evidence for the plural transfer pathways to the terminal emitters. Biochim Biophys Acta 973:153–162

13. Yamazaki I, Ohta N, Yoshinari S, Yamazaki T (1994) Site-selected excitation energy transport in Langmuir–Blodgett multilayer films. In: Masuhara H, DeSchryver FC, Kitamura N, Tamai N (eds) Microchemistry: spectroscopy and chemistry in small domains. North-Holland, Amsterdam, pp 431–440

14. Blankenship RE, Brune DC, Wittmershaus BP (1988) Chlorosome antennas in green photosynthetic bacteria. In: Stevens SE, Bryant D (eds) Light–energy transduction in photosynthesis: higher plants and bacterial models. American Society of Plant Physiologists

15. Sprague SG, Staehelin LA, DiBartolomeis MJ, Fuller RC (1981) Isolation and development of chlorosomes in the green bacterium *Chloroflexus aurantiacus*. J Bacteriol 147:1021–1031

16. Brune DC, Nozawa T, Blankenship RE (1987) Antenna organization in green photosynthetic bacteria. 1. Oligomeric bacteriochlorophyll *c* as a model for the 740 nm absorbing bacteriochlorophyll *c* in *Chloroflexus aurantiacus* chlorosomes. Biochemistry 26:8644–8652

17. Mimuro M, Nozawa T, Tamai N, Shimada K, Yamazaki I, Lin S, Knox RS, Wittmershaus BP, Brune DC, Blankenship RE (1988) Excitation energy flow in chlorosome antennas of green photosynthetic bacteria. J Phys Chem 93:7503–7509

18. Pierson BK, Thornber JP (1983) Isolation and spectral characterization of photochemical reaction centers from the thermophilic green bacterium *Chloroflexus aurantiacus* strain J-10-f1. Proc Natl Acad Sci USA 80:80–84

19. Karrasch S, Bullough P, Ghosh R (1995) EMBO J 14:631–638

20. McDermott G, Prince SM, Freer AA, Hawthornthwaite-Lawless AM, Papiz MZ, Cogdell RJ, Isaacs NW (1995) Crystal structure of an integral membrane light-harvesting complex from photosynthetic bacteria. Nature 374:517–521

21. Külbrandt W (1995) Many wheels make light work. Nature 374:497–498

22. Van Mourik F, Verwijst RR, Mulder JM, Van Grondelle R (1992) Excitation transfer dynamics and spectroscopic properties of the light-harvesting BChl *a* complex of *Prostecochloris aestuarii*. J Lumin 53:499–502

23. Du M, Xie X, Jia Y, Mets L, Fleming GR (1993) Direct observation of ultrafast energy transfer in PSI core antenna. Chem Phys Lett 201:535–542

24. Song PS (1984) Phytochrome. In: Wilkins M (ed) Advanced Plant Physiology. Pitman, London, pp 354–379

25. Furuya M (1989) Molecular properties and biogenesis of phytochrome I and II. Adv Biophys 25:133–167

26. Hershey HP, Barker RF, Idler KB, Lissemore JL, Quail PH (1985) Analysis of cloned cDNA and genomic sequences for phytochrome: complete amino acid sequences for two gene products expressed in etiolated *Avena*. Nucleic Acids Res 13:8543–8560

27. Romanowski M, Song PS (1992) Structural domains of phytochrome deduced from homologies in amino-acid sequences. J Protein Chem 11:139–155
28. Sommer D, Song PS (1990) Chromophore topography and secondary structure of 124-kilodalton *Avena* phytochrome probed by Zn^{2+}-induced chromophore modification. Biochemistry 29:1943–1948
29. Rudiger W, Thummler F, Cmiel E, Schneider S (1983) Chromophore structure of the physiologically active form (Pfr) of phytochrome. Proc Natl Acad Sci USA 80:6244–6248
30. Savakhin S, Wells T, Song PS, Struve WS (1993) Ultrafast pump–probe spectroscopy of native etiolated oat phytochrome. Biochemistry 32:7512–7518
31. Zhang CF, Farrens DL, Bjorling SC, Song PS, Kliger DS (1992) Time-resolved absorption studies of native etiolated oat phytochrome. J Am Chem Soc 114:4569–4580
32. Song PS, Hader DP, Poff KL (1980) Set-up photophobic response in the ciliated *Stentor coeruleus*. Arch Micorobiol 126:181–186
33. Kim IH, Rhee JS, Huh JW, Florell S, Faure B, Lee KW, Kahsai T, Song PS, Tamai N, Yamazaki T, Yamazaki I (1990) Structure and function of the photoreceptor stentorins in *Stentor coeruleus*. I. Partial characterization of the photoreceptor organelle and stentorins. Biochem Biophys Acta 1040:43–57
34. Tao N, Orlando M, Hyon JS, Gross M, Song PS (1993) A new photoreceptor molecule from *Stentor coeruleus*. J Am Chem Soc 115:2526–2528
35. Fabczak H, Park PB, Fabczak S, Song PS (1993) Photosensory transduction in ciliates. II. Possible role of G-protein and cGMP in *Stentor coeruleus*. Photochem Photobiol 57:702–706
36. Song PS, Kim IH, Florell S, Tamai N, Yamazaki T, Yamazaki I (1990) Structure and function of the photoreceptor stentorins in *Stentor coeruleus*. II. Primary photoprocess and picosecond time-resolved fluorescence. Biochim Biophys Acta 1040:58–65
37. Savikhin S, Tao N, Song PS, Struve W (1993) Ultrafast pump–probe spectroscopy of the photoreceptor stentorins from the ciliate *Stentor coeruleus*. J Phys Chem 97:12379–12386
38. Yamazaki I, Tamai N, Yamazaki T (1990) Electronic excitation transfer in organized molecular assemblies. J Phys Chem 94:516–525
39. Osuka A, Maruyama K, Mataga N, Asahi T, Yamazaki I, Tamai N (1990) Geometry dependence of intramolecular photoinduced electron transfer in synthetic zinc–ferric hybrid diporphyrins. J Am Chem Soc 112:4958–4959
40. Osuka A, Nakajima S, Maruyama K, Mataga N, Asahi T, Yamazaki I, Nishimura Y, Ohno T, Nozaki K (1993) 1,2-phenylene-bridge diporphyrin linked with porphyrin monomer and pyromellitimide as a model for a photosynthetic reaction center: synthesis and photoinduced charge separation. J Am Chem Soc 115:4577–4589
41. Osuka A, Yamada H, Maruyama K, Mataga N, Asahi T, Ohkouchi M, Okada T, Yamazaki I, Nishimura Y (1993) Synthesis and photoexcited-state dynamics of aromatic group-bridged carotenoid–porphyrin dyads and carotenoid–porphyrin–pyromellitimide triads. J Am Chem Soc 115:9439–9452
42. Segawa H, Nakayama N, Shimidzu T (1992) Electrochemical synthesis of one-dimensional donor–acceptor polymers containing oligothiophenes and phosphorus porphyrins. J Chem Soc, Chem Commun 784–786
43. Yonemura H, Nakamura H, Matsuo T (1989) External magnetic field effects on photoinduced electron transfer reactions in phenothiazine–viologen-linked systems complexed with cyclodextrins. Chem Phys Lett 155:157–161

44. Kuhn H, Möbius D, Bücher H (1972) Spectroscopy of monolayer assemblies. In: Weissberger A, Rossiter BW (eds) Techniques of chemistry, vol 1, part 3B. Wiley, New York

45. Ahuja R, Möbius D (1989) Photoinduced electron transfer in Langmuir–Blodgett films. Thin Solid Films 179:457–462

46. Mooney WF, Whitten DG (1986) Energy- and electron-transfer quenching of surfactant *trans*-stilbenes in supported n ltilayers: the use of hydrophobic substrate chromophores to determine short-range distance dependence in assemblies. J Am Chem Soc 108:5712–5719

47. Polymeropoulos EE, Möbius D, Kuhn H (1980) Monolayer assemblies with functional units of sensitizing and conducting molecular components: photovoltage, dark conduction and photoconduction in systems with aluminium and barium electrodes. Thin Solid Films 68:173–190

48. Tachibana H, Goto A, Nakamura T, Matsumoto M, Manda E, Niino H, Yabe A, Kawabata T (1989) Photoresponsive conductivity in Langmuir–Blodgett films. Thin Solid Films 179:207–213

49. Yamazaki I, Ohta N (1995) Photochemistry in LB films and its application to molecular switching devices. Pure Appl Chem 67:209–216

50. Förster Th (1960) Excitation transfer. In: Burton M, Kirby-Smith JS, Magee JL (eds) Comparative effects of radiation. Wiley, New York, pp 300–341

4.2 Photoelectrochemical Conversion

AKIRA FUJISHIMA and DONALD A. TRYK

Anticipating forthcoming changes in the 21st century, a century predicted to be the "age of light," we have been interested in light-related chemical phenomena. Utilization of light energy to induce chemical reactions is our main concern. The two main photoelectrochemical approaches that have actively been pursued are the production of hydrogen, a clean energy source, and the efficient reduction of carbon dioxide, which is implicated in global warming and which can be regarded as a readily available feedstock for the production of fuels. We have also been interested in the production of electrical energy from solar energy without attendant fuel production. Most recently, we have become interested in the use of light-activated photocatalysts for use in the control of air and water pollution, particularly indoor air pollution.

The solar energy impinging on the earth's surface is about 3×10^{24} J per year, or approximately 10^4 times the world-wide yearly consumption of energy. The search for an efficient way to convert solar energy into other useful forms of energy is, in view of the increasing anxiety over the exhaustion of energy resources, one of the most important challenges for future research and technological development.

In systems designed for the purpose of converting solar energy into electricity and/or chemicals (for fuel or other purposes), two principal criteria must be met. The first is the absorption, by some chemical substance, of solar irradiation, leading to the creation of electrons (i.e., reduced chemical moieties) and holes (i.e., oxidized chemical moieties). The second is an effective separation of these electron–hole pairs with little energy loss before they lose their input energy through recombination.

Plants capture the energy from sunlight and thus grow. During this process, they produce oxygen by oxidizing water and reducing carbon dioxide. In other words, the oxidation of water and the reduction of CO_2 are achieved with solar energy. By analogy with natural photosynthesis, we began to investigate the photoelectrolysis of water using light energy [1]. This approach involves what is essentially a photochemical battery making use of a photoexcited semiconductor (Fig. 4.2-1).

Such photoinduced charge separation can proceed effectively provided an electric field (potential gradient) has been established at the position where the

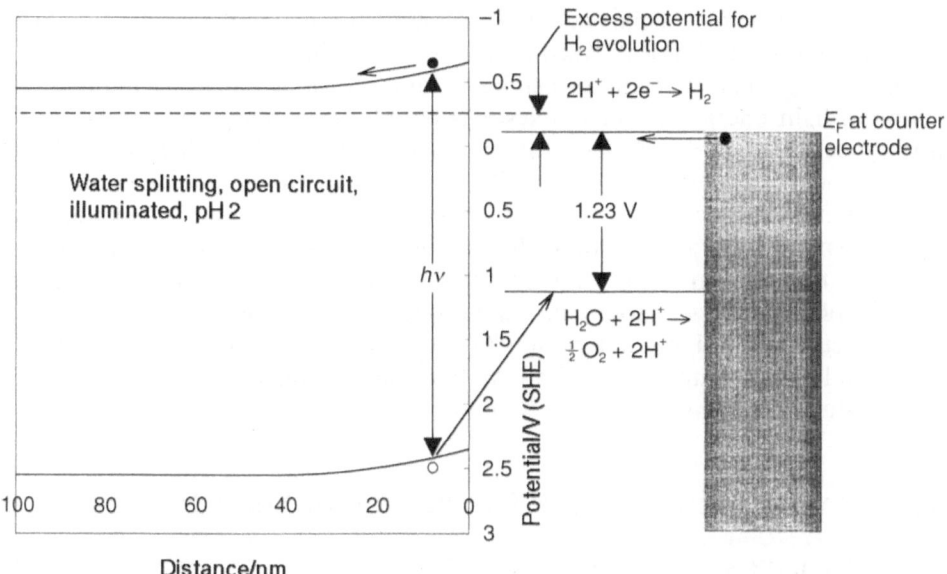

FIG. 4.2-1. Schematic representation of photoelectrochemical water electrolysis using an illuminated oxide semiconductor electrode. Open circuit (or small current), pH 2, illuminated conditions are shown for an oxide with an E_{CB} of $-0.65\,V$ (SHE) and an E_{VB} of $2.35\,V$ (SHE). With an open circuit, a small excess potential ($\sim 0.15\,V$) is available for H_2 evolution, assuming a reversible counter electrode

primary photoexcitation takes place. In general, a potential gradient can be produced at an interface between two different substances (or different phases). For example, a very thin (ca. 50 Å) lipid membrane separating two aqueous solutions inside the chloroplasts of green plants is believed to play the essential role in the process of photosynthesis, which is the cheapest and perhaps the most successful solar conversion system available.

Another well-known example is the photocell or solar photovoltaic cell, in which the photogenerated electron–hole pairs are driven efficiently in opposite directions by an electric field existing at the boundary (junction) of n- and p-type semiconductors (or at semiconductor/metal junctions). A potential gradient can also be created, by a process to be described later in more detail, at the interface of a semiconducting material and a liquid phase. Hence, if a semiconductor is used as an electrode which is connected to another (counter) electrode, photoexcitation of the semiconductor can generate electrical work through an external load and simultaneously drive chemical (redox) reactions on the surfaces of each electrode. On the other hand, in a system where semiconductor particles are suspended in a liquid solution, excitation of the semiconductor can lead to redox processes in the interfacial region around each particle. These types of systems have drawn the attention of a large number of investigators over the past 20 years, primarily in connection with solar energy conversion. During the last 5 years, the area of particulate semiconductors has also seen tremendous growth in terms of photocatalyzed air and water purification.

This section deals with the principles of, and recent advances in, the investigation of light energy conversion systems based on semiconductor/liquid junctions, focusing on fuel and electrical energy generation, CO_2 reduction, and the purification of air and water. Specific topics covered include: the TiO_2–Pt system for photoelectrochemical water electrolysis; compound semiconductor electrodes and their stabilization; the use of p-type semiconductors as photocathodes; the use of novel layered compounds as semiconductor electrodes; the use of photosensitizer-modified, nanocrystalline semiconductor electrodes; photoelectrochemical reduction of CO_2 at p-type semiconductor electrodes; photocatalytic decomposition of air and/or water pollutants with illuminated semiconductors.

4.2.1 Photoelectrochemical Conversion of Solar Energy

4.2.1.1 TiO$_2$–Pt System—Fundamentals

The possibility of solar photoelectrolysis was demonstrated for the first time with a system in which an n-type TiO_2 semiconductor electrode, which was connected through an electrical load to a platinum black counter electrode, was exposed to near-UV light (Fig. 4.2-2). When the surface of the TiO_2 electrode was irradiated with light consisting of wavelengths shorter than ~415 nm, a photocurrent flowed from the platinum counter electrode to the TiO_2 electrode through the external circuit. The direction of the current reveals that the oxidation reaction (oxygen evolution) occurs at the TiO_2 electrode and the reduction reaction (hydrogen evolution) at the Pt electrode. This fact shows that water can be decomposed, using UV-visible light, into oxygen and hydrogen, without the application of an external voltage, according to the following scheme:

$$TiO_2 + 2h\nu \rightarrow 2e^- + 2h^+ \qquad (4.2\text{-}1)$$

(excitation of TiO_2 by light)

$$2h^+ + H_2O \rightarrow 1/2O_2 + 2H^+ \qquad (4.2\text{-}2)$$

(at the TiO_2 electrode)

$$2e^- + 2H^+ \rightarrow H_2 \qquad (4.2\text{-}3)$$

(at the Pt electrode)
The overall reaction is

$$H_2O + 2h\nu \rightarrow 1/2O_2 + H_2 \qquad (4.2\text{-}4)$$

When a semiconductor electrode is in contact with an electrolyte solution, thermodynamic equilibration takes place at the interface. This results in the formation of a space charge layer within a thin surface region of the semiconduc-

FIG. 4.2-2. Schematic diagram of an electrochemical photocell. *1*, *n*-type TiO$_2$ electrode; *2*, platinum black counter electrode; *3*, ionically conducting separator; *4*, gas buret; *5*, load resistance; *6*, voltmeter

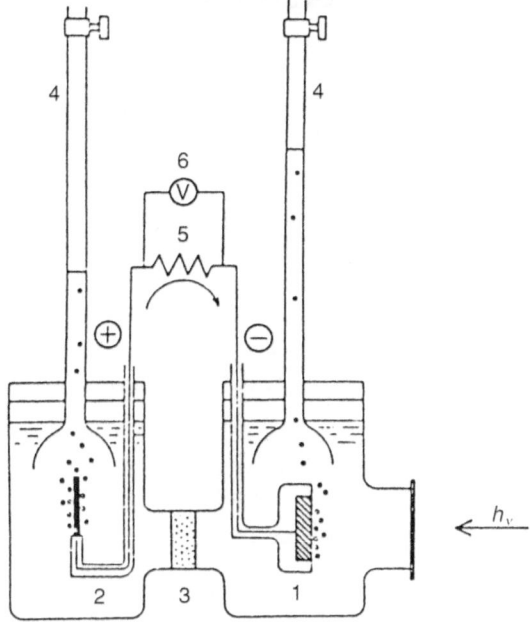

tor, in which the electronic energy bands are generally bent upwards or downwards in the cases of *n*- and *p*-type semiconductors, respectively. The thickness of the space charge layer is usually of the order of $1-10^3$ nm, depending on the carrier density and dielectric constant of the semiconductor. If this electrode receives photons with energies greater than that of the material's band gap, E_G, electron–hole pairs are generated within the space charge layer. In the case of an *n*-type semiconductor, the electric field existing across the space charge layer drives photogenerated holes toward the interfacial region (i.e., solid–liquid) and electrons toward the interior of the electrode and from there to the electrical connection to the external circuit. The reverse process takes place at a *p*-type semiconductor electrode. These fundamental processes have been discussed in a number of excellent reviews, including those of Gerischer [2], Nozik [3], Wrighton [4], and Heller [5], and the monograph by Morrison [6].

The flatband potential, E_{FB}, is an important parameter in semiconductor–electrolyte systems. When the semiconductor is polarized at E_{FB} in the dark, there is no space charge or electric field in the semiconductor. The flatband potentials are generally determined via capacitance measurements based on the Mott–Schottky equation (see Ref. [6]). In general, the flatband potential for an *n*-type semiconductor is close (0.1–0.2 eV) to the conduction band energy E_{CB}, while that for a *p*-type semiconductor is close to the valence band energy E_{VB} [6]. Figure 4.2-3 shows the energy levels of the conduction and valence bands for a number of semiconductors, including *n*-type oxide semiconductors. If the con-

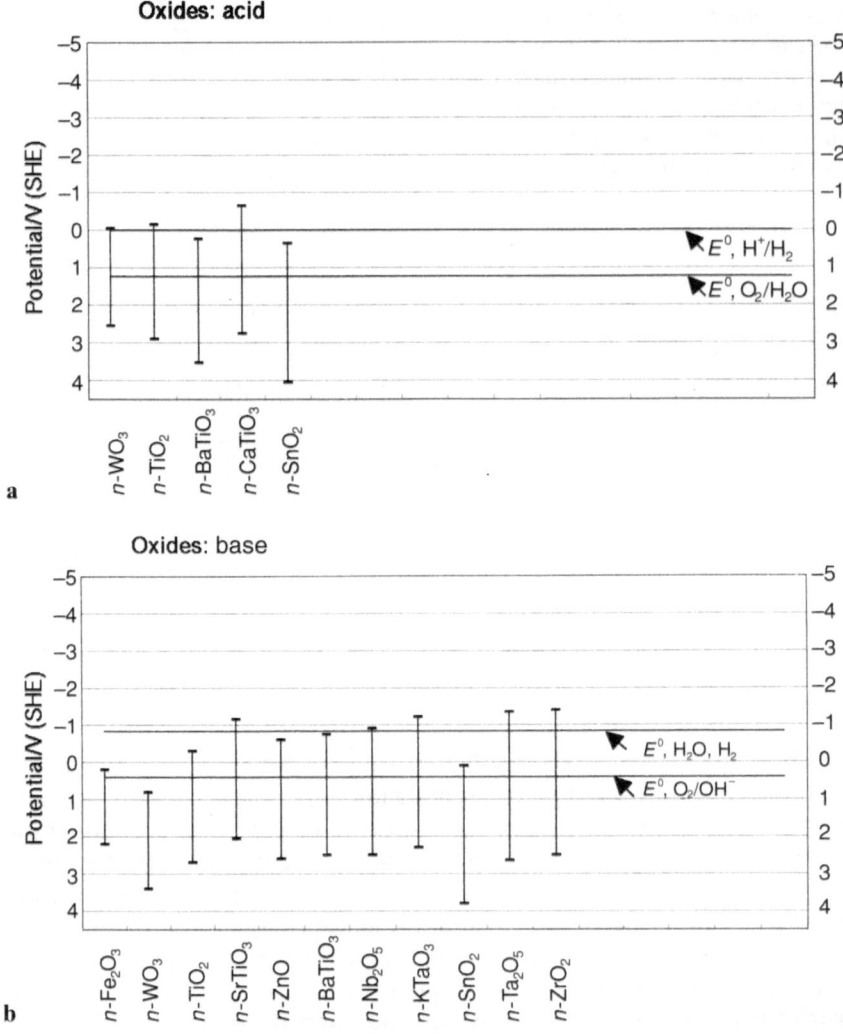

FIG. 4.2-3. Relative energy levels of selected semiconductor materials. **a** *n*-Type oxides in acid solution (0–2 pH). **b**, *n*-Type oxides in alkaline solution (12–14 pH). **c** Non-oxide semiconductors in acid solution (0–2 pH). **d** Non-oxide semiconductors in alkaline solution (12–14 pH)

duction band energy E_{CB} is higher (i.e., more negative) than the hydrogen evolution potential, photogenerated electrons can flow to the counter electrode and reduce protons, resulting in hydrogen gas evolution without an applied potential, although, as shown in Fig. 4.2-1, E_{CB} should be at least as negative as $-0.4\,V$ (SHE) in acid solution or $-1.2\,V$ (SHE) in alkaline solution. Among the oxide semiconductors, TiO_2 (acid), $SrTiO_3$, $CaTiO_3$, $KTaO_3$, Ta_2O_5, and ZrO_2 satisfy this requirement, as shown in Fig. 4.2-3. On the other hand, the employment of

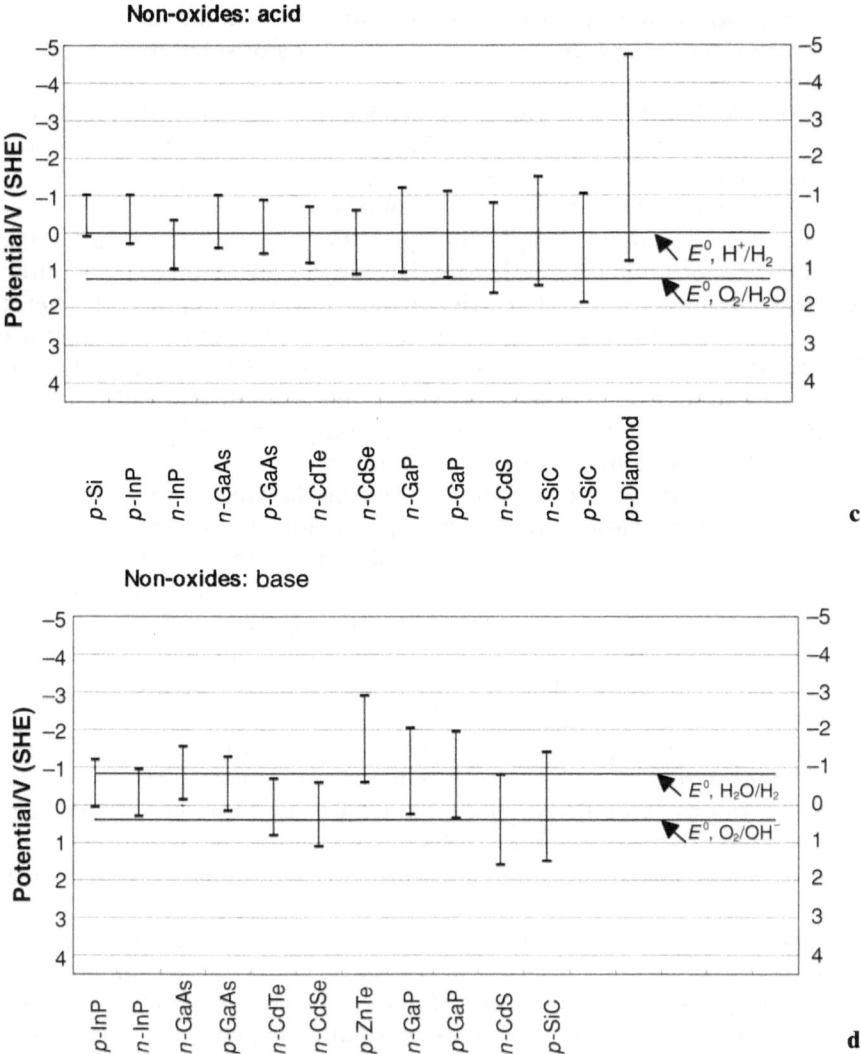

FIG. 4.2-3. *Continued*

an external bias or of a difference in pH between the anolyte and catholyte is required in the case of the other materials in order to achieve hydrogen evolution. For example, in early n-TiO_2-based photoelectrochemical cells used in our laboratory [7], the problem of E_{CB} being slightly lower (less negative) than that necessary to evolve hydrogen redox was circumvented by using an anolyte and catholyte with different pH values, higher in the former and lower in the latter. Effectively this decreases the equilibrium cell potential and thus there is excess overpotential with which the cell reactions can be driven to higher current densities. Our group [7] and others [4] also found that $SrTiO_3$, which has a sufficiently negative E_{CB}, was able to photoelectrolyze water without an additional voltage. With its very large band gap, however, the efficiency of solar energy conversion is very low.

It is desirable that the band gap of the semiconductor is near that for optimum utilization of solar energy, which would be ~1.35 eV for a solid-state *p–n* junction or for a photoelectrochemical cell (PEC) under ideal conditions [6, 8]. Even when photons are completely absorbed, excess photon energy ($E > E_{BG}$) cannot normally be utilized in a simple, single band-gap device, since vibrational relaxation occurs in the upper excited states before the charge transfer takes place, although there is some evidence for "hot" carriers under certain conditions [9]. Therefore, a fraction of the photon energy is dissipated as heat. When semiconductor electrodes are used as either photoanodes or photocathodes for water electrolysis, the band gap should be at least 1.23 eV (i.e., the equilibrium cell potential for water electrolysis at 25°C and 1 atm), particularly considering the existence of polarization losses due to, e.g., oxygen evolution.

The quantum efficiency (number of electrons flowing through the external circuit per absorbed photon) should be high, ideally 1.0, in order to obtain an efficient PEC. For the efficient utilization of light energy, the incident photons should be totally absorbed within the space charge layer. Since the thickness of the space charge layer depends on the carrier density, the latter must be adjusted in a number of cases. If the density of recombination centers in the space charge layer is high, the quantum efficiency will be lowered accordingly.

Bard and Fox [10] have recently reviewed the question of photoelectrochemical splitting of water in terms of a kind of "Holy Grail," i.e., an object of almost religious importance that is nearly unattainable. They define their Holy Grail as "an efficient and long-lived system for splitting water to H_2 and O_2 with light in the terrestrial (AM 1.5) solar spectrum at an intensity of one sun." (AM 1.5 refers to the solar spectrum corresponding to sunlight passing through the atmospheric air mass at an angle of ~42° from normal [8, 11].) They further define it to mean an energy efficiency of 10%, a lifetime exceeding 10 years, and production of hydrogen at a cost (on an energy basis) competitive with that of fossil fuels. They point out that, thus far at least, a single-junction semiconductor electrode has not been found that has both a negative enough conduction band energy to generate hydrogen and a small enough band gap to effectively utilize the solar spectrum. One of the best systems was developed by Heller and co-workers and involved *p*-type InP photocathodes with very thin (<100 Å) overlayers of Rh, Ru, or Pt as photocathodes for hydrogen generation [5]. Efficiencies of up to 12% were obtained, with a maximum power density of 10 mW cm^{-2}. *p*-Type materials will be discussed in greater detail later.

Bard and Fox [10] also reviewed a number of novel multicomponent approaches involving photosensitizers and catalysts, and concluded by speculating that types of heterogeneous array systems are needed which may involve multifunctional components, possibly including new types of multijunction semiconductors, multi-electron catalysts, and new classes of photosensitizers. Some of these approaches will also be discussed here. The requirement of water splitting makes all of this more challenging in terms of the stabilities of the components in the presence of the highly energetic electrons and holes needed for the electrode reactions. Approaches which avoid this requirement, i.e., using a single, well-placed redox couple for the generation of electrical power only, have shown much promise and are discussed in the next two sections.

4.2.1.2 Compound n-Type Semiconductor Electrodes and Their Stabilization

As mentioned above, stable photoanodes such as TiO_2 have relatively wide gaps (~3.0 eV) and hence can utilize only a small portion of the solar spectrum. Semiconductors with smaller band gaps have good visible light response and, in this sense, are suitable as photoelectrodes. The energy levels for some of these materials are shown in Fig. 4.2-3. However, many of these materials decompose easily through photoanodic dissolution of the electrode surface. For example, in the case of CdS, the material is oxidized by photogenerated holes so that Cd^{2+} ions dissolve and elemental sulfur deposits on the electrode surface. Thus the photocurrent at a CdS photoanode shows an abrupt decay with time due to the filtering effect of the deposited sulfur.

$$Cd + 2h^+ \rightarrow Cd^{2+} + S \qquad (4.2\text{-}5)$$

Such unfavorable phenomena must be alleviated in order for small band-gap semiconductors to be used in PECs. Much effort has been devoted to the problem of photocorrosion, particularly in the decade following the first report concerning PECs. Beginning with independent reports in 1976 by groups at Bell Labs [5], MIT [4], and the Weizmann Institute [12] that the aqueous S^{2-}/S_2^{2-} couple could be used to stabilize CdS and CdSe, it has become well established that stabilization of these electrodes can be achieved through the addition of suitable redox species to the electrolyte.

Early work in our laboratory involved the examination of competitive photoanodic oxidation at a CdS electrode in electrolytes containing various reducing agents using the rotating ring-disk electrode (RRDE) technique [13]. In the case of CdS, the competitive ratio for the oxidation of the added redox species becomes larger as its redox potential becomes more negative (i.e., more easily oxidizable). The efficiency of the electrode stabilization is controlled by the relationship between the dissolution potential E_D for the semiconductor and E^0 for the redox couple. Similar conclusions were reached by Gerischer [2], who also pointed out the importance of kinetics. Lewis [14] has discussed the fact that some of these electrochemical reactions are extremely rapid. In addition, the importance of strong adsorption of the redox reactant at the semiconductor electrode surface has been pointed out, particularly for aqueous electrolytes [2, 4, 5]. Unstable semiconductor electrode materials such as CdS, CdSe, CdTe, GaP, and GaAs have all been stabilized using chalcogenides [4, 5, 12]. Other redox couples have also been used, and some representative electrode/redox couple systems are listed in Table 4.2-1 [15–40].

TABLE 4.2-1. Representative semiconductor electrode/redox couple systems

Electrode material	Electrolyte	Redox couple	η (%)	i_{sc} (mA cm^{-2})	E_{OC} (V)	FF (%)	P_{max} (mW cm^{-2})	P_{imp} (mW cm^{-2})	Year	Ref.
n-type	*Aqueous*									
n-CdS	aq. KCl	$[Fe(CN)_6]^{3-/4-}$	9.4	6.0	1.0	70	4	40	1976	15
n-GaAs	aq. KOH	Se^{2-}/Se_2^{2-}	8.8	16.5	0.65	57	6.1	69	1977	16
n-CdSe	aq. NaOH	S^{2-}/S_2^{2-}	8.1	14	0.73	60	6.16	75.4	1977	17
n-CdSe[a]	aq. NaOH	S^{2-}/S_2^{2-}	6.9	10.6	0.765	55	4.4	64.6	1978	18
n-GaAs[b]	aq. KOH	Se^{2-}/Se_2^{2-}	12.0	25	0.70	65	11.4	95	1978	19
n-GaAs[b,c]	aq. KOH	Se^{2-}/Se_2^{2-}	4.8*	21.4	0.55	39	4.6	96	1980	20
n-GaAs[c,d]	aq. KOH	Se^{2-}/Se_2^{2-}	7.8*	11.3	0.68	55	4.2	53.7	1980	21
n-WSe$_2$	aq.	I_2/I^-	3.7	13.1	0.51	53	3.5	92.5	1980	22
n-CuInSe$_2$	aq. HI	I_2/I^-	9.5	46	0.41	50	9.5	99	1983	23
n-CuInSe$_2$	aq.	I_2/I^-	11.7	50	0.47	60	14.0	120	1984	24
n-CuInSeTe[e]	aq.	S^{2-}/S_2^{2-}	12.7	22.2	0.78	69	11.9	93.7	1985	25
n-WSe$_2$	aq.	I_2/I^-	14.3	28.6	0.63	62	11.1	77.6	1985	26
n-GaAs[c,f]	aq. KOH	Se^{2-}/Se_2^{2-}	16.0*	26	0.81	75	16	(100)	1987	27
n-CdSe	aq. KOH	$[KFe(CN)_6]^{2-/3-}$	16.4	18	1.2	63	14	84.4	1990	28
p-type	*Aqueous*									
p-InP	aq. HCl	V^{3+}/V^{2+}	9.4	24.8	0.66	63.5	10.4	110	1980	29
p-InP	aq. HCl	V^{3+}/V^{2+}	11.5	24.8	0.66	64	10.5	89.5	1981	30

			η	i_{SC}	E_{OC}	FF	P_{max}	P_{mp}		
n-type	*Nonaqueous*									
n-Si	LiClO$_4$/CH$_3$OH	(1-OH)EtFc$^{0/+}$	10.1	25	0.53	54	7.1	70	1983	31
n-Si	LiClO$_4$/CH$_3$OH	Me$_2$Fc$^{0/+}$	14	34	0.60	69	14	100	1984	32
n-Si^c	LiClO$_4$/CH$_3$OH	Me$_2$Fc$^{0/+}$	9.6*	27.4	0.53	66	9.6	100	1984	32
n-InP	LiClO$_4$/CH$_3$OH	Me$_2$Fc$^{0/+}$	7.0	15	0.61	47	4.3	62	1989	33
n-GaAs	LiClO$_4$/CH$_3$OH	Fc$^{0/+}$	11.0	20	0.83	58	9.6	88	1990	34
Dye-sensitizer nanocrystals										
TiO$_2$/sens.^{g}	CH$_3$CN, Li$^+$	I$_2$/I$^-$	10.0	18.2	0.72	73	9.6	96	1993	35
Solid state p–n junctions										
Si^c			17.8*	36	0.62	79.4	17.8	100	1990	36
CuInGaSe$_2^c$			15.9*	31.9	0.65	76.6	15.85	100	1994	37
GaInP/GaAs^c			29.5*	14	2.39	88.5	29.55	100	1994	38
Si			24.0	40.8	0.71	83.1	24	100	1995	39
Theoretical single junction			31	62	—	—	26.2	84.4	1980	40

[a] Se^0 added to electrolyte.

[b] Adsorbed Ru^{3+}.

[c] Polycrystalline (efficiencies are also marked*).

[d] Adsorbed Ru^{3+}, Pb^{2+}.

[e] Cu^{2+} added to electrolyte.

[f] Adsorbed Os^{3+}.

[g] Dye sensitizer was *cis*-di(thiocyanato)bis(2,2'-bipyridyl-4,4'-dicarboxylate)ruthenium(II).

η, efficiency; i_{SC}, short circuit current density; E_{OC}, open circuit cell voltage; FF, fill factor; P_{max}, maximum power density; P_{mp}, input power density.

In this type of PEC, the redox agents react reversibly at both electrodes.

$$\text{Reduced form} + h^+ \xrightarrow{\text{photoanode}} \text{oxidized form} \tag{4.2-6}$$

$$\text{Oxidized form} + e^- \xrightarrow{\text{cathode}} \text{reduced form} \tag{4.2-7}$$

Redox agents are regenerated via this process, and no overall chemical change occurs. Thus, such systems have been referred to as regenerative PECs, in which light energy is converted only to electricity. The driving force for this type of PEC is attributed to the underpotential developed for the oxidation reaction at the semiconductor photoanode. The open circuit cell voltage V_{oc} under illumination is an important quantity in determining the power characteristics of an attainable regenerative PEC. This is given by the following equation:

$$V_{\text{OC}} = \left| E_{\text{FL,semi.,OC,ill.}} - E^0 \right| \approx \left| E_{CB} - E^0 + 0.2 \right| \tag{4.2-8}$$

where $E_{\text{FL,semi.,OC,ill.}}$ is the Fermi level energy of the semiconductor at open circuit under illumination, E_{CB} is the conduction band edge energy, and E^0 is the standard redox potential of the redox couple. This is illustrated in Fig. 4.2-4. The V_{OC} value can thus be chosen by selecting various combinations of semiconductors and redox couples. For an n-type material, the maximum value is limited by the fact that the photogenerated holes must have sufficient oxidizing power to oxidize the redox couple, so that the E_{VB} at the electrode surface must be more positive than the E^0 of the redox couple. This means in essence that V_{OC} is limited by the bandgap of the semiconductor. Figure 4.2-4 also shows simplified band diagrams for conditions of maximum power and short circuit, together with the corresponding points on the I–V curve. The picture shown for the short circuit condition, which shows the maximum band bending, is also valid for the dark condition (Fig. 4.2-4a). The latter is like a chemical short circuit: the redox couple removes a small amount of negative charge from the bulk of the semiconductor, bringing down the Fermi level. Under illumination on open circuit, a photovoltage develops which is due to a balancing of the dark current with the photoassisted oxidation current (Fig. 4.2-4d). This situation can be described by the following equation, which is valid for solid-state p–n junction cells and Schottky barrier (semiconductor-metal junction) cells, as well as PECs [41]:

$$i = -i_{\text{s}} \left[\exp\left(-e\left(E - E^0_{\text{redox}} \right) / kT \right) - 1 \right] + i_{\text{L}} \tag{4.2-9}$$

where i_{s} is the diode saturation current density, and i_{L} is the current due to excitation of excess carriers by light. Under short circuit conditions, the first term (dark current) becomes zero, and the total current density becomes i_{L}. Under

Equilibrium, dark (also short circuit, illuminated)

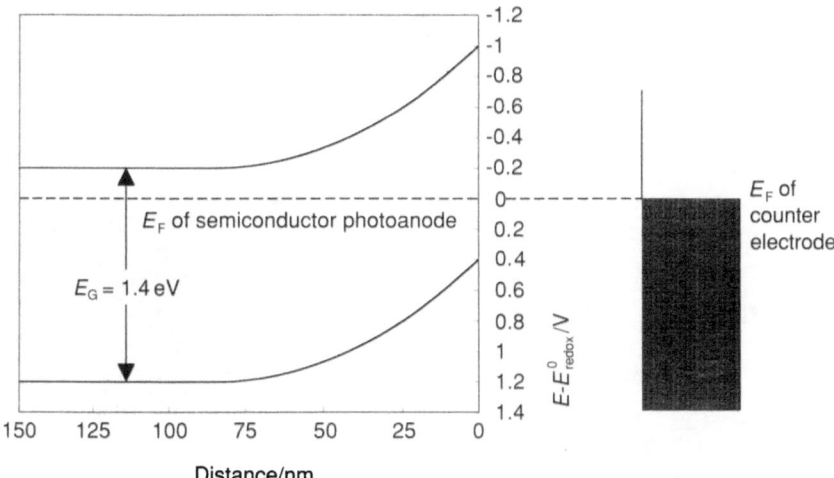

a

Open circuit, illuminated

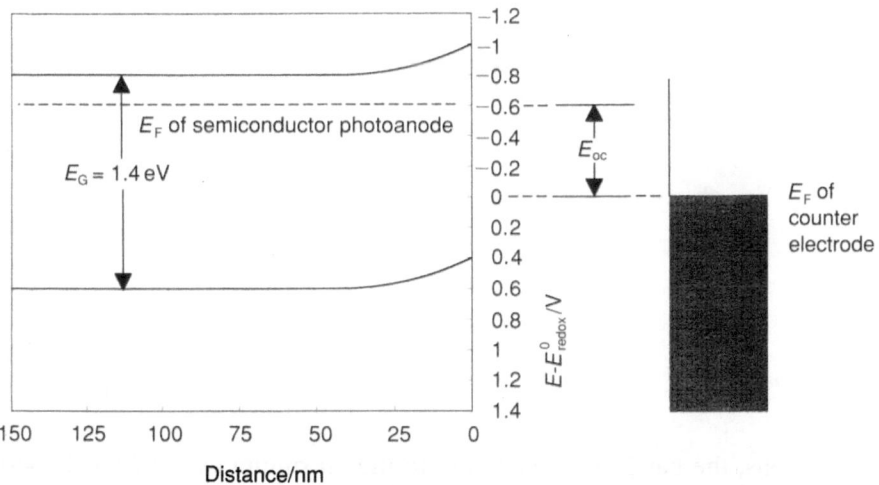

b

FIG. 4.2-4. Schematic diagram of the energy levels for an n-type semiconductor in a regenerative PEC. **a** Equilibrium, dark conditions, and also short circuit, illuminated condtions. **b** Open circuit, illuminated conditions. **c** Maximum power conditions. **d** Corresponding current–potential curve. The band-bending diagrams are based on Fig. 9.1 in [6], with a carrier density of $10^{17}\,cm^{-3}$ and a dielectric constant of eight. For the current–voltage curve, the photocurrent density and the reverse saturation current density are assumed to be $20\,mA\,cm^{-2}$ and $10^{-12}\,A\,cm^{-2}$, respectively

Maximum power, illuminated

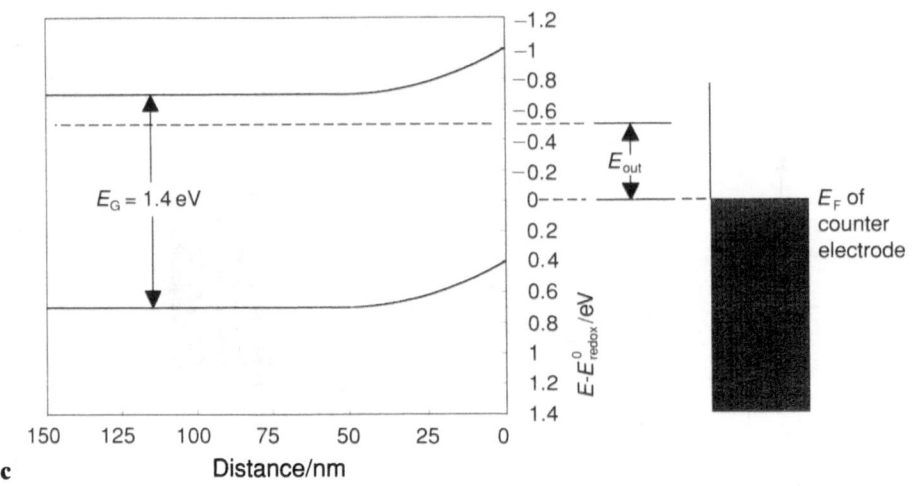

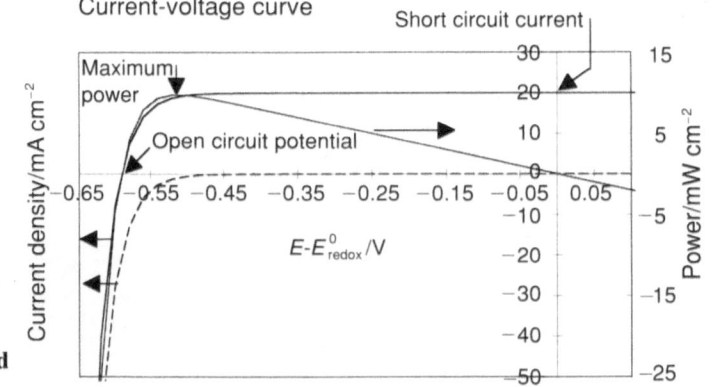

FIG. 4.2-4. *Continued*

these condtions, the bands are again bent to the maximum extent (Fig. 4.2-4a), leading to the maximum amount of electric field for charge separation. At the maximum power point, the bands are bent to an intermediate degree in order to maximize the product of photocurrent and photovoltage (Fig. 4.2-4c). The degree to which the *I–V* curve approximates a perfect rectangle is termed the fill factor.

$$\mathrm{FF}(\%) = \left[(i\,V)_{max}\right] \big/ \left(i_{SC}\,V_{OC}\right) \tag{4.2-10}$$

where i_{SC} is the short circuit photocurrent density. The maximum power divided by the illumination light power (for a specified spectral distribution) gives the efficiency. All these factors must be self-consistent.

Table 4.2-1 gives a list of representative regenerative PEC systems. The incident light intensities vary over a significant range, often with different spectral distributions, which complicates comparisons. For example, the first value given, 9.4% at 40 mW cm^{-2} incident light power, obtained with a xenon lamp, is equivalent to 5.5% efficiency with a typical ground-level solar spectrum and intensity, i.e., 80–100 mW cm^{-2} for direct sunlight (for discussion of light source standards, see [11] and [42]). Aside from this minor complication, it can be seen that great progress has been made in the last 20 years in the area of regenerative PECs, with the highest efficiency, 16.4%, being realized for n-CdSe in an aqueous electrolyte by Licht and co-workers [28]. One of the most impressive results is the 16.0% efficiency for polycrystalline n-GaAs with adsorbed Os^{3+}, which was obtained by Lewis and co-workers [43]. This value approaches the best value reported for a polycrystalline solid-state p–n junction cell, which is 17.8%.

These improvements in efficiency have resulted from a long series of efforts in which many different aspects of semiconductor–electrolyte systems were examined, often from a fundamental viewpoint. Improving the quality of the semiconductor surfaces in order to decrease recombination rates has proved to be important in several cases.

Heller and co-workers [5] have made improvements in the efficiency of CdSe-based cells by improving surface preparation in order to remove trapping and recombination sites, and have also improved the efficiency of GaAs-based cells by decreasing recombination rates via adsorbed transition metal ions such as Ru^{3+} and Pb^{2+} (Table 4.2-1). Further work on surface modification via adsorbed metal ions by Lewis and co-workers [43], however, led to the conclusion that instead of decreasing the recombination rate, these ions were actually increasing the hole capture rate in the electrolyte. After examining a number of different adsorbates, they found that Os^{3+} adsorbed on n-GaAs improved the efficiency to 16% (see Table 4.2-1).

Even with the great improvements in semiconductor stability brought about by the presence of very efficient redox couples, stability remains a very important issue. It was recognized by Bard and Wrighton in 1977 (see [4]) that semiconductor stability could be improved with the use of non-aqueous solvents, thus avoiding the presence of water as an oxygen donor for the oxidation process. The efficiencies were low in the early work with nonaqueous systems, but Lewis and co-workers [43] were later able to improve them by adjusting the semiconductor quality in order to increase the quantum efficiency and decrease the recombination rates, and also changing the conductivity of the electrolyte. The Caltech group [43] developed a high-efficiency system (14%) based on n-Si in a methanol electrolyte with dimethylferrocene/dimethylferricinium as the redox couple (see Table 4.2-1).

4.2.1.3 Stable p-Type Semiconductor Electrodes

Another important advance in achieving stable semiconductor electrodes was the use of p-type semiconductor electrodes [5]. The basic idea is as follows. Instead of the active, minority carrier species at the electrode surface being the hole, which has sufficient oxidizing power, particularly in aqueous electrolytes, to photocorrode most semiconductor materials, the active species is the electron. Most semiconductor materials are much less vulnerable to cathodic corrosion, so that this strategy should lead to improved stability. Using p-InP with a thin oxide layer, the Bell Labs group was initially able to achieve 9.3% efficiency and later, with improved surface preparation, 11.5% [5] (see Table 4.2-1). Figure 4.2-5 shows the relative positions of the band energies and the redox potential for a regenerative PEC based on a p-type semiconductor photocathode.

More recently, it has been found that high-quality p-type semiconducting and even conducting chemical vapor-deposited (CVD) diamond films can be prepared using boron as a dopant [44, 45]. Diamond has some unique properties that make it highly attractive as an electrode material, including excellent chemical inertness, an extremely wide potential window, extreme hardness, excellent transparency, and high thermal conductance. Our work has focused on the use of diamond as both a highly conductive electrode material [45] and a semiconducting photocathode. In very recent work, we have found that, using extremely pure, highly crystalline boron-doped films, the potential window in an aqueous acid solution extends from −1.0 V to +1.5 V vs. SCE. We have also found that the flat band potential, which approximates E_{VB}, is around +0.5 V vs. SCE. Since the band gap is approximately 5.5 eV, this puts E_{CB} at −5.0 V, i.e., above the vacuum level! Therefore, our photoelectrochemical work has involved the use of excimer lasers to exceed this band gap energy. Some very interesting experiments can be carried out in which extremely hot electrons are produced.

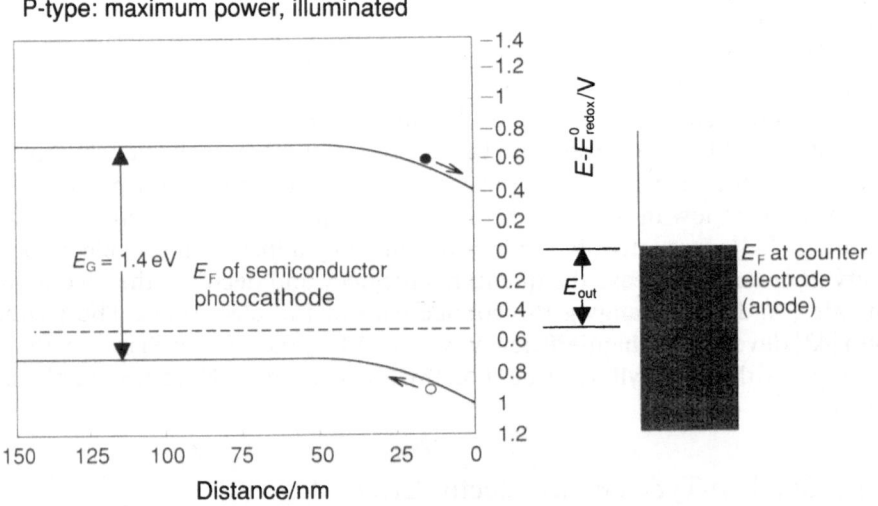

FIG. 4.2-5. Energy level diagram for a photoelectrochemical cell employing a p-InP photocathode, using similar assumptions to those in Fig. 4.2-4

In summary, an efficient regenerative PEC should possess the following characteristics:

1. high quantum efficiency;
2. sufficient absorption of solar radiation;
3. sufficient stability for the photoanode;
4. large output photovoltage based on good reversibility of redox couples in the solution and significant band-bending for the semiconductor electrode;
5. sufficiently small ohmic resistance of the overall semiconductor–electrolyte system;
6. negligible absorption of light by the electrolyte and redox couple.

For the design and construction of such regenerative PECs, the following points must also be taken into account:

1. simple construction;
2. selection of inexpensive materials for the electrodes;
3. ease of manufacture of large-area electrodes;
4. low toxicity for electrode materials, electrolyte components, and redox couples.

Of the large number of systems that have been reported in the literature, many of them meet at least some of these criteria, but to date none meets all of the requirements.

4.2.1.4 Novel Types of Semiconductors

Semiconductors with unusual structural and physical properties have been of interest in photoelectrochemical energy conversion since the late 1970s, when Tributsch first reported results on layered transition metal chalcogenides [46]. These were initially examined because it was thought that the presence of transition metal d-bands might facilitate sluggish electrode reactions such as oxygen evolution, but it became apparent that one of their most interesting aspects was their great lack of interaction with water or other electrolyte components. Another aspect of great interest was the high light-to-electrical energy conversion efficiency of some of these compounds (Table 4.2-1). Several researchers noticed significant sample-to-sample variations in performance [42], and it was not until several years later that it was clearly demonstrated that specific types of surface preparation methods were needed in order to remove recombination sites. When this was done, for example, by using photoelectrochemical etching of the surface, high efficiencies could routinely be reached, particularly for WSe_2 [47].

While it was initially thought that the stability of the layer-type transition metal chalcogenides such as MoX_2 (X = S, Se) was due to the presence of a van der Waals (vdW) surface, it was later shown by Tributsch and co-workers that the key factor was the metal d-band character of the valence band [46]. This makes it possible for strongly coordinating redox species such as I^- to capture photogenerated holes efficiently. One strong piece of evidence for this is that transition metal chalcogenides with pyrite structures, such as FeS_2 and RuS_2, which do not have vdW surfaces but do have metal d-band-based valence bands, are almost completely stable in the presence of the I_3^-/I^- redox couple. Partially for this reason, there was interest in FeS_2 in terms of practical applications in solar energy conversion, but its principal disadvantage was found to be that, due to the relatively high dark current, the open circuit photopotential was limited to only 20–30% of the band-gap energy, as compared with up to 60% for the Mo and W layer-type compounds. Due to the vdW surfaces of the latter, the dark currents are typically very low.

Therefore there is continuing interest in the layer-type compounds, especially if methods could be developed to manufacture high-quality thin films. A number of approaches have been tried, with varying degrees of success. Molecular beam epitaxy (MBE) has the ability to produce high-quality films [48], but some development work would be needed in order to lower the costs for large-scale application.

An interesting fundamental aspect of the MX_2–I_3^-/I^- system is the fact that, even though the interaction between the redox couple species and the vdW surface is so weak, it is enough to make the photocurrent significantly greater than that available with an outer-sphere redox species such as Fe^{2+}. Jaegermann and co-workers [49] have shown in UHV studies that I_2 adsorbs on the vdW surface of WSe_2 only at relatively low temperatures ($\leq 100 K$), and is completely desorbed in its intact form by annealing at room temperature. In later work, Jaegermann [50] carried out further UHV simulations of the semiconductor/electrolyte interface, for example, simulating the presence of a Br_2/Br^- redox couple on the WSe_2 vdW surface.

In the area of layered semiconductor materials, there has also been some very intersting work done during the past 10 years on layered transition metal oxides in the form of particle suspensions. Domen and co-workers [51, 52] have focused on light-assisted water-splitting reactions, while Mallouk and co-workers [53] have focused on the light-assisted production of hydrogen and triiodide from an aqueous HI solution.

Domen and co-workers [51] found that the layered niobate $K_4Nb_6O_{17}$ has photoactivity for water splitting, and that the reaction can be catalyzed by the presence of Ni in the form of extremely small particles (~5 Å) in one of the two types of interlayer spaces (Type I). These particles appear to catalyze hydrogen generation on one side of a given niobate sheet, while oxygen is generated on the opposite side of the sheet, with both reactions being driven by a photoinduced electric field gradient across the layer (Fig. 4.2-6). However, the exact reason for the existence of such a field gradient is still unclear.

These workers also examined the same layered niobate material with very small Pt particles contained in interlyer I to catalyze hydrogen generation [52]. Two interesting results were obtained: (1) Pt could be ion-exchanged using the cationic $[Pt(NH_3)_4]^{2+}$ and not the anionic $[PtCl_6]^{2-}$, providing evidence that the Pt was being ion-exchanged at interior, anionic (i.e., hydroxide) sites; (2) it was found to be important to remove Pt particles from the outsides of the particles

FIG. 4.2-6. Diagram of layered semiconductor material (from Fig. 7 in Kudo et al. [51], with permission)

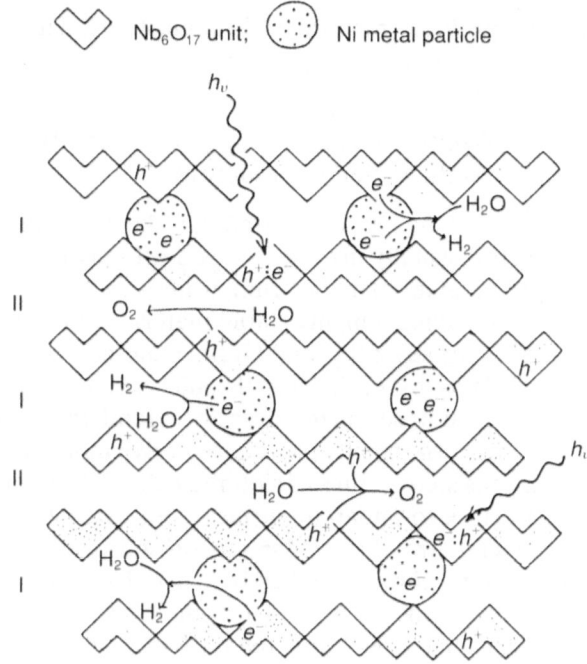

using *aqua regia*, because such particles can catalyze the recombination reaction efficiently.

Mallouk and co-workers [53] examined a series of several different types of layered oxides, including titanates, niobates, and titanoniobates. For all but two, photoassisted HI decomposition was obtained, with the highest activity being found for the niobate $K_4Nb_6O_{17}$. All were loaded with Pt and were photosensitized using a tris bipyridyl ruthenium complex. As with the work on water-splitting, it was found that the presence of Pt particles on the outsides of the particles can catalyze the recombination reaction.

In terms of practical applications, one serious problem with these catalysts is their lifetime, which is about 200 h in the case of the water-splitting reaction, due to the gradual exposure of Pt particles on the outsides of particles and to oxide degradation. Another problem which will have to be addressed is how to immobilize the particles in coatings for easier handling. In fact, it may be possible to solve both problems simultaneously, as follows. One scenario for the exposure of Pt particles is the exfoliation of niobate layers. This problem could be greatly alleviated by trapping the particles in a thin film of eleastomeric, hydrophilic polymer, using techniques that have been worked out to stabilize colloidal suspensions [54].

In connection with the water-splitting system, it is interesting to note that there already exists a novel type of thin-film fuel cell which can directly utilize a gas mixture of hydrogen and oxygen [55], thus conveniently solving a potentially serious problem for practical applications. The quantitative separation of gas mixtures is not a trivial problem.

4.2.1.5 Approaches Involving Photosensitizer-Modified, Nanocrystalline Semiconductors

Many researchers have realized that the conventional approach using bulk semiconductors usually involves a trade-off between capturing a reasonable portion of the available solar spectrum, using a material with a band gap in the range 1–2 eV, and having a highly stable material which could have a reasonable chance of lasting for several years. Grätzel and co-workers [35, 56] have tried to avoid this dilemma by opting for the ultrastable semiconductor TiO_2 together with adsorbed photosensitizer dyes to extend the spectral absorption. In order to make up for the small absorbance of dye monolayers, they used photoelectrodes which consisted of films of sintered colloidal TiO_2 particles, so that the roughness factors (true-to-apparent surface area) are of the order of 1000 for films that are several μm thick. Many photosensitizer dyes are relatively unstable in contact with photocatalytically active TiO_2, but Grätzel and co-workers found that ruthenium bis-bipyridyl complexes are reasonably stable. One of the dyes that these workers favor, $RuL_2(NCS)_2$, where L = 2,2'-bipyridyl-4,4'-dicarboxylate, has high absorbance and quantum yield over a wide spectral range (Fig. 4.2-7). As shown, the incident photon-to-current conversion efficiency reaches a maximum of ~70% at 510 nm, and, after correction for reflection and scattering losses, the quantum yield approaches unity. This dye is also able to inject electrons into the TiO_2 conduction band very rapidly (<7 ps). The carboxy-

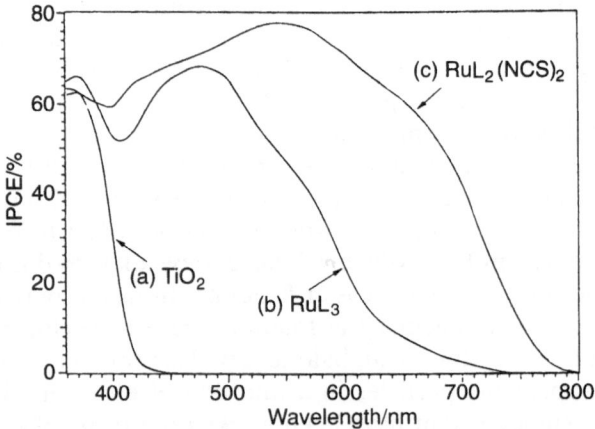

FIG. 4.2-7. Photocurrent action spectrum obtained with a nanocrystalline TiO_2 film supported on conducting glass and derivatized with a monolayer of the Ru complex [Ru(4,4'-Me_2-2,2'-bipy)(4'-PO_3H-terpy)(NCS)] (reprinted with permission (Fig. 11) from Hagfeldt and Grätzel [56]); copyright (1995) American Chemical Society

late groups on the bipyridyl ligands adsorb on the TiO_2 surface and assist in the electronic coupling.

These workers [35] have been able to achieve a relatively high performance for regenerative PECs based on this technology, particularly in view of the polycrystalline nature of the electrode, reaching ~10% efficiency (Table 4.2-1). These cells were based on the I_3^-/I^- redox couple in a nonaqueous electrolyte. The stability of the performance was found to be good over 10 months of testing. The reduction of the oxidized dye by I^- was found to be very fast, while the reaction of photoinjected electrons with I_3^- was found to be slow.

An interesting method to measure the flat band potential of nanocrystalline TiO_2 films was developed using optical absorption spectroscopy as a function of potential [56]. The E_{FB} was found to depend upon the pH of the aqueous electrolyte, upon the protic nature of the nonaqueous solvent, and upon the nature of the cation used with the aprotic nonaqueous solvents. These results are consistent with the electrochromic properties of the TiO_2 films [57] and with the charge storage properties [58]. However, the precise nature of the interactions of both protons and alkali metal cations such as Li^+ with electrochemically or photochemically reduced TiO_2 remains unclear.

4.2.2 Photoelectrochemical Reduction of CO_2

Halmann [59] has stated that, after efforts have been made to limit the consumption of fossil fuels via conservation, at some point in the future it may also become necessary to take steps to chemically fix the combustion product of fossil fuels, i.e., carbon dioxide. Certainly everyone is now becoming more aware of the possible problems associated with increased levels of CO_2 in the atmosphere and global warming. It is also recognized that CO_2 in the atmosphere provides a reservoir for inorganic carbon that could be used as a feedstock for the production of fuels and chemicals.

The use of solar energy to reduce CO_2 photoelectrochemically is an appealing concept and has been studied almost as long as the photoelectrochemical water-splitting reaction. Halmann (see [60]) reported the first work on the reduction of CO_2 at an illuminated p-GaP photocathode in 1978. Our group reported the first work on the use of semiconductor powder dispersions in 1979 [61]. This area has recently been covered in reviews by Halmann [59] and by Lewis and Shreve [62]. Electrode materials that have been examined include p-GaP, p-GaAs, p-CdTe, p-InP, and p-Si. Semiconductors used in the form of powders have included TiO_2, ZnO, CdS, GaP, SiC, WO_3, ZnS, and CdSe. Various types of metal coatings have been examined, including Au, Zn, Pb, Cu, In, Pd, and noble metal alloys. Both aqueous and nonaqueous electrolytes have been examined.

Some pertinent electrode reactions are shown below, together with their redox potentials [63].

				E^0, V (SHE)	
$CO_2 + 2H^+ + 2e^-$	=	$CO + H_2O$		-0.11	(4.2-11)
$CO_2 + 2H^+ + 2e^-$	=	$HCOOH$		-0.20	(4.2-12)
$CO_2 + H_2O + 2e^-$	=	$HCOO^- + OH^-$		-0.72	(4.2-13)
$CO_2 + e^-$	=	CO_2^-		-1.67[a]	(4.2-14)

[a] Approximated by the polarographic half-wave potential.

In aqueous acid solutions, the redox potentials for CO_2 reduction are only slightly negative of the hydrogen potential, so that, on electrode materials which have high overpotentials for hydrogen evolution, CO_2 reduction can be competitive. In aqueous alkaline solutions (Eq. 4.2-13), the potential for CO_2 reduction to formate is actually 0.11 V positive of the hydrogen potential (-0.828 V). The very negative potential for the one-electron reduction of CO_2 to its radical anion is one of the disadvantages of using a nonaqueous solvent. On the other hand, the hydrogen evolution reaction is suppressed in the absence of water. In an aqueous electrolyte, although the redox potentials are less negative, the overall reactions are slow and require substantial overpotentials.

During CO_2 reduction at a p-type semiconductor electrode, the photogenerated minority carriers (electrons) are swept to the electrode surface at close to E_{CB}. Thus the E_{CB} of a particular semiconductor should be negative enough to reach the thermodynamic redox potential of the predominant CO_2 reduction reaction plus any kinetic overpotential. In a dry nonaqueous electrolyte, in the absence of any species that could react with the CO_2^- radical anion or catalyze its disproportionation, E_{CB} should be more negative than -1.67 V (SHE). Figure 4.2-8 shows a band diagram with the band edge positions of p-InP as a typical example, in relation to the redox potentials for CO_2 reduction. It should also be noted here that, even though the potentials required to reduce CO_2 in nonaqueous electrolytes are more negative than those needed in an aqueous electrolyte, an advantage of the former is that many of the semiconductor electrode materials are more stable. This advantage, which has already been mentioned in relation to n-type photoanodes, is also true in the case of p-type photocathodes.

In recent work carried out in our laboratory, p-InP photocathodes were used to reduce CO_2 in a highly concentrated, high pressure (40 atm) CO_2–methanol solvent, with a tetraalkylammonium perchlorate supporting electrolyte. This solvent system has been extensively examined using various metal electrodes [64]. Both high current densities (200 mA cm^{-2}) and high current efficiencies (~95%) were achieved for CO_2 reduction to CO (Fig. 4.2-9). However, the electrode potentials were very negative (~−4 V vs. SHE), due to the redox potential for Eq. 4.2-14, plus the resistances of the electrode and electrolyte (Fig. 4.2-10).

Instead of trying to reduce CO_2 directly at a photocathode, Lewis and Shreve [62] argued that it may actually be more efficient to separate the light-to-electrical energy conversion and the CO_2 reduction reaction, i.e., to use separate systems that are optimized for each. This comment is certainly valid and should be kept in mind in evaluating systems in which the two functions are combined.

FIG. 4.2-8. Energy-
level diagram for
CO_2 photoreduction
at a p-type
semiconductor

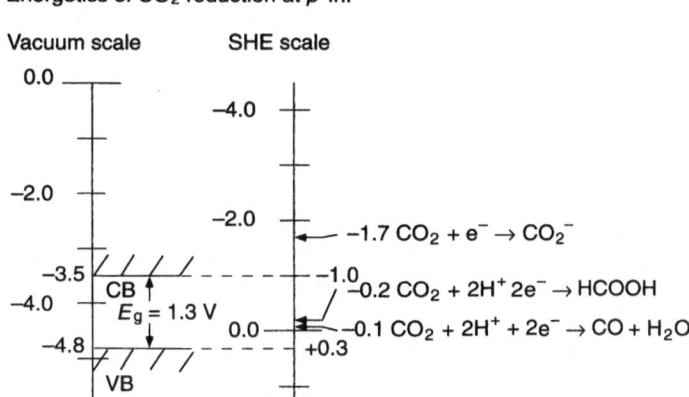

Energetics of CO_2 reduction at p-InP

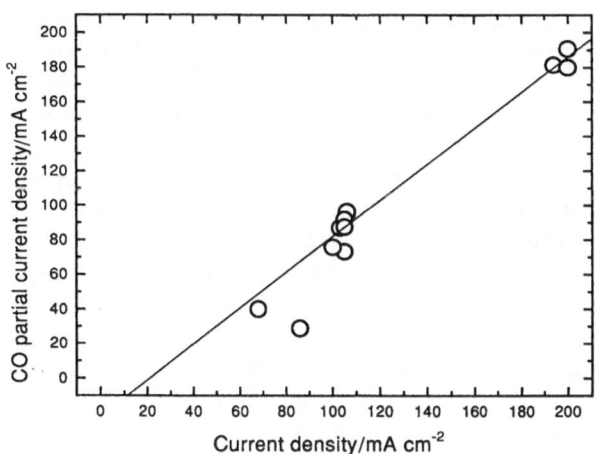

FIG. 4.2-9. Correlation of equivalent current density for CO production with total current density for a p-InP photocathode in CO_2-saturated methanol at 40 atm at 25°C (Hirota K, Tryk DA, Hashimoto K, Fujishima A, unpublished results)

In fact, an analogous comment was made a number of years ago in relation to water-splitting, where the electrode reactions, particularly oxygen evolution, are kinetically demanding [2].

4.2.3 Photocatalytic Decomposition of Air and Water Pollutants

Much of the scientific and technological body of knowledge that has been built up in the area of semiconductor/electrolyte interfaces during the past three decades

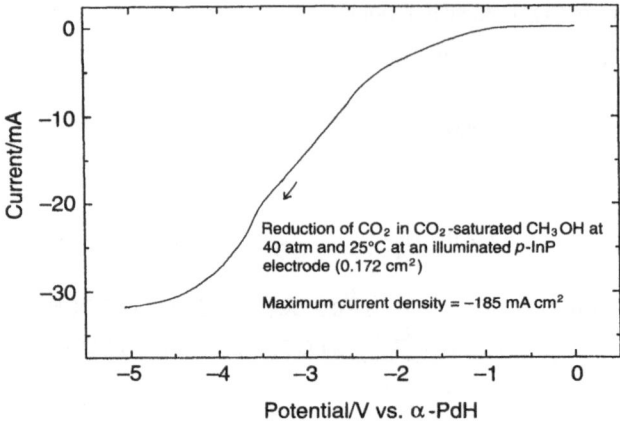

F_{IG}. 4.2-10. Current–potential curve for CO_2 photoreduction at a p-type semiconductor (Hirota K, Tryk DA, Hashimoto K, Fujishima A, unpublished results)

is now being applied in the area of semiconductor-based photocatalysis of various reactions. One of the main driving forces in this development has been the appealing idea of using solar energy in the clean-up of the environment. This area has seen explosive growth, particularly during the past 5 years [65–69]. The list of substances, particularly organics, that can be completely mineralized (i.e., to inorganic products) is impressive. One of the particularly noteworthy applications of semiconductor photocatalysis is the development of TiO_2-coated glass microbubbles for use in the photodegradation of oil and chemical slicks on water, as investigated by the Heller group [68].

Recently, our research group has focused attention on four aspects of semiconductor photocatalysis: (1) development of photoactive TiO_2 films, paints, and paper; (2) analysis of the kinetics of photodegradation of various organic compounds on TiO_2 films; (3) selective monitoring of important reactive species that are photolytically produced, using microelectrodes, on TiO_2 surfaces; (4) photo-assisted selective destruction of cancer cells.

In collaboration with Ishihara Sangyo Kaisha, Ltd., we have developed transparent coatings that can be used on glass and ceramic tiles [70]. These coatings have been found to be able to photodegrade various noxious, malodorous chemicals. They were also able to photodegrade tobacco smoke residues and cooking oil residues. Surprisingly, it has been found that even low-intensity interior lighting is sufficient to induce photocatalysis. Thus it has become conceivable to use such TiO_2 coatings to create self-cleaning, deodorizing surfaces for indoor environments. The possible applications also include public restrooms, which are notorious for foul odors, and lighting fixtures used in highway tunnels, which are easily fouled by automobile exhaust. In addition, these coatings have been found to be antibacterial, and are even effective against methicillin-resistant *Staphylococcus aureus*, which has increasingly become a problem in hospitals.

The development of photoactive TiO_2-based paints is ongoing and promises to widen the field of possible applications considerably, so that any surface that can be painted can be endowed with photocatalytic properties, i.e., self-cleaning,

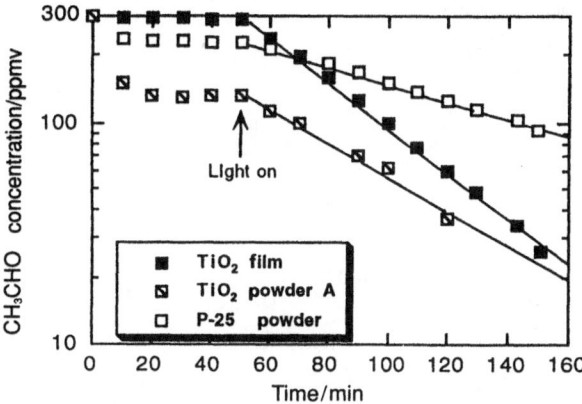

FIG. 4.2-11. Acetaldehyde photodegradation at a film-type TiO$_2$ photocatalyst (Sopyan I, Watanabe M, Murusawa S, Hashimoto K, Fujishima A, unpublished results). TiO$_2$ type-A powder is the starting material used to prepare the TiO$_2$ film-type photocatalyst

deodorizing, and antibacterial. The development of TiO$_2$-containing paper has been motivated by the desire to purify and deodorize room air using small, fan-driven air cleaners fitted with paper filters, or paper blow-by surfaces [71].

The kinetics of photodegradation of various noxious, malodorous compounds, such as acetaldehyde [72], ammonia, and hydrogen sulfide, have been examined on TiO$_2$ thin films. The dark adsorption process was found to be Langmuirian, and the photodegradation was found to follow Langmuir–Hinshelwood kinetics. The TiO$_2$ films were found to exhibit higher photoactivity than the widely used Degussa P-25 powder when the latter was spread over the same size area. Figure 4.2-11 shows a plot of gas-phase acetaldehyde concentration in ambient air as a function of time in contact with an illuminated TiO$_2$ film, and shows the high photoactivity of the film compared with P-25 powder.

Our group has also been trying to gain insight into the reactions occurring on illuminated TiO$_2$ surfaces by monitoring the photogenerated products very close ($\leq 50\,\mu$m) to the surface using microelectrodes [73]. We have been able to measure peroxide production above both TiO$_2$-covered and bare indium–tin oxide surfaces using a horseradish peroxidase-modified microelectrode. We found that peroxide was produced at much higher rates above the bare ITO sections, demonstrating that, like a TiO$_2$ particle, the surface is a short-circuited photoelectrochemical cell. Photoinjected electrons are drawn toward areas on which O$_2$ reduction takes place, which in some cases could be covered with a metallic catalyst such as Pd, and holes are produced on other areas of the TiO$_2$ itself. This type of scheme is shown in Fig. 4.2-12 specifically for the case of acetaldehyde photooxidation, which, as pointed out by Heller, can be mediated via photogenerated holes alone, in contrast to other types of organics [68].

The remaining area our group has been involved in is that of the selective destruction of cancer cells using illuminated TiO$_2$ [74]. We have found that cancer

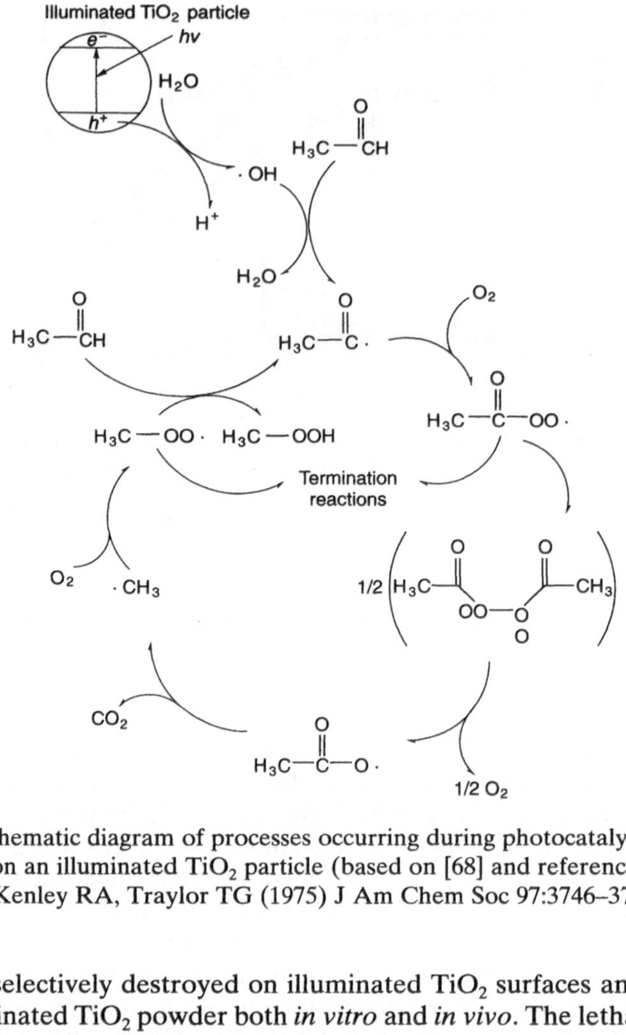

Illuminated TiO$_2$ particle

hv

e^-

h^+

H$_2$O

·OH

H$^+$

H$_2$O

H$_3$C—CH (with =O)

H$_3$C—CH (with =O)

H$_3$C—C· (with =O)

O$_2$

H$_3$C—C—OO· (with =O)

H$_3$C—OO·   H$_3$C—OOH

Termination reactions

O$_2$ ·CH$_3$

1/2 (H$_3$C—C(=O)—OO—O—O—C(=O)—CH$_3$)

CO$_2$

H$_3$C—C—O· (with =O)

1/2 O$_2$

FIG. 4.2-12. Schematic diagram of processes occurring during photocatalytic oxidation of acetaldehyde on an illuminated TiO$_2$ particle (based on [68] and references cited therein (Clinton NA, Kenley RA, Traylor TG (1975) J Am Chem Soc 97:3746–3751, 3752–3757)

cells can be selectively destroyed on illuminated TiO$_2$ surfaces and in the presence of illuminated TiO$_2$ powder both *in vitro* and *in vivo*. The lethal effects have been ascribed to the photoproduction of both hydroxyl radicals and hydrogen peroxide. Thus, it is possible to envision cancer treatment methodologies involving illuminated TiO$_2$ particles, for example, using fiber optic probes.

4.2.4 Concluding Remarks

Even though photoelectrochemical cells have not come into widespread use in solar energy conversion, for various reasons, the scientific and technological knowledge base that has been accumulated has been extremely useful in a number of other areas, particularly in the area of semiconductor-based photocatalysis. Other areas that might be mentioned include the photoelectrochemical characterization of semiconductor surfaces [75] and the photoelectrochemical processing of semiconductor surfaces, such as for electronics applications [76]. It is also important to continue to explore the possibilites for solar energy conversion, to explore the fundamental aspects of the semiconductor/electrolyte interface, and to explore new, as yet unimagined, applications.

References

1. Fujishima A, Honda K (1972) Electrochemical photolysis of water at a semiconductor electrode. Nature 238:37–38
2. Gerischer H (1979) Solar photoelectrolysis with semiconductor electrodes. In: Seraphin BO (ed) Solar energy conversion; solid-state physics aspects. Springer, Berlin, Chap. 4
3. Nozik A (1978) Photoelectrochemistry: applications to solar energy conversion. In: Rabinowitch BS, Schurr JM, Strauss HL (eds) Annual review of physical chemistry, vol 29. Annual Reviews, Palo Alto, pp 189–222
4. Wrighton MS (1979) Photoelectrochemical conversion of optical energy to electricity and fuels. Acc Chem Res 12:303–310
5. Heller A (1981) Conversion of sunlight into electrical power and photoassisted electrolysis of water in photoelectrochemical cells. Acc Chem Res 14:154–162
6. Morrison SR (1980) Electrochemistry at semiconductor and oxidized metal electrodes. Plenum, New York
7. Watanabe T, Fujishima A, Honda K (1976) Photoelectrochemical reactions at $SrTiO_3$ single crystal electrode. Bull Chem Soc Jpn 49:355–358
8. Sze SM (1981) Physics of semiconductor devices, 2nd edn. Wiley, New York, Chap. 14
9. Kolodinski S, Werner JH, Wittchen T, Queisser HJ (1993) Quantum efficiencies ex-ceeding unity due to impact ionization in silicon solar cells. Appl Phys Lett 63:2405–2407
10. Bard AJ, Fox MA (1995) Artificial photosynthesis: solar splitting of water to hydrogen and oxygen. Acc Chem Res 28:141–145
11. Wang A, Zhao J, Green MA (1990) 24% efficient silicon solar cells. Appl Phys Lett 57:602–604
12. Hodes G, Manassen J, Cahen D (1985) Photoelectrochemical energy conversion and storage using polycrystalline chalcogenide electrodes. Nature 261:403–404
13. Inoue T, Watanabe T, Fujishima A, Honda K, Kohayakawa K (1979) Suppression of surface dissolution of CdS photoanode by reducing agents. J Electrochem Soc 124:719–722
14. Lewis N (1991) An analysis of charge transfer rate constants for semiconductor/liquid interfaces. In: Strauss HL, Babcock GT, Leone SR (eds) Annual review of physical chemistry, vol 42. Annual Reviews, Palo Alto, pp 543–580
15. Gerischer H, Gobrecht (1976) On the power characteristics of electrochemical solar cells. Z Phys Chem 80:327–330
16. Chang KC, Heller A, Schwartz B, Menezes S, Miller B (1977) Stable semiconductor liquid junction cell with 9 percent solar-to-electrical conversion efficiency. Science 196:1097–1099
17. Heller A, Chang KC, Miller B (1977) Spectral response and efficiency relations in semiconductor liquid junction solar cells. J Electrochem Soc 124:697–700
18. Heller A, Schwartz GP, Vadimsky RG, Menezes S, Miller B (1978) Output stability of n-CdSe/Na_2S-S-NaOH/C solar cells. J Electrochem Soc 125:1156–1160
19. Parkinson BA, Heller A, Miller B (1978) Enhanced photoelectrochemical solar-energy conversion by gallium arsenide surface modification. Appl Phys Lett 33:521–533

20. Johnston WD Jr, Leamy HJ, Parkinson BA, Heller A, Miller B (1980) Effect of ruthenium ions on grain boundaries in gallium arsenide thin film photovoltaic devices. J Electrochem Soc 127:90–95
21. Heller A, Lewerenz HJ, Miller B (1980) Combined ruthenium lead surface treatment of gallium arsenide photoanodes. Ber Bunsenges Phys Chem 84:592–595
22. Lewerenz HJ, Heller A, DiSalvo FJ (1980) Relationship between surface morphology and solar conversion of WSe_2 photoanodes. J Am Chem Soc 102:1877–1880
23. Menezes S, Lewerenz HJ, Bachmann KJ (1983) Efficient and stable solar cell by interfacial film formation. Nature 305:615–616
24. Cahen D, Chen YW (1984) n-$CuInSe_2$ based photoelectrochemical cells: improved, stable performance in aqueous polyiodide through rational surface and solution modification. Appl Phys Lett 45:746–748
25. Licht S, Tenne R, Dagan G, Hodes G, Manassen J, Cahen D, Triboulet R, Rioux J, Levy-Clement C (1985) High efficiency n-$Cd(Se,Te)/S^=$ photoelectrochemical cell resulting from solution chemistry control. Appl Phys Lett 46:608–610
26. Tenne R, Wold A (1985) Passivation of recombination centers in n-WSe_2 yields high efficiency (>14%) photoelectrochemical cell. Appl Phys Lett 47:707–709
27. Tufts BJ, Abrahams IL, Santangelo PG, Ryba GN, Casagrande LG, Lewis NS (1987) Chemical modification of n-GaAs electrodes with Os^{3+} gives a 15% efficient solar cell. Nature 326:861–863
28. Licht S, Peramunage D (1990) Efficient photoelectrochemical solar cells from electrolyte modification. Nature 345:330–333
29. Heller A, Miller B, Lewerenz HJ, Bachmann KJ (1980) An efficient photocathode for semiconductor liquid junction cells: 9.4% solar conversion efficiency with p-InP/VCl_3-VCl_2-HCl/C. J Am Chem Soc 102:6555–6556
30. Heller A, Miller B, Thiel FA (1981) 11.5% solar conversion efficiency in the photocathodically protected p-InP/V^{3+}-V^{2+}-HCl/C semiconductor liquid junction cell. Appl Phys Lett 38:282–284
31. Gronet CM, Lewis NS, Cogan G, Gibbons J (1983) n-Type silicon photoelectrochemistry in methanol: design of a 10.1% efficient semiconductor/liquid junction solar cell. Proc Natl Acad Sci USA 80:1152–1156
32. Gibbons JW, Cogan GW, Gronet CM, Lewis NS (1984) A 14% efficient nonaqueous semiconductor/liquid junction solar cell. Appl Phys Lett 45:1095–1097
33. Heben MJ, Kumar A, Zheng C, Lewis NS (1989) Efficient photovoltaic devices for InP semiconductor/liquid junctions. Nature 340:621–623
34. Tufts BJ, Casagrande LG, Lewis NS, Grunthaner FJ (1990) Erratum: correlations between the interfacial chemistry and current-voltage behavior of n-GaAs/liquid junctions [Appl Phys Lett 57, 1242 (1990)]. Appl Phys Lett 57:2262–2264
35. Nazeeruddin MK, Kay A, Rodicio I, Humphry-Baker R, Müller E, Liska P, Vlachopoulos N, Grätzel M (1993) Conversion of light to electricity by cis-X_2bis(2,2′-bieyridyl-4,4′dicarboxylate)ruthenium(II) charge transfer sensitizers ($X = Cl^-$, Br^-, I^-, CN^- and SCN^-) on nanocrystalline TiO_2 electrodes. J Am Chem Soc 115:6382–6390
36. Narayanan S, Wenham SR, Green MA (1990) 17.8-percent efficiency polycrystalline silicon solar cells. IEEE Trans Electron Dev 37:382–384
37. Gabor AM, Tuttle JR, Albin DS, Contreras MA, Noufi R, Hermann AM (1994) High-efficiency $CuIn_xGa_{1-x}Se_2$ solar cells made from $(In_x,Ga_{1-x})_2Se_3$ precursor films. Appl Phys Lett 65:198–200

38. Bertness KA, Kurtz SR, Friedman DJ, Kibbler AE, Kramer C, Olson JM (1994) 29.5%-efficient GaInP/GaAs tandem solar cells. Appl Phys Lett 65:989–991
39. Zhao J, Wang A, Altermatt P, Green MA (1995) Twenty-four percent efficient silicon solar cells with double layer antireflection coatings and reduced resistance loss. Appl Phys Lett 66:3636–3638
40. Henry CH (1980) Limiting efficiencies of ideal single and multiple energy gap terrestrial solar cells. J Appl Phys 51:4494–4500
41. Schefold J, Vetter M (1994) Solar energy conversion at the p-InP/vanadium$^{3+/2+}$ semiconductor/electrolyte contact. J Electrochem Soc 141:2040–2048
42. Parkinson B (1984) On the efficiency and stability of photoelectrochemical devices. Acc Chem Res 17:431–437
43. Lewis NS (1990) Mechanistic studies of light-induced charge separation at semiconductor/liquid interfaces. Acc Chem Res 23:176–183
44. Swain GM (1994) The use of CVD diamond thin films in electrochemical systems. Adv Mater 6:388–392
45. Reuben C, Galun E, Cohen H, Tenne R, Kalish R, Muraki Y, Hashimoto K, Fujishima A, Butler JM, Levy-Clement C (1995) Efficient reduction of nitrite and nitrate to ammonia using thin-film B-doped diamoned electrodes. J Electroanal Chem 396:233–239
46. Tributsch H (1992) Electronic structure, coordination photoelectrochemical pathways and quantum energy conversion by layered transition metal dichalcogenides. In: Aruchamy A (ed) Photoelectrochemistry and photovoltaics of layered semiconductors. Kluwer, Dordrecht
47. Levy-Clement C, Tenne R (1992) Modification of surface properties of layered compounds by chemical and (photo) electrochemical procedures. In: Aruchamy A (ed) Photoelectrochemistry and photovoltaics of layered semiconductors. Kluwer, Dordrecht
48. Ueno K, Shimada T, Saiki K, Koma A (1990) Heteroepitaxial growth of layered transition metal dichalcogenides on sulfur-terminated GaAs (111) surfaces. Appl Phys Lett 56:327–329
49. Mayer T, Lehmann J, Pettenkofer C, Jaegermann W (1992) Coadsorption of Na and Br$_2$ on WS$_2$ (0001). Creating a surface redox couple? Chem Phys Lett 198:621
50. Jaegermann W (1996) The semiconductor/electrolyte interface: a surface science aproach. In: White BE et al. (eds) Modern aspects of electrochemistry, vol. 30. Plenum, New York, Chap 1
51. Kudo A, Sayama K, Tanaka A, Asakura K, Domen K, Maruya K, Onishi T (1989) Nickel-loaded K$_4$Nb$_6$O$_{17}$ photocatalyst in the decomposition of H$_2$O into H$_2$ and O$_2$: structure and reaction mechanism. J Catal 120:337–352
52. Sayama K, Tanaka A, Domen K, Maruya K, Onishi T (1991) Photocatalytic decomposition of water over platinum-intercalated K$_4$Nb$_6$O$_{17}$. J Phys Chem 95:1345–1348
53. Kim YI, Atherton SJ, Brigham ES, Mallouk TE (1993) Sensitized layered metal oxide semiconductor particles for photochemical hydrogen evolution from nonsacrificial electron donors. J Phys Chem 97:11802–11810
54. Grätzel M, Kalyanasundaram K, Kiwi J (1982) Visible light induced cleavage of water into hydrogen and oxygen in colloidal and microheterogeneous systems. In: Clarke MJ et al. (eds) Structure and bonding 49: solar energy materials. Springer, Berlin, pp 37–125
55. Dyer CK (1990) A novel thin-film electrochemical device for energy conversion. Nature 343:547–548

56. Hagfeldt A, Grätzel M (1995) Light-induced redox reactions in nanocrystalline systems. Chem Rev 95:49–68
57. Hagfeldt A, Vlachopoulos N, Grätzel M (1994) Fast electrochromic switching with nanocrystalline oxide semiconductor films. J Electrochem Soc 141:L82–L84
58. Huang SY, Kavan L, Exnar I, Grätzel M (1995) Rocking chair lithium battery based on nanocrystalline TiO_2 (anatase). J Electrochem Soc 142:L142–L144
59. Halmann MM (1978) Photoelectrochemical reduction of aqueous carbon dioxide on p-type gallium phosphide in liquid junction solar cells. Nature 275:115–116
60. Halmann MM (1993) Chemical fixation of carbon dioxide—methods for recycling CO_2 into useful products. CRC Press, Boca Raton, Chaps. 1 and 8
61. Inoue T, Fujishima A, Konishi S, Honda K (1979) Photoelectrochemical reduction of carbon dioxide in aqueous suspensions of semiconductor powders. Nature 277:637–638
62. Lewis NS, Shreve GA (1993) Photochemical and photoelectrochemical reduction of carbon dioxide. In: Sullivan BP et al. (eds) Electrochemical and electrocatalytic reactions of carbon dioxide. Elsevier, Amsterdam, Chap. 8
63. Randin JP (1976) Carbon. In: Bard AJ (ed) Encyclopedia of electrochemistry of the elements, vol VII. Marcel Dekker, New York, Chap. VII-1
64. Saeki T, Hashimoto K, Fujishima A, Kimura N, Omata K (1995) Electrochemical reduction of CO_2 with high current density in a CO_2–methanol medium. J Phys Chem 99:8440–8446
65. Ollis DF, Al-Ekabi, H (eds) (1993) Photocatalytic purification and treatment of water and air. Elsevier, Amsterdam
66. Fox MA, Duly MT (1993) Heterogeneous photocatalysis. Chem Rev 93:341–357
67. Hoffman MR, Martin ST, Choi W, Bahnemann DW (1995) Environmental applications of semiconductor photocatalysis. Chem Rev 95:69–96
68. Heller A (1995) Chemistry and applications of photocatalytic oxidation of thin organic films. Acc Chem Res 28:503–508
69. Linsebigler AL, Lu G, Yates JT Jr (1995) Photocatalysis on TiO_2 surfaces: principles, mechanisms and selected results. Chem Rev 95:735–758
70. Negishi N, Iyoda T, Hashimoto K, Fujishima A (1995) Preparation of transparent TiO_2 thin film photocatalyst and its photocatalytic activity. Chem Lett 841–842
71. Matsubara H, Takada M, Koyama S, Hashimoto K, Fujishima A (1995) Photoactive TiO_2 containing paper: preparation and its photocatalytic activity under weak UV light illumination. Chem Lett 767–768
72. Sopyan I, Murasawa S, Hashimoto K, Fujishima A (1994) Highly efficient TiO_2 film photocatalyst: degradation of gaseous acetaldehyde. Chem Lett 723–726
73. Sakai H, Baba R, Hashimoto K, Fujishima A, Heller A (1995) Local detection of photoelectrochemically produced H_2O_2 with a "wired" horseradish peroxidase microsensor. J Phys Chem 99:11896–11900
74. Kubota Y, Shuin T, Kawasaki C, Hosaka M, Kitamura H, Cai R, Sakai H, Hashimoto K, Fujishima A (1994) Photokilling of T-24 human bladder cancer cells with titanium dioxide. Br J Cancer 70:1107–1111
75. Tomkiewicz M (1992) Photoelectrochemical characterization. In: McHardy J, Ludwig F (eds) Electrochemistry of semiconductors and electronics: processes and devices. Noyes Publications, Park Ridge, Chap. 5
76. Rauh RD (1992) Photoelectrochemical processing of semiconductors. In: McHardy J, Ludwig F (eds) Electrochemistry of semiconductors and electronics: processes and devices. Noyes Publications, Park Ridge, Chap. 4

4.3 Mechanoactive Molecular Systems

Masahiro Irie

Shape memory polymers show a mechanoactive effect. The polymers directly convert heat, chemical or light energy into mechanical work. One polymer which performs this effect is rubber. Rubber shows an elastic property at room temperature. However, at temperatures below the glass transition temperature, T_g, the elasticity is lost. When the expanded rubber is cooled below T_g, the elongated shape is fixed and does not return to its original shape as long as the temperature is kept below T_g. However, the deformed shape returns to its original shape when the rubber is heated above T_g. The strain is released and the recovery stress can be used for mechanical work. Such a mechanoactive effect is induced not only by heating, but also by chemicals or light.

The mechanisms of the mechanoactive effect of polymers are classified as follows.

Case 1. The original shape is formed from the polymer powder or pellets by melt-molding. If necessary, the polymer is crosslinked by the addition of crosslinking agents or by radiation. Then the shape is deformed under stress at a temperature near or above T_g, or melting temperature, T_m. The deformed shape is fixed by cooling below T_g or T_m. The deformed form reverts to the original shape by heating the sample above T_g or T_m.

$$\text{Polymer powder, pellets} \xrightarrow{\substack{\text{Crosslinking} \\ \text{Melt-molding}}} L_0 \xrightarrow{T > T_g, T_m} L_0 + \Delta L$$

$$\xrightarrow{T < T_g, T_m} L_0 + \Delta L \xrightarrow{T > T_g, T_m} L_0$$

Case 2. The polymer shape is reversibly controlled by chemical reactions of the polymer. The original shape is formed by crosslinking the polymer chains as Case 1.

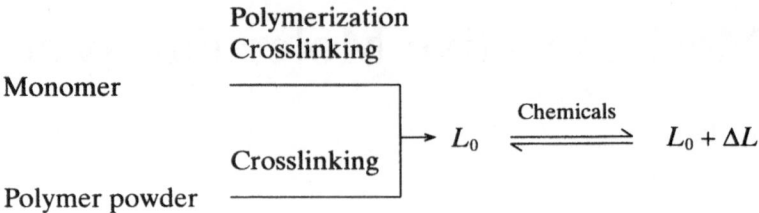

Case 3. The polymer shape is reversibly controlled by photo- or electrochemical reactions of the polymer. The original shape is formed by crosslinking the polymer chains as Case 1.

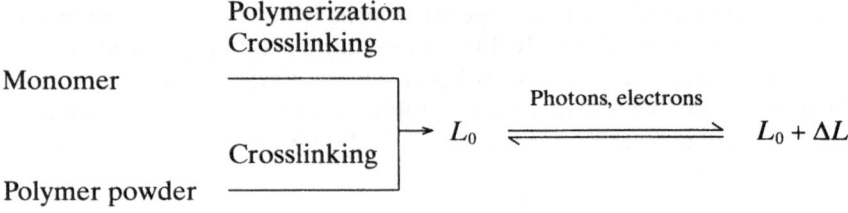

4.3.1 Thermomechanical Effect

Table 4.3-1 shows various types of thermomechanoactive polymers, along with the physical interactions used for memorizing the original and transient shapes of the polymers. To maintain a stable shape, polymer chains should have three-dimensional networks. Interpolymer chain interactions which are useful for constructing polymer networks are crystal, aggregate, or glassy-state formation, chemical crosslinking, and chain entanglement. The latter two interactions are permanent, and are used for constructing the original shape. The other interactions are thermally reversible, and are used for maintaining the transient shapes. Two examples are described below.

TABLE 4.3-1. Thermomechanoactive polymers and mechanism

Interchain interaction	Polynorbornene	*trans*-polyisoprene	Styrene–butadiene copolymer	Polyethylene
Entanglement	O			
Crosslinking		O		O
Microcrystals		T	O, T	T
Glassy state	T			

O, used for memorizing the original shape.
T, used for maintaining the transient shape.

4.3.1.1 Polyisoprene

Polyisoprene has four microstructures, as shown below.

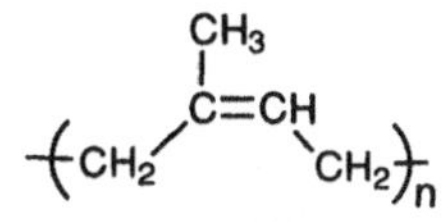

$$CH_3$$
$$C=CH$$
$$-(CH_2 \qquad CH_2)_n-$$

CSIP

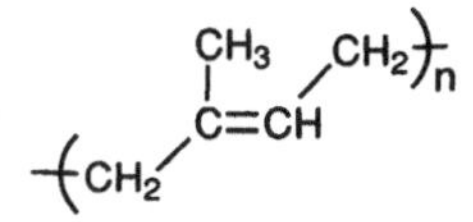

$$CH_3 \quad CH_2)_n$$
$$C=CH$$
$$-(CH_2$$

TSIP

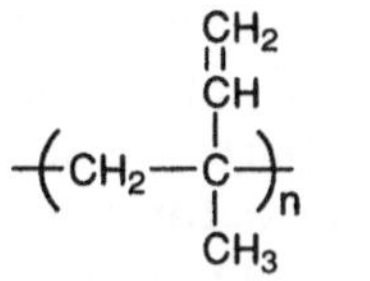

$$CH_2$$
$$CH$$
$$-(CH_2-C)_n-$$
$$CH_3$$

IP12

$$CH_2$$
$$C-CH_3$$
$$-(CH_2-CH)_n-$$

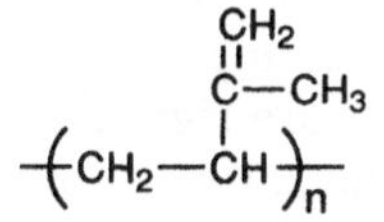

IP34

Among these, *trans*-form polyisoprene (TSIP) has a T_m of 67°C, and the degree of crystallinity is around 40%. The polymer can be chemically crosslinked by peroxides. Below T_m, the TSIP polymer has a chain network whose links are connected to each other by both chemical bonds and microcrystal parts of the polymer. Above T_m, the microcrystal phase disappears and only the chemical bonds remain as crosslinking points. The polymer has an elasticity similar to rubber at temperatures above T_m.

The original shape is formed by heating the polymer powder or pellets with crosslinking agents around 145°C for 30 min, and then cooling it to room temperature. The crosslinked networks construct the original shape. Subsequent deformation can be done by heating the polymer to around 80°C. The transient shape is fixed by microcrystal parts which are formed in the polymer during the cooling process. The deformed shape returns to the original one upon heating above 80°C.

The TSIP polymer has a relatively large recovery or shrinking stress (10–30 Kg cm^{-2}) and can be used for mechanical work. The stress increases in proportion to the stretching ratio (200–400%). The disadvantage of this polymer is its lack of durability because of the presence of reactive diene groups in the polymer main chain.

4.3.1.2 Styrene–Butadiene Copolymer

This polymer has the chemical structure shown below. It is made of polystyrene and polybutadiene parts, and these are connected in a chain.

$$\left(CH_2-\underset{\underset{\bigcirc}{|}}{CH}\right)_n \left(CH=CH-CH=CH\right)_m$$

SBC

The mechanoactive behavior of this copolymer is illustrated in Fig. 4.3-1 [2]. Above 120°C the copolymer melts and flows. Below 120°C the polystyrene parts aggregate to form the initial shape. The aggregate or glassy-state formation of polystyrene parts is used to memorize the original shape. The deformation is carried out by heating the sample to around 80°C, at which temperature the polystyrene parts are rigid but the polybutadiene parts are flexible. Below 40°C the polybutadiene parts become crystallized and the deformed shape is fixed. The shape returns to the original one upon heating to around 80°C, at which temperature the microcrystals in the polybutadiene parts melt. Microcrystal formation in the polybutadiene parts is used to fix the transient deformed shape.

4.3.2 Chemomechanical Effect

The shape of polymers can also be changed by chemicals when the polymers contain reactive pendant groups. The polymers convert chemical energy into mechanical work. The most extensively studied polymers are polyelectrolytes. Figure 4.3-2 shows several polymers which dissociate into ions by changing the pH. When these polymers are connected to each other by crosslinking agents, the polymer networks make a gel or film which changes shape on the addition of acid or alkaline in water and performs mechanical work.

Figure 4.3-3 shows the reversible shape-change of a poly(acrylic acid) film by alternate additions of acid and alkaline [3]. The film (5 cm in length with a load of 3.5 N) shrinks as much as 1 cm by the addition of 0.02 N HCl and expands again to the initial shape by the addition of 0.02 N NaOH. The cycles can be repeated as many as 2000 times.

If we apply the gels to mechanical work, the contraction force should be large. In order to achieve a strong contraction force, the gels must be constructed with a large number of crosslinked networks or with rigid chemical structures. One example of such a gel is a PANG gel [4].

The gel is prepared by peroxidation and subsequent hydrolysis of polyacrylonitrile fiber. The gel showed considerable swelling in an alkaline solution and collapsed in an acidic solution at 750 g cm^{-2} load. The change in length with swelling was about 80% and the contraction/elongation response time was less than 20s. The maximum contraction force was ca. 12 Kg cm^{-2}. The contraction force is comparable to that of living skeletal muscle.

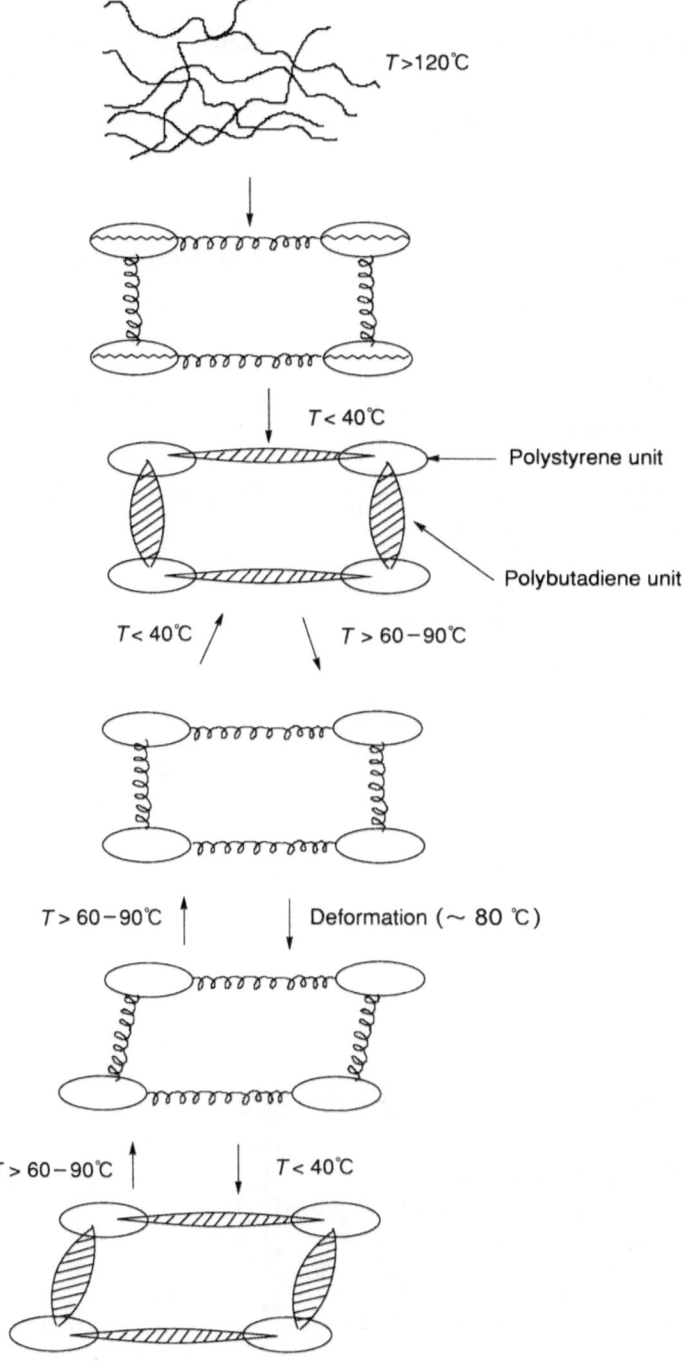

FIG. 4.3-1. Schematic illustration of the shape memory effect of styrene–butadiene copolymer

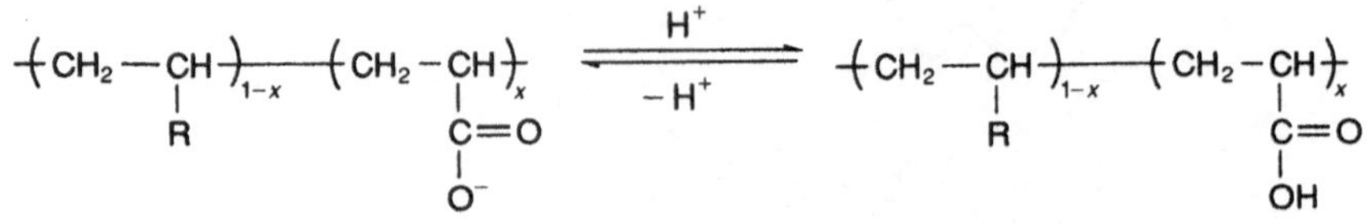

$$\left\{ CH_2-CH \right\}_{1-x} \left(CH_2-CH \right)_x \overset{H^+}{\underset{-H^+}{\rightleftharpoons}} \left\{ CH_2-CH \right\}_{1-x} \left(CH_2-CH \right)_x$$

PAAC

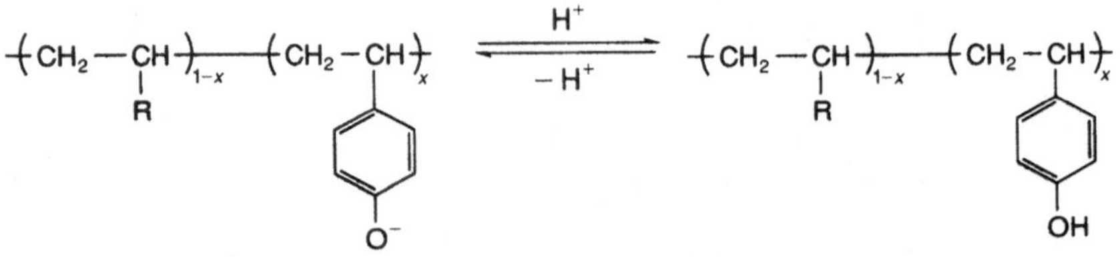

PPHA

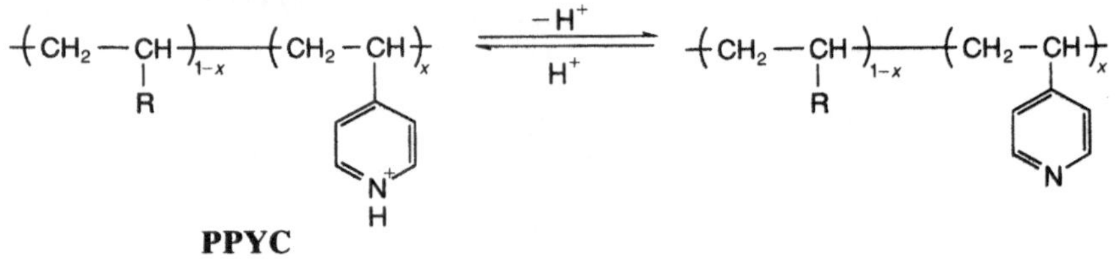

PPYC

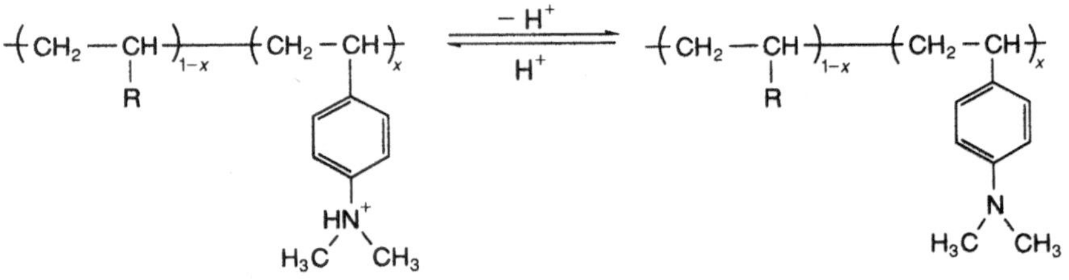

PDAC

FIG. 4.3-2. pH-sensitive polymers

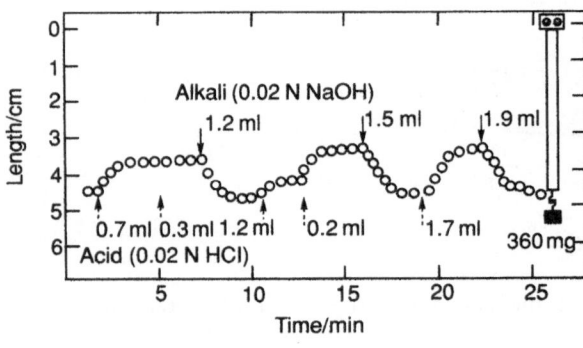

FIG. 4.3-3. Shape changes of a poly(acrylic acid) film by alternative addition of acid (0.02 N HCl) and alkali (0.02 N NaOH) in water. Initial length, 5 cm; weight, 350 mg

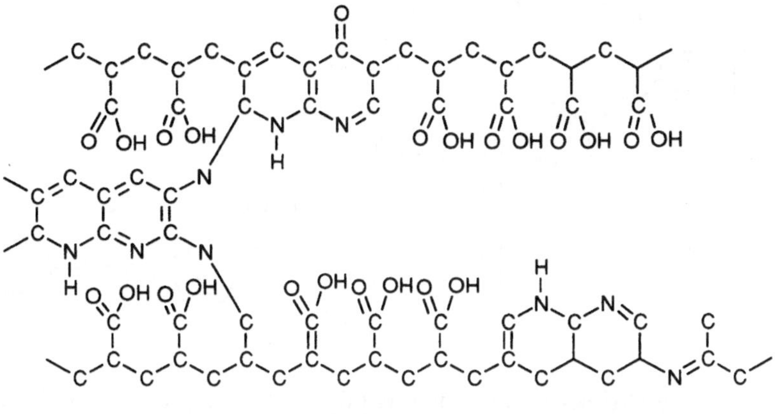

PANG

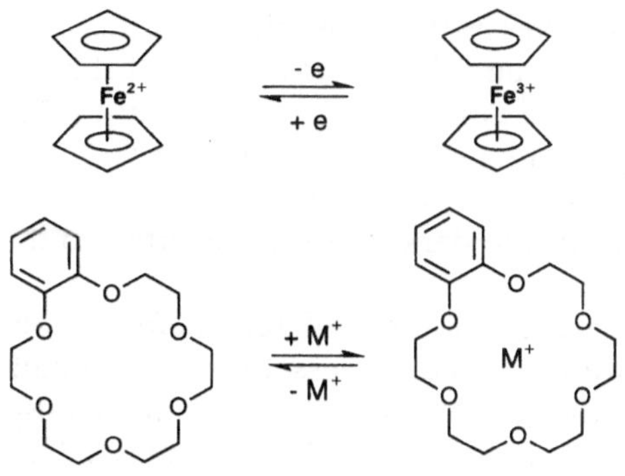

FIG. 4.3-4. Chemical responsive chromophores

Chemically responsive gels can also be constructed by introducing ferrocene or crown ether groups into polymers, as shown in Fig. 4.3-4 [5]. The gels change shape by oxidation/reduction reactions or by the addition of metal ions.

4.3.3 Photomechanical Effect

When a mechanoactive function can be switched on or off by light, the materials involved could have various applications. It is well known that many photochromic molecules are transformed under photoirradiation to another isomer, which returns to the initial isomer either thermally or photochemically (see Sect. 3.2.1). These changes induce shape changes in the polymers into which the chromophores are incorporated.

The use of structural changes in photochromic molecules to produce a size-change in a polymer solid was proposed for the first time by Merian [6]. He studied a nylon filament fabric, 6 cm wide and 30 cm long, dyed with 15 mg g^{-1} azo-

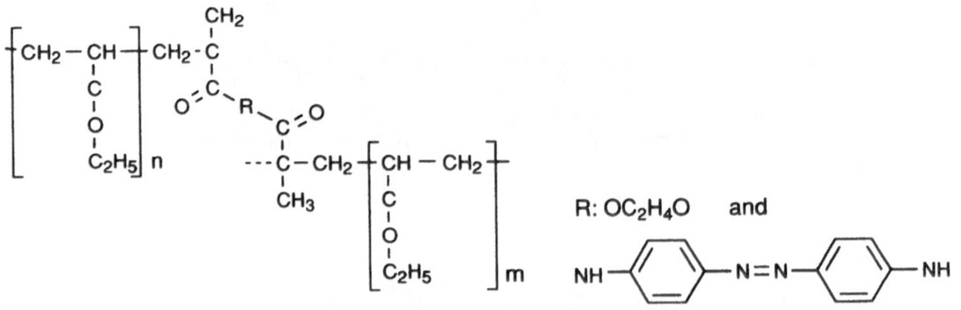

PAEN

dye. The azo-dye shrinks by as much as 60% with the photochromic reaction. After exposure to a xenon lamp, he found that the dyed fabric shrank 0.33 mm. Since this finding, many materials have been reported to exhibit photo-stimulated deformation.

A typical example is a polymer having azo-dyes as the crosslinking agents. The polymer PAEN shown below was synthesized, and the photostimulated shape-change was measured [7]. When the azo-dye was converted from the *trans* to the *cis* form by ultraviolet irradiation, the polymer film shrank by as much as 0.27%, while it expanded by as much as 0.16% with visible irradiation, which isomerizes the azo-dye from the *cis* to the *trans* form. In the dark, the film reverted to its initial length.

Although many polymer systems showing photostimulated deformation have been reported, the deformations were limited to less than 5%. Small deformations make it difficult to judge whether the effect is due to photochemistry or just to photo-heating. If the deformation is larger than 20%, we may safely say that it is due to the photochemical effect.

Polyacrylamide gels PTAN containing a small amount of triphenyl-methane leucohydroxide or leucocyanide groups were prepared [8]. The leucohydroxide or leucocyanide groups dissociate into ion pairs by photoirradiation, and recombine with each other in the dark.

A disk-shaped gel (10 mm diameter and 2 mm thickness) having 3.7 mol% triphenylmethane leucohydroxide residues showed photostimulated dilation in water, as shown in Fig. 4.3-5. Upon ultraviolet irradiation the gel swelled by as much as three times its original weight in 1 h. In the dark, the gel contracted to its initial weight in 20 h. The cycles of dilation and contraction could be repeated several times. When the leucohydroxide residues were replaced with leucocyanide residues, the gel swelled as much as ten times in weight and to twice its length.

This gel expansion is interpreted as follows. Upon photo-irradiation, the leuco-derivatives in the polymer chains dissociate into ions. The formation of ions, fixed cations, and free anions generates an osmotic pressure difference between the inside of the gel and the outer solution, and this osmotic effect is considered to be responsible for the gel expansion. The polyacrylamide gel is the first example

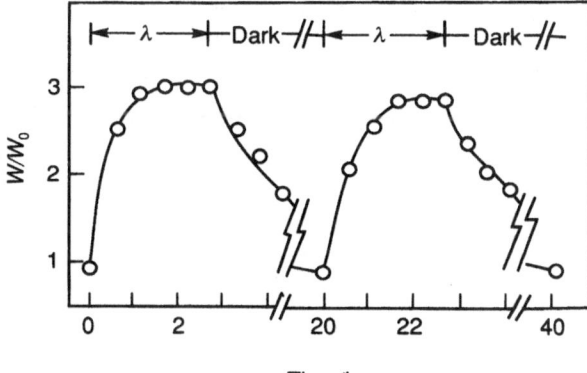

PTAN

FIG. 4.3-5. Photostimulated reversible shape changes of a polyacrylamide gel having pendant triphenylmethane leucohydroxide groups in water at 25°C. $\lambda > 270\,nm$

showing a reversible deformation of more than 100%. The electric field can also control the shape of polymers, although the mechanism is not yet clear [9, 10].

References

1. Japan Kokai (1987) Shape-memory materials and their application. JP 62-192440, Kuraray
2. Japan Kokai (1988) Shape-memory polymer materials. JP 63-179955, Asahi Chemical
3. Kuhn W, Katchalsky A, Eisenberg H (1950) Reversible dilation and contraction by changing the state of ionization of high polymer acid networks. Nature 165:514–516

4. Okui N, Umemoto S (1993) Contraction behavior of poly(acrylo-nitrile) gel fibers. Application to an artificial muscle. In: Tsuruta T, Doyama M, Seno M, Imanishi Y (eds) New functionality materials. Elsevier, Amsterdam, p 165
5. Irie M, Misumi Y, Tanaka T (1993) Stimuli-responsive polymers: chemical-induced reversible phase separation of an aqueous solution of poly(N-isopropylacrylamide) with pendant crown ether groups. Polymer 34:4531–4535
6. Merian E (1966) Organic fiber-formation research. Text Res J 36:612–615
7. Eisenbach CD (1982) Isomerization of aromatic azo chromophores in poly(ethyl acrylate) networks and photomechanical effect. Polymer 21:1175–1178
8. Irie M, Kungwatchakun D (1986) Photoresponsive polymers. 8. Reversible photostimulated dilation of polyacrylamide gels having triphenylmethane leuco derivatives. Macromolecules 19:2476–2480
9. Irie M (1990) Photoresponsive polymers. Adv Polym Sci 94:27–65
10. Irie M (1993) Stimuli-responsive poly(N-isopropylacrylamide). Photo- and chemical-induced phase transitions. Adv Polym Sci 110:49–65

5. Molecular Devices and Molecular Systems —Present Status and the Future

Takeo Shimidzu and Kenichi Honda

The extremely rapid advance of modern science and engineering has taken many by surprise. In particular, solid state devices have played a very important role in the progress of electronics and information engineering, and the lives of modern human beings have been greatly improved by this progress.

However, a living system such as a human being can be regarded as the most advanced system in nature, and in such a system there are no solid state devices, but only molecular systems. Therefore, it is expected that molecular systems will be the most important functional artificial systems of the future.

This book has given an overview of many possible future applications of molecular systems, and recent developments in each field have been introduced. However, accessing molecular systems still remains a problem, and much research is needed into the manner in which molecules and/or molecular systems can be connected in order to establish effective input and output terminals.

Figure 5-1 [1] shows an attempt to establish an interconnection between a chromophore molecule and a substrate electrode. Here the chromophores are impregnated in multilayers of amphiphiles, and the hydrophilic parts (circles) of the amphiphile molecule are crosslinked to become electro-conductive. In this way, the electric response of the photoexcited chromophore leads to the substrate electrode.

In an ideal case, each molecule should be independently accessed in order to obtain the highest efficiency in terms of molecular function. Recent progress indicates that nanotechnology is poised to solve such problems.

Multifunctionalization is a distinguishing characteristic of molecules which makes the accumulation of functions possible. Various types of function-integrated molecules and devices can be designed with combinations of intra- and intermolecular functional units. The role of the connector is important for both multifunctionalization and accumulation. The connector, for example, the "molecular wire", governs the properties of the multifunctionalized molecule. The function of the resulting molecularly connected functional molecular units is controlled via the connector. The function depends on the nature of the connector, for example, conducting vs. insulating, or long vs. short. Energetic, electronic, and/or geometric considerations are important in order to realize the desired

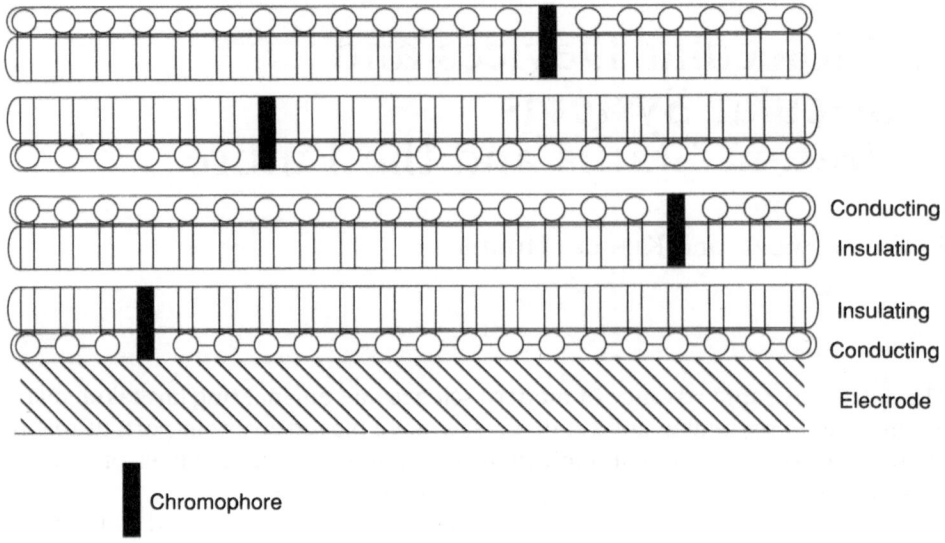

Chromophore

FIG. 5-1. Attempt to establish access to a molecular device

function. The connector can be compared to an interconverter in drawing upon the intrinsic quantum states of the functional unit in order to optimize its functions. The connector plays the role of a barrier, in a broad sense, to control the multifunctionality and/or the novel function of the resulting functionally connected molecule. Whether the functions are independent or coupled depends on the properties of the barrier. A donor–sensitizer–acceptor molecule, a multimode chemical transducer molecule, and a photoswitching molecule are examples of unique systems resulting from different types of barriers. An appropriate barrier will facilitate control of the functions of photoinduced etectron transfer, leading to the possibilities of, for example, either photoinformation storage, or 2^n functionalities where n is the number of functional groups in a molecule.

Three-dimensional construction can also increase the density of accumulated functions. Recent developments in molecular assembly fabrication methods, including techniques involving Langmuir–Blodgett films and liposome manipulation, etc., make three-dimensional construction of molecular systems possible.

Recent developments in organic chemistry have also led to improvements in the stability of organic molecules, so that long-lived practical devices can be envisioned.

The systematization of various molecules promises to provide many interesting and effective molecular systems. Recently, many concepts have been presented concerning the construction of functional molecular materials, molecular devices, supramolecules, integrated chemical systems, accumulated reaction fields, and intelligent molecular systems. In particular, proposals for a molecular electronic

FIG. 5-2. Basic processes of a molecular device and a solid state device

Molecular device $\left\{ \begin{array}{l} \text{Electronic process} \\ \text{Ionic process} \\ \text{Molecular process} \end{array} \right\}$ Reaction
Transport
Energy interconversion
etc.

(Identity of respective molecules)

Solid state device ········· Electronic process

device [2] and supramolecular chemical systems [3] should be noted. The impact of the idea of molecular devices has been strong. Such devices are characterized by intramolecular systematization of the functional unit. Optimized systematization is very important.

The themes described in this book cannot cover all the possible functional molecular materials, but the basic ideas and fundamentals of molecular functions and systematized molecular materials are discussed. Certainly, the molecule is the smallest unit of functional material and is the smallest possible molecular device. We are sure that when nanotechnology is more fully developed, molecules will immediately be considered to be the principal players on the nano-stage, since only molecules can act as freestanding nano-sized functional materials.

However, all the descriptions given above do not mean that solid state devices will be replaced by molecular systems. With the advancement of molecular systems, a hybrid system will be developed where solid state devices are partly replaced by molecular systems. In such cases, a way to combine solid state devices and molecular devices is obviously necessary.

It is useful to compare the essential features of both molecular devices (systems) and solid state devices. Figure 5-2 gives a simple description of the fundamental processes of both devices. While the solid state device depends solely on the electronic processes in the solid, the molecular device depends on all electronic, ionic, and molecular processes. Here the electronic process means the energy level transition and electron transfer from a specific molecule to another specific molecule.

One of these electronic, ionic, or molecular processes, or any combination of these processes, leads to a reaction, transport, energy conversion, information conversion, etc. Two typical examples of information conversion are chromogenesis and morphogenesis. In all cases, the response of the molecular device to the input signal represents the identity of the functional molecule itself.

Figure 5-3 shows the relative advantages and disadvantages of both devices. A circle indicates an advantage.

From the point of view of size, i.e., the degree of integration, it is clear that a molecular device is superior because the minimum scale is that of a molecule. From the viewpoint of access time, a solid state device is excellent because the electronic process is much faster than that of an ionic or molecular device by several orders of magnitude.

	Solid state device	Molecular device
Size (integration)		◯
Access time	◯	
Feasibility of access	◯	
Energy consumption Unit function	◯	
Integrated function		◯
Multiple functions (dynamic response)		◯
Life	◯	

FIG. 5-3. Comparison of various features of a solid state device and a molecular device

As mentioned above, the problem of the interconnection between a molecular system and the outside system is yet to be solved, while no problem exists with a solid state device.

Energy consumption is an important consideration for the future of molecular systems. In general, the energy requirement of an electronic process is lower than that of an ionic or molecular process. Consequently, when based upon unit elementary function, the energy consumption of a solid state device is lower than that of a molecular device.

However, taking the degree of integration into consideration, the molecular system seems to be superior in overall function. The human brain is a good example. It consumes very little energy considering its unbelievable capabilities.

In terms of multiplicity of function, the molecular device has great potential possibilities because of its variety of profiles. However, in terms of the lifetime or stability of the device, the solid state device is far better than the molecular device at present.

In the future, it is probable that the development of other functions such as recognition, learning, self-repair, and so on, will improve the lifetime of molecular systems. We look forward to the day when artificial molecular devices play a significant role in daily life as truly functional materials.

References

1. Honda K (1988) Photoresponsive molecular systems. Proceedings or the MRS International Meeting on Advanced Materials, vol 12 pp 131–137
2. Carter FL, Siatkowski RE, Wohltjen H (1988) Molecular electronic devices. North-Holland, Amsterdam
3. Balzani V, Scandola F (1991) Supramolecular photochemistry. Elis Horwood, London

Index